TRP-Mediated Signaling

The field of transient receptor potential (TRP) channels has gained momentum in recent years not only because of the 2021 Nobel Prize in Physiology or Medicine awarded to David Julius and Ardem Patapoutian for their discoveries of receptors for temperature and touch, but also because of the growing appreciation of the diverse and important physiological and pathophysiological functions of this diverse family of cation channels. In the past decade, there have been important discoveries in the TRP field: the resolution of the 3-D structures of major subfamilies using single-particle cryogenic electron microscopy and X-ray crystallography, identification and development of selective agonists and antagonists for most TRP subfamily members, discovery of new human diseases associated with TRP gene mutations, as well as improved understanding of TRP channel function and regulation in various systems. These new discoveries are addressed in *TRP-Mediated Signaling*, a volume in the Methods in Signal Transduction book series.

Key Features

- Summarizes recent advances in the study of TRP channels
- Written by a team of leading international researchers
- Reviews physiological and pathophysiological functions of TRP channels
- Identifies diseases associated with TRP gene mutations
- Examines the evolution of temperature-sensing TRP channels

Methods in Signal Transduction

Series Editors: Michael X. Zhu and Harpreet Singh

The overall theme of this series continues to be the presentation of the wealth of up-to-date research methods applied to the many facets of signal transduction. Each volume is assembled by one or more editors who are pre-eminent in their specialty. In turn, the guiding principle for editors is to recruit chapter authors who will describe concepts, procedures and protocols with which they are intimately familiar in a reader-friendly format. The intent is to assure that each volume will be of maximum practical value to a broad audience, including students and researchers just entering an area, as well as seasoned investigators.

Gap Junction Channels and Hemichannels
Donglin Bai and Juan C. Sáez

Signaling Mechanisms Regulating T Cell Diversity and Function
Jonathan Soboloff and Dietmar J. Kappes

Lipid-Mediated Signaling Transduction, Second Edition
Eric Murphy, Thad Rosenberger, and Mikhail Golovko

Calcium Entry Channels in Non-Excitable Cells
Juliusz Ashot Kozak and James W. Putney, Jr.

Autophagy and Signaling
Esther Wong

Signal Transduction and Smooth Muscle
Mohamed Trebak and Scott Earley

Polycystic Kidney Disease
Jinghua Hu and Yong Yu

New Techniques for Studying Biomembranes
Qiu-Xing Jiang

Ion and Molecule Transport in Lysosomes
Bruno Gasnier and Michael X. Zhu

Neuron Signaling in Metabolic Regulation
Qingchun Tong

Non-Classical Ion Channels in the Nervous System
Tian-Le Xu and Long-Jun Wu

Immune cells, Inflammation and Cardiovascular Diseases
Shyam S. Bansal

Opsin-free Optogenetics: Technology and Applications
Yubin Zhou and Kai Zhang

Lung Biology and Pathophysiology
Yutong Zhao

TRP-Mediated Signaling
Michael Xi Zhu

For more information about this series, please visit: https://www.routledge.com/Methods-in-Signal-Transduction-Series/book-series/CRCMETSIGTRA

TRP-Mediated Signaling

Edited by
Michael Xi Zhu

CRC Press
Taylor & Francis Group
Boca Raton London New York

CRC Press is an imprint of the
Taylor & Francis Group, an **informa** business

Designed cover image: Jie Zheng

First edition published 2025
by CRC Press

2385 NW Executive Center Drive, Suite 320, Boca Raton FL 33431

and by CRC Press

4 Park Square, Milton Park, Abingdon, Oxon, OX14 4RN

CRC Press is an imprint of Taylor & Francis Group, LLC

ISBN: 9781032515335 (hbk)
ISBN: 9781032515366 (pbk)
ISBN: 9781003402732 (ebk)

DOI: 10.1201/9781003402732

Typeset in Times
by KnowledgeWorks Global Ltd.

Contents

Preface

The term transient receptor potential (TRP) originates from a mutant phenotype of the fruit fly (*Drosophila melanogaster*), which exhibited a short-lived response to a pulse of light, as detected by electroretinography—a method measuring the eye's electrical response to light. In wild-type flies, the response persists as long as the light pulse continues, but the *trp* mutant displays only a brief, transient response regardless of the pulse duration (Cosens and Manning, 1969; Montell, 2005). The mutant gene was later found to encode a transmembrane protein with sequence similarities to voltage-gated ion channels (Montell and Rubin, 1989; Wong et al., 1989), and functional studies on *Drosophila* photoreceptor cells confirmed that the *trp* mutant lacks a component of light-induced currents (Hardie and Minke, 1992). The field gained momentum at the turn of the century when numerous genes homologous to the *Drosophila trp* were cloned from mammals and other animals (Montell et al., 2002). In mammals, there are 28 nonallelic TRP genes divided into six subfamilies: TRPA (ankyrin), TRPC (canonical), TRPM (melastatin), TRPML (mucolipin), TRPP (polycystin), and TRPV (vanilloid). In invertebrates, there is an additional subfamily called TRPN (nompC). Humans have 27 TRP genes because TRPC2 became a pseudogene.

The TRP field has advanced significantly over the past two decades, witnessing a deeper and deeper understanding of the regulation, function, physiological significance, and disease relevance of individual TRP subtypes. We now know that all TRP channels are tetramers of identical or homologous transmembrane proteins, each containing six transmembrane α-helices, with the ion-conducting pore formed by the fifth and sixth helices from all four subunits. The other four transmembrane helices form a regulatory domain that contributes to channel gating, together with parts of the cytoplasmic amino and carboxyl termini. Most of these channels are cation nonselective with permeability to calcium ions (Ca^{2+}) and are localized to the plasma membrane, where their main functions are to cause membrane depolarization and an intracellular Ca^{2+} rise in response to environmental changes both inside and outside of the cell (Clapham, 2003). However, some TRP channels are also found on the membranes of intracellular organelles, such as TRPML1 in lysosomes (Dong et al., 2008), or on both the plasma membrane and intracellular organelles, like TRPC3, TRPM2, and TRPP2 (Feng et al., 2013; Köttgen et al., 2005; Lange et al., 2009). The strong health relevance of TRP channels is underscored by findings that genetic mutations in more than half of the mammalian TRP channels, encompassing all subfamilies, lead to various diseases (or channelopathies) (Zhao and Rohacs, 2021).

Recent progress in the TRP field has significantly advanced our knowledge of signal transduction, impacting broad areas of life sciences, including biophysics, structural biology, pharmacology, physiology, cell biology, and evolution. This volume of the CRC Methods in Signal Transduction series features chapters written by experts using cutting-edge technology to investigate the structure, regulation, and function of TRP channels. These include structural resolutions of TRP channels at atomic levels by cryogenic electron microscopy in different states—both ligand-bound and unbound—and their functional implications; new tools applied to study lipid regulation of TRP channels; the physiological and pathophysiological functions of TRP channels in various tissue systems; and the evolution of temperature-sensing TRP channels. These materials should be of great interest and value to a broad audience, including seasoned investigators and students or new investigators considering entering the TRP field.

I would like to thank all of the contributors for their timely and enthusiastic responses in preparing the chapters. The materials included in this volume will be an excellent resource for all investigators in the TRP and related fields.

REFERENCES

Clapham, D. E. TRP channels as cellular sensors. *Nature*, 2003, 426(6966): 517–524. doi: 10.1038/nature02196.

Cosens, D. J., Manning, A. Abnormal electroretinogram from a Drosophila mutant. *Nature*, 1969, 224(5216): 285–287. doi: 10.1038/224285a0.

Dong, X. P., Cheng, X., Mills, E., Delling, M., Wang, F., Kurz, T., Xu, H. The type IV mucolipidosis-associated protein TRPML1 is an endolysosomal iron release channel. *Nature*, 2008, 455(7215): 992–996. doi: 10.1038/nature07311.

Feng, S., Li, H., Tai, Y., Huang, J., Su, Y., Abramowitz, J., Zhu, M. X., Birnbaumer, L., Wang, Y. Canonical transient receptor potential 3 channels regulate mitochondrial calcium uptake. *Proc Natl Acad Sci U S A*, 2013, 110(27): 11011–11016. doi: 10.1073/pnas.1309531110.

Hardie, R. C., Minke, B. The *trp* gene is essential for a light-activated Ca^{2+} channel in *Drosophila* photoreceptors. *Neuron*, 1992, 8(4): 643–651. doi: 10.1016/0896-6273(92)90086-s.

Köttgen, M., Benzing, T., Simmen, T., Tauber, R., Buchholz, B., Feliciangeli, S., Huber, T. B., Schermer, B., Kramer-Zucker, A., Höpker, K., Simmen, K. C., Tschucke, C. C., Sandford, R., Kim, E., Thomas, G., Walz, G. Trafficking of TRPP2 by PACS proteins represents a novel mechanism of ion channel regulation. *EMBO J*, 2005, 24(4): 705–716. doi: 10.1038/sj.emboj.7600566.

Lange, I., Yamamoto, S., Partida-Sanchez, S., Mori, Y., Fleig, A., Penner, R. TRPM2 functions as a lysosomal Ca^{2+}-release channel in beta cells. *Sci Signal*, 2009, 2(71): ra23. doi: 10.1126/scisignal.2000278.

Montell, C. Drosophila TRP channels. *Pflugers Arch*, 2005, 451(1): 19–28. doi: 10.1007/s00424-005-1426-2.

Montell, C., Rubin G. M. Molecular characterization of the Drosophila trp locus: A putative integral membrane protein required for phototransduction. *Neuron*, 1989, 2(4): 1313–23. doi: 10.1016/0896-6273(89)90069-x.

Montell, C., Birnbaumer, L., Flockerzi, V., Bindels, R. J., Bruford, E. A., Caterina, M. J., Clapham, D. E., Harteneck, C., Heller, S., Julius, D., Kojima, I., Mori, Y., Penner, R., Prawitt, D., Scharenberg, A. M., Schultz, G., Shimizu, N., Zhu, M. X. A unified nomenclature for the superfamily of TRP cation channels. *Mol Cell*, 2002, 9(2): 229–231. doi: 10.1016/s1097-2765(02)00448-3.

Wong, F, Schaefer, E. L., Roop, B. C., LaMendola, J. N., Johnson-Seaton, D., Shao, D. Proper function of the Drosophila trp gene product during pupal development is important for normal visual transduction in the adult. *Neuron*, 1989, 3(1): 81–94. doi: 10.1016/0896-6273(89)90117-7.

Zhao, S., Rohacs, T. The newest TRP channelopathy: Gain of function TRPM3 mutations cause epilepsy and intellectual disability. *Channels (Austin)*, 2021, 15(1): 386–397. doi: 10.1080/19336950.2021.1908781.

About the Editor

Michael Xi Zhu is a professor in the Department of Integrative Biology and Pharmacology, McGovern Medical School, The University of Texas Health Science Center at Houston, Houston, Texas. He received his BS from Fudan University, Shanghai, China, in 1984, and his MS and PhD degrees from the University of Houston in 1988–1991, respectively. He was a postdoctoral fellow in the Department of Cell Biology at Baylor College of Medicine, Houston, Texas, from 1991–1994 and an assistant researcher in the Department of Anesthesiology, University of California at Los Angeles, from 1994–1997. In 1997, he became an assistant professor in the Department of Pharmacology and Neurobiotechnology Center at The Ohio State University. In 2000, he transferred his appointment to the Department of Neuroscience and Center for Molecular Neurobiology at The Ohio State University, where he was promoted to associate professor in 2003 and to professor in 2010. In 2010, he moved to his current position. Dr. Zhu's research interests include many aspects of cell signaling, especially those that involve heterotrimeric G proteins and ion channels that affect Ca^{2+} signaling. Dr. Zhu's main contributions include the identification and characterization of multiple TRPC channels in mammalian species and the determination of the molecular identity of endolysosomal Ca^{2+} release channels activated by the Ca^{2+} mobilizing messenger, nicotinic acid adenine dinucleotide phosphate. He has published more than 200 research papers, reviews, and book chapters and delivered lectures at many international conferences and symposia. Dr. Zhu has served on the editorial boards of *Journal of Cellular Physiology, Pflügers Archiv - European Journal of Physiology, Biophysics Reports, Cells, and Molecular Pharmacology.* He has been serving as a series editor of the CRC Methods in Signal Transduction book series since 2013. Dr. Zhu has chaired or co-chaired many international conferences, including the 17th International Symposium on Ca^{2+}-binding proteins and Ca^{2+} function in Health and Disease, Beijing, 2011, and the 2nd Gordon Research Conference on Organellar Channels and Transporters, Vermont, 2017.

Contributors

Indu S. Ambudkar
National Institutes of Health
Bethesda, MD, USA

Lei Chen
Peking University
Beijing, China

Elisa Dai
Dalhousie University
Halifax, Nova Scotia, Canada

Rose Yang Dong
Dalhousie University
Halifax, Nova Scotia, Canada

Xian-Ping Dong
Dalhousie University
Halifax, Nova Scotia, Canada

Li Geng
Jiangnan University
Wuxi, Jiangsu, China

Klaus Groschner
Gottfried Schatz Research Center
Medical University of Graz
Graz, Austria

Wenjun Guo
Peking University
Beijing, China

Huifang Hou
Xinxiang Medical University
Xinxiang, Henan, China

Ning Hua
Zhejiang University
Hangzhou, Zhejiang, China

Christina Humer
Johannes Kepler University Linz
Linz, Austria

Lin-Hua Jiang
Xinxiang Medical University
Xinxiang, Henan, China

Denis Krivic
Gottfried Schatz Research Center
Medical University of Graz
Graz, Austria

Lin Lu
Zhejiang University
Hangzhou, Zhejiang, China

Yanping Luo
Zhejiang University
Hangzhou, Zhejiang, China

Xin Ma
Jiangnan University
Wuxi, Jiangsu, China

Ericka Martínez Martínez
National Institutes of Health
Bethesda, MD, USA

Li Meng
Xinxiang Medical University
Xinxiang, Henan, China

Hwei Ling Ong
National Institutes of Health
Bethesda, MD, USA

Sebastian Roger
University of Tours
Tours, France

Christoph Romanin
Johannes Kepler University Linz
Linz, Austria

Shigeru Saito
Nagahama Institute of Bio-science
and Technology
Nagahama, Shiga, Japan

Makoto Tominaga
National Institutes of Natural Sciences
Okazaki, Aichi, Japan

Simon Vu
University of California Davis
Davis, CA, USA

Lu Wang
The First Affiliated Hospital of Xinxiang Medical University
Xinxiang, Henan, China

Pingping Wang
Dalhousie University
Halifax, Nova Scotia, Canada

Yang Wei
Zhejiang University
Hangzhou, Zhejiang, China

Mengnan Xu
Dalhousie University
Halifax, Nova Scotia, Canada

Jing Yao
Zhejiang University
Hangzhou, Zhejiang, China

Xiaoqiang Yao
The Chinese University of Hong Kong
Hong Kong, China

Luyu Ye
Zhejiang University
Hangzhou, Zhejiang, China

Yaling Yin
Xinxiang Medical University
Xinxiang, Henan, China

Lixia Yue
University of Connecticut School of Medicine (UConn Health)
Farmington, CT, USA

Jun Zhang
The First Affiliated Hospital of Harbin Medical University
Harbin, Heilongjiang, China

Kaihua Zhang
Fudan University
Shanghai, China

Jie Zheng
University of California Davis, School of Medicine
Davis, CA, USA

Pengyu Zong
University of Connecticut School of Medicine (UConn Health)
Farmington, CT, USA

Series Preface

The concept of signal transduction is now long established as a central tenet of biological sciences. Since the inception of the field 50 years ago, the number and variety of signal transduction pathways, cascades, and networks have steadily increased and now constitute what is often regarded as a bewildering array of mechanisms by which cells sense and respond to extracellular and intracellular environmental stimuli. It is not an exaggeration to state that virtually every cell function is dependent on the detection, amplification, and integration of these signals. Moreover, there is increasing appreciation that in many disease states, aspects of signal transduction are critically perturbed.

Our knowledge of how information is conveyed and processed through these cellular molecular circuits and biochemical switches has increased enormously in scope and complexity since this series was initiated twenty-six years ago. Such advances would not have been possible without the supplementation of older technologies, drawn chiefly from cell and molecular biology, biochemistry, physiology, and pharmacology, with newer methods that make use of sophisticated genetic approaches as well as structural biology, imaging, bioinformatics, and systems biology analysis.

The overall theme of this series continues to be the presentation of the wealth of up-to-date research concepts and methods applied to the many facets of signal transduction. Each volume is assembled by one or more editors who are pre-eminent in their specialty. In turn, the guiding principle for editors is to recruit chapter authors who will describe procedures and protocols with which they are intimately familiar in a reader-friendly format. The intent is to assure that each volume will be of maximum practical value to a broad audience, including students and researchers just entering an area, as well as seasoned investigators.

It is hoped that the information contained in the books of this series will constitute a useful resource to the life sciences research community well into the future.

Michael Xi Zhu
Harpreet Singh
Series Editors

1 Recent Advances in the Structural Pharmacology of TRPC Channels

Wenjun Guo and Lei Chen

1.1 INTRODUCTION

Transient receptor potential canonical (TRPC) channels are activated by diacylglycerol (DAG) produced by phospholipase C (PLC) [1]. Based on sequence similarity and channel characteristics, TRPC channels are classified into two subgroups: TRPC3/6/7 and TRPC1/4/5. Several high-resolution structures of homomeric TRPC channels have been elucidated [2–14]. These structures reveal that TRPC channels consist of tetramers with a two-layer architecture, including the intracellular domain (ICD) and the transmembrane domain (TMD) [15] (Fig. 1.1).

The ICD of the TRPC protein is formed by folding both the N-terminal amino acids and the C-terminal amino acids together. The N-terminal amino acids contribute to the formation of an unstructured N-terminal region, an ankyrin repeat domain (Ank1-4, ARD), and a linker helices domain. On the other hand, the C-terminal amino acids include a TRP helix, a flexible linker, two C-terminal helices (CH), and an unstructured region at the C-terminus (Fig. 1.1C). ICD harbors the binding sites of ions and interacting proteins, as described later.

Like other TRP channels, the TMD of TRPC channels consists of the voltage sensor-like domain (VSLD) and the pore domain, which are interconnected by the S4-S5 linker in a domain-swapped arrangement (Fig. 1.1). In addition to these common structural elements, all TRPC proteins also feature a pre-S1 elbow and a TRP re-entrant loop in the TMD, structural

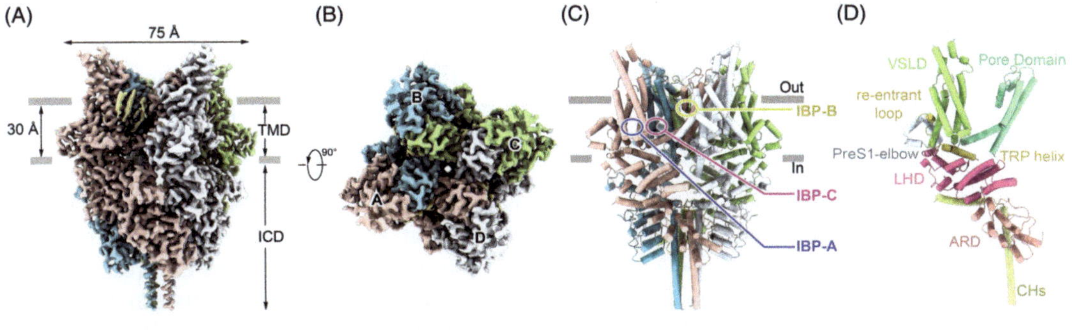

FIGURE 1.1 The architecture of the homotetrameric TRPC channels. (A) The cryo-EM density map of the full-length SAR7334-bound hTRPC6 structure (PDB: 7DXG) is shown in a side view. The structure consists of four subunits labeled as A, B, C, and D, arranged in a clockwise manner. The subunits are colored in pink, blue, green, and gray, respectively. The ligands SAR7334 are colored the same as the subunits they belong to. Lipids are colored in yellow. The approximate boundary of the cell membrane is indicated by gray lines. The structure has a transmembrane domain (TMD) and an intracellular domain (ICD). (B) The top view of the full-length SAR7334-bound hTRPC6 structure. (C) The cartoon model of SAR7334-bound hTRPC6 is depicted using cylinders and colored in the same manner as (A) and (B). The three drug binding pockets (IBP-A-C) in TRPC are clearly circled. (D) A single subunit of hTRPC6 is shown in cartoons with different domains in different colors. ARD, ankyrin repeats domain; LHD, linker helices domain; and CHs, C-terminal helices.

DOI: 10.1201/9781003402732-1

characteristics closely resembling those of No Mechanoreceptor Potential C (NOMPC) [16] and melastatin-like TRP (TRPM) channels [17–22]. In contrast, the TRPA [23] and TRPV [24] channels differ from TRPC in the fact that they do not have a pre-S1 elbow or a TRP re-entrant loop motif.

All the identified binding sites for synthetic small molecular compounds targeting TRPC channels are located within the TMD. These binding sites include inhibitors specific to TRPC3/6 such as SAR7334 [12], AM-1473 [8], and BTDM [2, 12]; inhibitors specific to TRPC5 including piperazinone derivatives GFB-8438, GFB-8749, and GFB-9289 [9], clemizole [11], and methylxanthine derivatives HC-070 [11] and HC-608 [10], as well as the TRPC5 selective activator riluzole [13], the TRPC6 selective activator AM-0883 [8], and the endogenous ligand DAG [11]. Previous literature has classified these binding sites into three groups according to their locations: inhibitor-binding pocket A (IBP-A), IBP-B, and IBP-C [11] (Fig. 1.1C). In the following section, we will explore in greater depth the detailed ligand binding poses and interactions between the ligands and TRPC channels.

1.2 LIGAND BINDING IN IBP-A

IBP-A is a location in the VSLD. Resolved structures have shown that seven compounds, including SAR7334, AM-1473 (an analog of SAR7334), GFB-8438, GFB-9289, GFB-8749, clemizole, and riluzole, bind in this pocket. SAR7334 was discovered during the optimization of aminoindanol derivatives as inhibitors of TRPC6 [25]. It is one of the most potent TRPC6 channel inhibitors, with an IC_{50} below 10 nM [25]. SAR7334 has also demonstrated favorable oral pharmacokinetics and reasonable selectivity compared to TRPC3 (IC_{50} approximately 282 nM) and TRPC7 (IC_{50} approximately 226 nM) channels [25]. The structure of TRPC6 in complex with SAR7334 has been resolved to a resolution of 2.9 Å (PDB: 7DXG) [12]. This structure reveals that the SAR7334 binding pocket is surrounded by several amino acids, including Lys442 and His446 on S1, Glu509 and Glu512 on S2, Asn527 and Asp530 on S3, Arg609, Tyr612, and Ile613 on S4, Ser752 and Tyr753 on the TRP helix, and Arg758 on the re-entrant loop [12] (Fig. 1.2A-C). The piperidine ring of SAR7334 forms polar contacts with Glu512 on S2, Asn527 on S3, and Arg609 on S4 (Fig. 1.2C). Additionally, the primary amine group of the piperidine ring participates in hydrogen-bond interactions with Glu509 on S2 and Asp530 on S3 (Fig. 1.2C). The benzonitrile group of SAR7334 engages in cation–π interactions with Arg758 on the re-entrant loop and aromatic-stacking interactions with His446 on S1 and Tyr753 on the TRP helix (Fig. 1.2C). The chlorine atom of the chlorobenzonitrile motif exhibits polar contact with Ser752 on the TRP helix (Fig. 1.2C). Finally, the indene double ring of SAR7334 contributes to van der Waal's contacts with Tyr612 and Ile613 on S4 [12] (Fig. 1.2C).

AM-1473 is an antagonist of TRPC6 with an IC_{50} of 200 nM, and it possesses a similar aminoindanol structure as SAR7334 [8]. The only difference between the two compounds is that the chlorine atom on the benzonitrile group in SAR7334 is substituted with a hydrogen atom in AM-1473. Because of their similar molecular structures, these two compounds exhibit almost identical binding pockets and binding poses, as evidenced by the resolved structure (PDB: 6UZA) [8].

The residues involved in the interaction with AM-1473 are nearly identical between TRPC3 and TRPC6, with the exception of a substitution of Arg758 in TRPC6 with a lysine (Lys689) in TRPC3. In SAR7334, the cation–π interaction between Arg758 and the benzonitrile group is critical. The R758K mutation in TRPC6 reduces the potency of AM-1473 by approximately 5-fold, while in TRPC3, the K689R mutation increases potency by 3.5-fold compared to wild-type TRPC3 [8]. Furthermore, the R758A mutation in TRPC6 significantly decreases the binding affinity of AM-1473 by approximately 300-fold [8]. These findings highlight the significance of the cation–π interaction between Arg758 and the benzonitrile group, which also explains the selectivity of SAR7334 for TRPC6 over TRPC3. Additionally, the E509A and D530A mutations in TRPC6 largely abolish the inhibition of AM-1473 [8], underscoring the importance of negatively charged side chains in ligand recognition and binding.

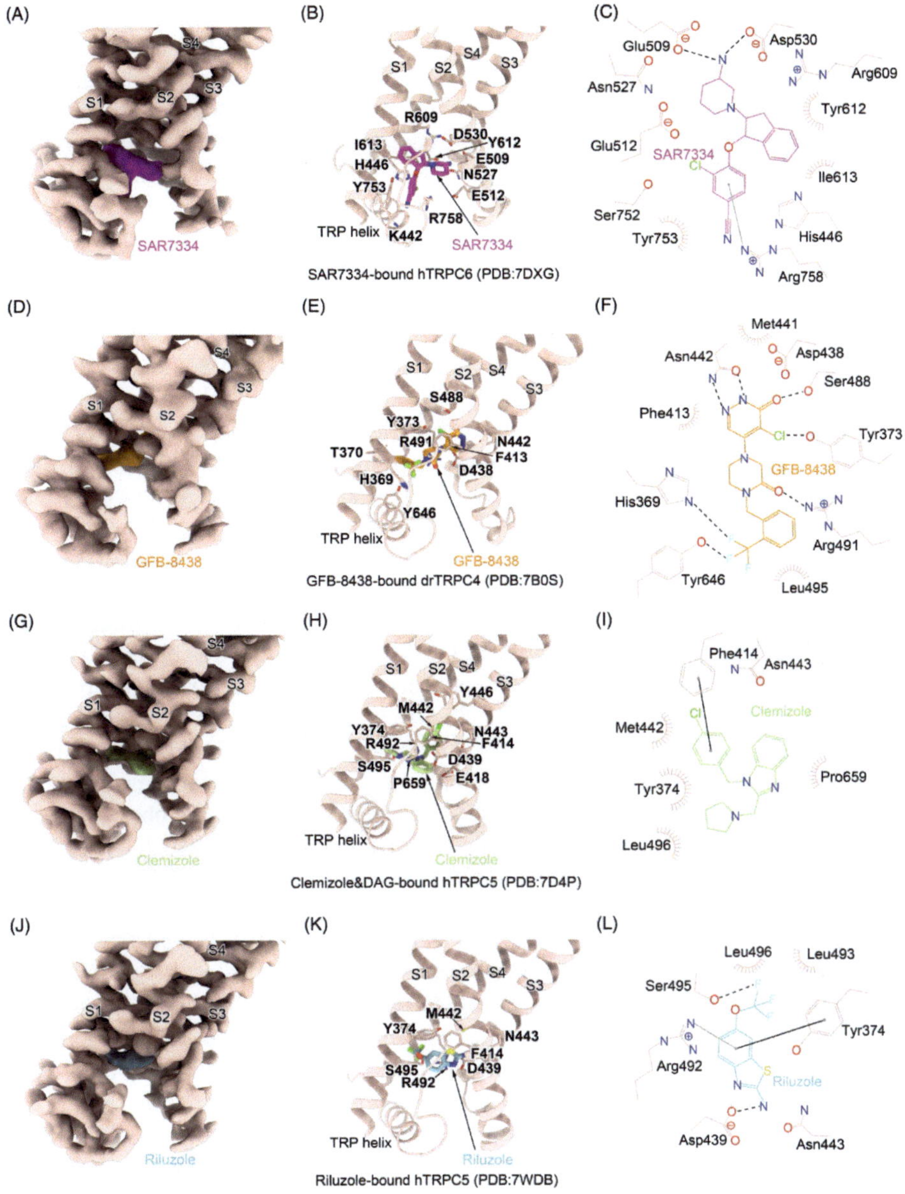

FIGURE 1.2 Ligands binding in IBP-A of TRPC channels. Among the seven compounds bound in IBP-A, SAR7334 and AM-1473 share a similar structure, while GFB-9289, GFB-8749, and GFB-8438 share a similar core structure as well. Therefore, SAR7334 and GFB-8438 are selected as examples to demonstrate the ligand binding modes in IBP-A. (A) Close-up view of the density map of the SAR7334-bound TRPC6 channel (PDB:7DXG). The densities of SAR7334 and TRPC6 are colored in magenta and pink, respectively. (B) Close-up view of the SAR7334 binding site in TRPC6 shown in cartoon. TRPC6 and SAR7334 are colored the same as (A). Residues of TRPC6 within 4 Å from SAR7334 are shown as sticks and colored by elements. (C) The TRPC6-SAR7334 interactions analyzed using LigPlot. Hydrogen bonds, cation-π interaction, and hydrophobic interactions are indicated with dashed lines, gray line, and spoked arcs, respectively. Side chains of hydrophilic residues are shown as sticks. (D–F) Close-up view of the GFB-8438 binding site in TRPC4 channel (PDB: 7B0S). GFB-8438 is colored in orange. (G, H) Close-up view of the clemizole binding site in TRPC5 channel (PDB: 7D4P). Clemizole is colored in green. (I) The clemizole binding mode. π-π interaction is indicated with the black line. (J–L) Close-up view of the riluzole binding site in TRPC5 channel (PDB: 7WDB). Riluzole is colored in blue.

GFB-8438 is a selective inhibitor of TRPC4/5 and belongs to the class of piperazinone derivatives. It was identified through high-throughput screening using the FLIPR assay [26]. In vitro studies have shown that GFB-8438 is capable of protecting mouse podocytes from injury induced by protamine sulfate [26]. Additionally, in an in vivo study using a hypertensive deoxycorticosterone acetate-salt rat model of FSGS, GFB-8438 significantly reduced both total protein and albumin concentrations in urine [26]. Furthermore, an analog of GFB-8438 called GFB-887 has been developed by Goldfinch Bio. GFB-887 is currently in a phase 1 clinical trial for the treatment of kidney disease, with the clinical trial identifier NCT03970122.

The resolution of the structure of TRPC4 with bound GFB-8438 was determined to be 3.6 Å [9]. Based on the map, GFB-8438 binds in the IBP-A region (Fig. 1.2D). The binding pocket of GFB-8438 is surrounded by several residues, including His369 and Tyr373 on S1, Phe413 on S2, Met441, Asn442, and Asp438 on S3, Ser488, Arg491, and Leu495 on S4, and Tyr646 on the TRP helix [9] (Fig. 1.2E). The chemical structure of GFB-8438 consists of three six-membered rings. It has a pyridazinone ring, a piperazinone ring in the middle, and a trifluoromethyl benzyl group at opposing ends. The nitrogen atom and the oxo group of the pyridazinone ring form hydrogen bonds with Asn442 on S3 and Ser488 on S4, respectively (Fig. 1.2F). Additionally, the chlorine atom in the pyridazinone ring forms halogen bonds with Tyr373 on S1. The oxo group of the piperazinone ring is involved in a hydrogen bond with Arg491 on S4 (Fig. 1.2F). The fluorine atom of the benzyl group forms hydrogen bonds with His369 on S1 and Tyr646 on the TRP helix [9] (Fig. 1.2F).

GFB-9289 is an analog of GFB-8438 in which the trifluoromethyl benzyl group is replaced by a cyclohexyl ring. In the 3.2 Å TRPC4 map with GFB-9289 bound, it was observed that GFB-9289 also binds in the IBP-A region [9]. The binding pocket of GFB-9289 is surrounded by several amino acid residues, including Thr370, Tyr373 on S1, Phe413 on S2, Met 441, Asn442, and Asp438 on S3, and Ser488, Arg491 on S4 [9]. Similar to GFB-8438, the pyridazinone ring of GFB-9289 forms hydrogen bonds with Asn442, and its oxo group in the middle piperazinone ring engages in a hydrogen bond with Arg491 on S4. However, due to the reduced size of the cyclohexyl group in the tail part of GFB-9289, there is reduced interaction with surrounding residues such as His369 on S2 and Tyr646 on the TRP helix [9]. This decreased interaction between GFB-9289 and TRPC4 residues may account for its lower potency compared to GFB-8438.

GFB-8749 has a common chemical scaffold shared with GFB-8438 and GFB-9289. In GFB-8749, the trifluoromethyl benzyl group of GFB-8438 is replaced by a difluoro cyclohexyl ring. According to the 3.8 Å TRPC4 map with GFB-8749 bound, GFB-8749 binds in the VSLD region [9]. Several residues in TRPC4 are in close proximity to the GFB-8749 molecule, including Phe366, Tyr373 on S1, Phe413 on S2, Met441, Asn442, and Asp438 on S3, Ser488, Arg491, Leu495 on S4, and Tyr646 on the TRP helix [9]. The interactions between the pyridazinone and piperazinone rings of GFB-8749 and the surrounding residues are almost identical to those of GFB-8438. However, since the size of the terminal group in GFB-8749 is intermediate between that of GFB-9289 and GFB-8438, the hydrogen bond between the fluorine atom and Tyr646 on the TRP helix, as well as the hydrophobic contacts between the cyclohexyl ring and Leu495 on S4, could potentially contribute to the improved binding affinity of GFB-8749 [9].

The residues that interact with the piperazinone scaffold of inhibitors GFB-8438, GFB-9289, and GFB-8749 are conserved between TRPC4 and TRPC5, suggesting a similar ligand binding mode for both channels. In the automated patch clamp assay, GFB-8438 has shown selectivity for TRPC4/5 compared to other TRPC channels [26]. Upon analyzing the sequence alignment among TRPC channels, it has been observed that the Asn442 and Phe413 residues in TRPC4/5 are substituted with Leu534 and Met505 in TRPC6. These substitutions may weaken the interactions between TRPC6 and the pyridazinone ring of the inhibitors. These critical alterations in the ligand binding pocket help to explain the significantly reduced inhibitory effect of GFB-8438 against TRPC6.

Clemizole, M084, and AC1903 are TRPC4/5 inhibitors that share a common benzimidazole scaffold. Clemizole has been found to efficiently inhibit TRPC5 currents and calcium entry, with an IC_{50} of approximately 1 μM [27]. Fluorometric intracellular calcium measurements have shown that clemizole exhibits around sixfold selectivity for TRPC5 over TRPC4β and nearly 10-fold selectivity over TRPC3 and TRPC6 [27]. In 2021, the structure of the TRPC5-clemizole complex was resolved to a resolution of 2.7 Å [11]. The binding pose of clemizole in the VSLD region was easily discernible from the electron density map (Fig. 1.2G). The binding pocket of clemizole is surrounded by several amino acid residues, including Tyr374 on S1, Phe414, Gly417, and Glu418 on S2, Asp439, Met442, Asn443, and Tyr446 on S3, Arg492, Ser495, and Leu496 on S4, and Pro659 on the TRP re-entrant loop [11] (Fig. 1.2H). The chlorophenyl ring of the clemizole forms a π–π stacking interaction with Phe414 (Fig. 1.2I). The three components of the clemizole molecule, namely the chlorophenyl ring, pyrrolidin ring, and benzimidazole group, exhibit hydrophobic interactions with specific side chains of amino acid residues (Fig. 1.2I). The chlorophenyl ring interacts hydrophobically with Tyr374, the pyrrolidin ring interacts with Met442, and the benzimidazole group interacts with Leu496 and Pro659 [11] (Fig. 1.2I). These hydrophobic interactions contribute to the stabilization of the clemizole binding within the TRPC5 protein. The potency of clemizole and the role of specific interacting residues were assessed using the FLIPR calcium assay [11]. Three residues, Phe414, Asn443, and Ser495, were mutated to the corresponding residues found in TRPC6 (Met, Leu, and Ala, respectively). The results of the FLIPR calcium assay showed that while the mutations F414M and S495A had minimal effect, the N443L mutation led to a decrease in the potency of clemizole [11]. This suggests that the Van der Waals interactions between clemizole and the side chain of Asn443 play a crucial role in the subtype selectivity of clemizole.

M084 was identified through high-throughput screening [28]. It exhibits selectivity for TRPC4 and TRPC5 over TRPC3 and TRPC6. Fluorescence membrane potential measurements revealed that the IC_{50} of M084 for TRPC5 is approximately 8 μM, with an almost sevenfold selectivity over TRPC3 and TRPC6 [28]. AC1903 was identified through a series of molecules synthesized based on ML204 and clemizole [29]. Patch-clamp electrophysiology recording in TRPC5-, TRPC4-, or TRPC6- overexpressing HEK293 cells revealed that AC1903 is TRPC5-selective [29]. Because of the similar chemical structures of these three compounds, they probably share a common binding site at IBP-A.

Riluzole is a benzothiazole compound that is currently approved as the only drug for the treatment of amyotrophic lateral sclerosis [30]. Additionally, riluzole has shown promise for the treatment of psychiatric disorders [31, 32]. At the molecular level, riluzole has been found to affect the activity of various channels and transporters [33].

In 2014, riluzole was discovered to be a novel activator of TRPC5 [34]. It can activate both the heterologously overexpressed TRPC5 in HEK293 cells and the endogenously expressed TRPC5 in the U-87 glioblastoma cell line [34]. Concentration-response curves obtained through fluorometric calcium imaging measurements demonstrated that the EC_{50} of riluzole on TRPC5 is approximately 9.2 μM [34]. Patch clamp recordings further revealed that riluzole-induced TRPC5 currents could be potentiated by lanthanum ions (La^{3+}) [34], indicating that riluzole and La^{3+} activate TRPC5 through different mechanisms.

In 2022, the structure of the TRPC5-riluzole complex was resolved, achieving a global resolution of 2.4 Å [13]. The structure confirms that riluzole also binds within VSLD. Riluzole interacts with several nearby residues in the TRPC5 structure (Fig. 1.2J). These residues include Tyr374 on S1, Phe414, Glu418, and Glu421 on S2, Asp439, Met442, and Asn443 on S3, and Arg492, Leu493, Ser495, and Leu496 on S4 [13] (Fig. 1.2K). Specifically, the benzothiazole ring of riluzole forms a π–π stacking interaction with Tyr374 on S1 and a cation–π interaction with Arg492 on S4 (Fig. 1.2L). The trifluoromethoxy group and opposite amine of riluzole form hydrogen bonds with Ser495 on S4 and Asp439 on S3, respectively [13] (Fig. 1.2L). To understand the roles of key interacting residues in riluzole selectivity, mutations were analyzed. The mutations F414M (corresponding to Met in human (h)TRPC1/3/6/7) and N443L (corresponding to Leu in hTRPC1/3/6/7) were found to decrease the potency of riluzole towards hTRPC5 to varying extents [13]. This is consistent with the observed selectivity of riluzole for hTRPC5 over hTRPC3/6/7.

1.3 LIGAND BINDING IN IBP-B

IBP-B is surrounded by S5 and the pore helix from one subunit, and S6 of the adjacent subunit. Inhibitors AM-0883 and HC-070, Pico145 (an analog of HC-070), as well as the endogenous activator DAG, bind to IBP-B.

AM-0883 is a novel TRPC6 agonist with an EC_{50} of approximately 45 nM [8]. It was identified through small-molecule high-throughput screening [8]. Both AM-0883 and 1-oleoyl-2-acetyl-sn-glycerol (OAG, a DAG analog) induce significantly higher activity in the V867T/L868T mutant TRPC6 channel (deletion of amino acids 2–72) compared to the wild-type TRPC6 channel [8]. Therefore, the V867T/L868T TRPC6 channel was used in structural studies in complex with the small molecule agonist AM-0883, and the structure was determined at a resolution of 3.1 Å [8], revealing that AM-0883 binds to IBP-B (Fig. 1.3A-B). AM-0883 consists of a chloro-indole, a piperidine, and a benzodioxin. The main hydrophobic interactions occur between AM-0883 and nearby residues, including Phe675, Phe679, and TRP680 on the pore helix, as well as Tyr705, Val706, and Gly709 on the S6 helix [8] (Fig. 1.3C). In addition, the indole ring forms hydrophilic interactions with Glu672 on the pore helix and Asn702 on the S6 helix [8] (Fig. 1.3C). The dose-response activation of several TRPC6 mutants by the agonist AM-0883 was measured using the FLIPR calcium assay. According to the results, the expression levels of all these mutants on the cell membrane were similar to those of the wild-type TRPC6 channel. However, the mutations F675A, W680A, N702A, Y705A, and V710A almost completely abolished the channel activity elicited by the agonist AM-0883 [8]. On the other hand, the mutations E672A and V706A showed a 35-fold and 20-fold increase in agonist EC_{50}, respectively, suggesting the functional importance of these residues [8].

In the map of TRPC5 with clemizole bound, several putative lipid densities were observed in the TMD [11]. One lipid without the phosphate group is specifically located in the IBP-B region (Fig. 1.3D). Through HPLC-MS data and electron density analysis, this lipid was identified as a DAG molecule [11] (Fig. 1.3E). The free hydroxyl group on the head portion of the DAG molecule forms a polar contact with Gln573 (Fig. 1.3F). The two hydrophobic tails of the DAG molecule engage in hydrophobic interactions with Tyr524 and Leu528 on S5, Phe576, Trp577, and Val579 on the pore helix, as well as Val610, Ile611, and Leu616 on S6 of the adjacent subunit [11] (Fig. 1.3F). In mouse TRPC5, the double mutations F576A and W577A, which correspond to the F576A and W577A mutations in hTRPC5, abolished the response to Englerin A (a TRPC4/5 agonist) activation [6]. It is worth noting that mutations in the IBP-B region also influence the activity of DAG-activated TRPC3/6 channels [8, 35, 36]. The mutants affecting the IBP-B of TRPC6, namely E672A, F675A, W680A, N702A, Y705A, V706A, and V710A, correspond to F569, L572, W577, F599, A602, and T603, respectively, in hTRPC5. These mutations significantly impact the activation of the TRPC6 channel by OAG [8]. Additionally, the G652A mutation on S6 in hTRPC3, which corresponds to G606 in hTRPC5, results in both altered responses to different DAG species in TRPC3 and a modified pattern of photoactivation of TRPC3 and TRPC6 (corresponding to G709A) through a photoswitchable DAG analog, OptoDArG [35]. Interestingly, the G652A mutant of TRPC3 still maintains sensitivity to the non-lipid activator GSK1702934A [36]. These findings collectively demonstrate that the DAG binding pocket observed in hTRPC5 is also functionally conserved in TRPC3/6.

HC-070 is a highly potent small-molecule antagonist that inhibits both homomeric TRPC4 and TRPC5 channels, as well as TRPC1/C5 or TRPC1/C4 heterotetramers, with nanomolar potency [37]. According to the calcium influx assay, HC-070 demonstrates an IC_{50} of 9.3 nM for TRPC5 [37]. Treatment with HC-070 has also been shown to attenuate the anxiogenic effect when orally administered to mice [37]. HC-070 is a methylxanthine derivative with a hydroxypropyl tail, a chlorophenyl ring, and a chlorophenoxy ring. In 2021, the cryo-EM structure of TRPC5 with HC-070 bound was determined to have a resolution of 2.7 Å [11]. The high-resolution map revealed the unambiguous binding pose of HC-070 (Fig. 1.3G). The methylxanthine core of HC-070 forms a π–π stacking interaction with the side chain of Phe576 on the pore helix, while the hydroxypropyl tail moiety forms polar contacts with Arg557 on the linker and Gln573 on the pore helix [11] (Fig. 1.3H-I).

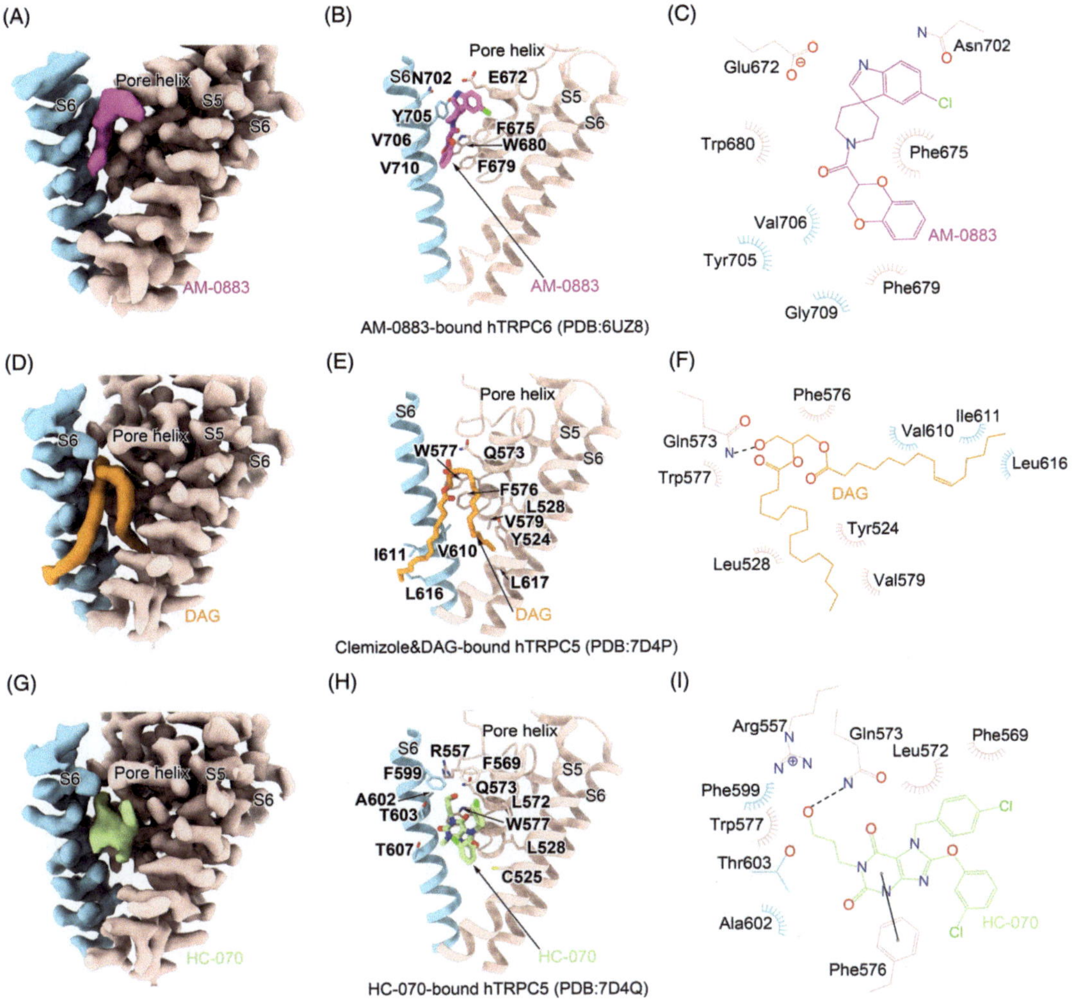

FIGURE 1.3 Ligands binding in IBP-B of TRPC channels. Among the four ligands bound in IBP-B, Pico145 (HC-608) and HC-070 share a similar core structure. Therefore, the binding site of HC-070 is shown in detail. (A) Close-up view of the density map of the activator AM-0883-bound TRPC6 channel (PDB:6UZ8). The density of AM-0883 is colored in magenta. One subunit of TRPC6 is colored in pink and the adjacent subunit is colored in blue. (B) Close-up view of the AM-0883 binding site in TRPC6 shown in cartoon. TRPC6 and AM-0883 are colored the same as (A). AM-0883 surrounding residues are shown as sticks and colored by elements. (C) The TRPC6-AM-0883 interactions analyzed using LigPlot. Hydrophobic interactions are indicated with spoked arcs. Side chains of hydrophilic residues are shown as sticks. (D–F) Close-up view of the DAG binding site in TRPC5 channel (PDB: 7D4P). DAG is colored in orange. Hydrogen bonds are indicated with dashed lines. (G–I) Close-up view of the HC-070 binding site in TRPC5 channel (PDB: 7D4Q). HC-070 is colored in green. The π-π interaction between HC-070 and TRPC5 molecules is indicated with the black line.

The hydroxypropyl tail of HC-070 also engages in hydrophobic interactions with Trp577 on the pore helix, Phe599 and Ala602 on S6 of the adjacent subunit. Additionally, Thr603 contributes to the stabilization of the hydroxypropyl tail of HC-070 (Fig. 1.3H-I). The chlorophenyl ring of HC-070 interacts hydrophobically with Phe569 and Leu572 on the pore helix [11] (Fig. 1.3H-I). Interestingly, the chlorophenoxy ring is only observed at low contour levels, indicating a weaker interaction with Leu521 and Cys525 on S5 [11]. Sequence alignment reveals that the residues involved in HC-070 binding are identical between hTRPC4 and hTRPC5, but only share 20% similarity with TRPC3/6/7 and 46%

similarity with TRPC1. These findings are consistent with the high selectivity of HC-070 towards TRPC4/5 compared to TRPC3/6/7, as well as its lower potency in the heterotetrameric hTRPC1/C5 and hTRPC1/C4 complexes [37]. In a fluorometric calcium assay, alanine mutations were introduced at various key residues, including R557A, F569A, L572A, Q573A, F599A, and T603A. These mutations resulted in decreased potency, with Q573A showing the most significant decrease, highlighting the crucial role of the hydrogen bond between Gln573 and the hydroxypropyl group of HC-070 [11]. Notably, as TRPC3/6 have Asn instead of Phe at position 599, the IC_{50} of HC-070 on the F599N mutant was measured, and it exhibited a markedly reduced potency as well [11].

Pico145 (HC-608) is an analog of HC-070 that exhibits higher potency and selective inhibition of TRPC1/4/5 [38]. It can inhibit homotetrameric TRPC4 at picomolar concentrations and homotetrameric TRPC5 at low nanomolar concentrations [38]. Compared to HC-070, Pico145 has a trifluoromethoxy group substitution for the chloride atom in the chlorophenoxy ring. In 2020, the cryo-EM structure of TRPC5 in complex with Pico145 was determined to have a resolution of 3.0 Å [10]. The binding pose of Pico145 is similar to HC-070, except for the presence of a trifluoromethoxy group on the chlorophenoxy ring, which forms hydrophobic interactions with Leu521 and Tyr524 on S5, as well as polar contacts with Cys525 [10]. The additional interactions with surrounding residues likely contribute to the higher potency of Pico145 compared to HC-070.

AM237 is also a member of the xanthine class, and it activates TRPC5 homotetramers with an EC_{50} of approximately 15–20 nM in fluorometric calcium assays [39]. Unlike Pico145, AM237 has an additional chlorine atom on its phenoxy ring. To confirm the activity of AM237, mutants related to IBP-B were tested using fluorometric intracellular calcium measurements [10]. The potency of AM237 was found to be lower for the Q573T mutant compared to the wild-type TRPC5 channel. Furthermore, neither the TRPC5 F576A mutant nor its W577A variant was responsive to AM237 for activation [10]. These functional data suggest that AM237 also binds to the same pocket in IBP-B and highlight the importance of the π–stack interaction between Phe576 and the xanthine core, as well as the hydrogen bond between Trp577 and the 3-hydroxypropyl group, for the activity of AM237 [10]. Due to the larger size of the phenoxy group in AM237 and the possible clash between its chloride atom and Leu521 on S5, it is presumed that AM237 might push Leu521 on S5 through its chloride-phenoxy group, thereby activating hTRPC5.

1.4 LIGAND BINDING IN IBP-C

IBP-C is membrane-embedded and situated between the VSLD and the pore domain. Currently, only one inhibitor, known as BTDM, has been identified to bind to IBP-C. BTDM exhibits high-affinity inhibition towards the hTRPC3/6 channels, effectively suppressing its activity with an IC_{50} of approximately 10 nM [2]. Additionally, BTDM can inhibit gain-of-function mutants of the hTRPC6 channel associated with focal segmental glomerulosclerosis (FSGS), such as P112Q, R895C, and E897K, with comparable potency [2]. In 2022, the cryo-EM structure of the TRPC6 channel bound to BTDM was resolved to a resolution of 2.9 Å [12]. The binding pose of BTDM within the channel was unambiguously identified (Fig. 1.4A). The binding pocket of BTDM is surrounded by residues Phe607, Ser608, and Ala611 on S4, as well as Gln624 and Leu627 on S5 from one subunit, and Ile640, Met643, Val644, and Phe648 on S5, as well as Thr714, Leu719, and Met722 on S6 from an adjacent subunit [12] (Fig. 1.4B).

BTDM consists of three chemical moieties: a benzodioxole group, a thiazole ring, and a dimethylpiperidine tail. The benzodioxole group forms hydrophilic contacts with Ser608 on S4 and Thr714 on S6 of the adjacent subunit (Fig. 1.4C). Furthermore, hydrophobic interactions between the benzodioxole group and Phe648 and Val644 contribute to the stabilization of the BTDM binding (Fig. 1.4C). The thiazole ring, situated in the middle, contacts the hydrophobic residues Ile610 and Met643 (Fig. 1.4C). Additionally, the carbonyl oxygen situated between the thiazole ring and dimethylpiperidine tail forms weak polar contacts with Gln624 on S5 (Fig. 1.4C). Despite several residues surrounding the dimethylpiperidine tail, including Leu627 and Val631 on S5 from one

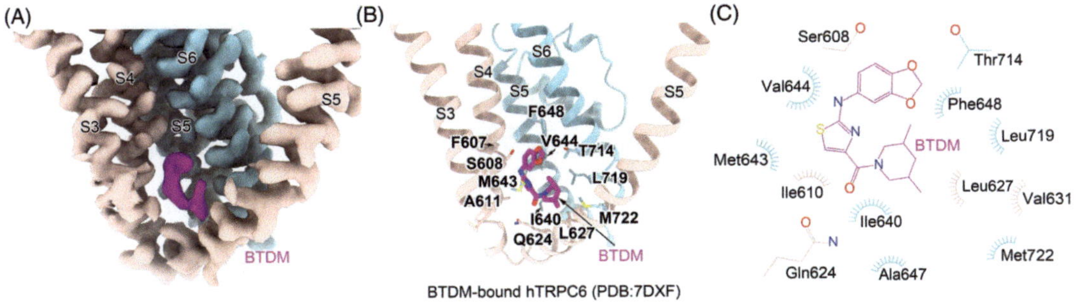

FIGURE 1.4 (A) Close-up view of the density map of the inhibitor BTDM-bound TRPC6 channel (PDB:7DXF). The density of BTDM is colored in magenta. One subunit of TRPC6 is colored in pink and the adjacent subunit is colored in blue. (B) Close-up view of the BTDM binding site in TRPC6 shown in cartoon. TRPC6 and BTDM are colored the same as (A). Residues of TRPC6 close to BTDM are shown as sticks and colored by elements. (C) The TRPC6-BTDM interactions analyzed using LigPlot. Hydrophobic interactions are indicated with spoked arcs, and side chains of hydrophilic residues are shown as sticks.

subunit, as well as Ile640, Ala647, Leu719, and Met722 from the adjacent subunit, they only establish weak hydrophobic contacts with the dimethylpiperidine tail [12] (Fig. 1.4C). This observation is consistent with the lack of density for the BTDM dimethylpiperidine tail in the BTDM-bound state TRPC6 map at lower resolution (3.8 Å) [2].

To evaluate the functional significance of these BTDM-interacting residues, they were subjected to mutational analysis. Through membrane potential FLIPR assays, it was observed that mutations such as S608L, V644F, and T714F dramatically reduced the affinity of BTDM [2]. This highlights the importance of the interactions between Ser608, Val644, and Thr714 residues with the benzodioxole group of BTDM. While only BTDM has been identified as binding to IBP-C in TRPC channels, it is worth mentioning that in TRPV1, other ligands have been found to bind to IBP-C. These include the vanilloid agonist resiniferatoxin, the vanilloid antagonist capsazepine, and an endogenous phosphoinositide [40, 41].

In TRPC4/5 channels, mutation of G503S/G504S (corresponding to Gly621 in TRPC6) on the S4-S5 linker results in constitutive channel opening [42]. Compared with other inhibitor-bound structures of TRPC6, the BTDM-bound TRPC6 has a π hinge around Ile717 in the S6 helix instead of a continuous helix. This structural difference may be attributed to the interactions between BTDM and residues Thr714, Leu719, and Met722 on S6. Importantly, the transition from a helix to a π hinge in the S6 helix has been shown to be essential for the gating of TRPV channels [43]. Collectively, these findings suggest that IBP-C plays a crucial role in the gating mechanism of TRP channels.

1.5 BINDING SITES OF IONS AND INTERACTING PROTEINS OF TRPC

1.5.1 Calcium and Calmodulin

The calcium ion is a universal second messenger [44] that modulates various ion channels, including TRPV, TRPA, TRPM, TRPP [45], and TRPC channels [46–57]. Intracellular calcium has been reported to regulate TRPC channels. Based on the high-resolution structures of TRPC3 in high-calcium and low-calcium conditions, three calcium binding sites (CBS1-3) were observed [12] (Fig. 1.5A-D). CBS1 is located between ankyrin repeat 2 (Ank 2) and CH2 and has two negatively charged residues. The calcium ion is surrounded by the main-chain carbonyl group and the side-chain carboxyl group of Glu73 on AR2, as well as the side-chain carboxyl group of Asp798 on CH2 (Fig. 1.5B). The mutation D798A in CBS1 significantly diminishes the inhibitory effect of intracellular calcium, suggesting that CBS1 plays an inhibitory role [12]. The residues of CBS1

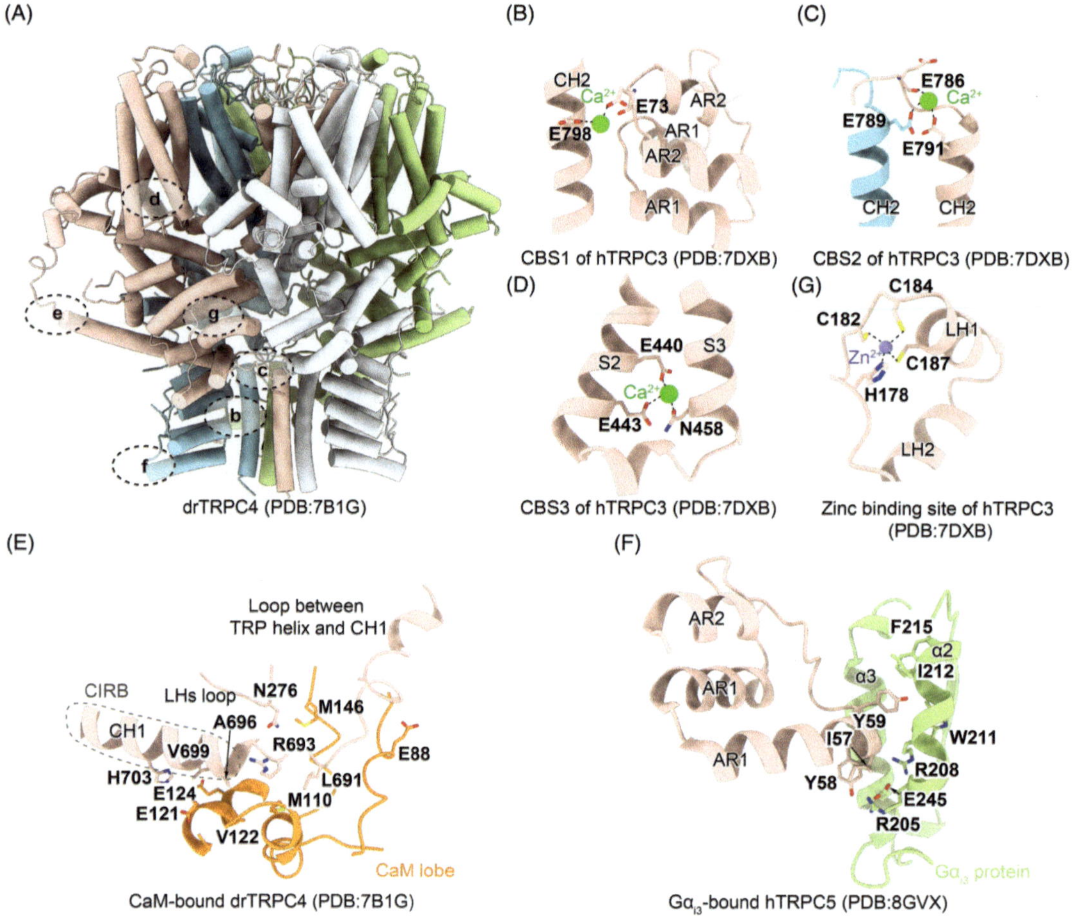

FIGURE 1.5 The binding sites of ions and interacting proteins of TRPC channels. (A) Overview of the binding sites for ions and interacting proteins in TRPC channels, exemplified by CaM-bound drTRPC4. The cartoon model of drTRPC4 in cylinders is shown and colored the same as in Fig. 1C. Ank1-3 of subunit A are omitted for better illustration. (B–D) Close-up view of the calcium-binding sites (CBS) 1-3 of hTRPC3 circled in (A). The calcium ion is colored in green, and the side chains of surrounding amino acids are shown as sticks and colored by elements. (E) Close-up view of the CaM binding region. CaM is colored in orange, and residues within 4 Å between CaM and TRPC4 are shown as sticks and colored by elements. The CIRB motif of TRPC4 is marked by a dashed line. LHs, linker helices. (F) Close-up view of $G\alpha_{i3}$ protein binding site in hTRPC5. $G\alpha_{i3}$ protein is colored in green, and interacting residues are shown as sticks and colored by elements. (G) Close-up view of the zinc binding site in hTRPC3. Zinc ion is colored in purple, and the side chains of surrounding amino acids are shown as sticks and colored by elements.

are conserved among drosophila TRP channels and human TRPC channels, highlighting its functional importance. CBS2 is located at the top of CH2 and is formed by the side chain of Glu791 on CH2 and the main chain of Glu786 on the linker between CH1 and CH2 of one subunit, plus the side chain of Glu789 of the adjacent subunit (Fig. 1.5C). However, when the key residue Glu789 was mutated to alanine, the mutant channel still exhibits robust calcium inhibition, suggesting that CBS2 may play a minor role in calcium regulation [12]. CBS3 is located within the VSLD of TMD, close to IBP-A, and is formed by the side chains of Glu440 and Glu443 on S2, as well as Asn458 on S3 (Fig. 1.5D). The mutation E440A in CBS3 abolishes the activation of the gain-of-function mutation R803C of TRPC3 by calcium in inside-out mode recording, confirming that CBS3 plays an activating role in calcium regulation [12]. This function of CBS3 is also conserved in TRPC5. The TRPC5 CBS3 triple mutant almost completely abolishes the activation by extracellular calcium [11]. CBS3 also plays a synergistic effect with riluzole on the activation of TRPC5 [13]. The functions of CBS1-3 are conserved between TRPC3 and TRPC6.

The conserved calmodulin (CaM)-inositol trisphosphate receptor (IP_3R)-binding (CIRB) motif has been reported to be critical during CaM-dependent calcium regulation among TRPC channels [58]. Under high calcium conditions, CaM competes with IP_3R for the same binding site to inhibit the channel. To understand the mechanism of CaM inhibition, the structure of TRPC4 in complex with CaM was solved to 3.3 Å resolution [9]. By improving the local resolution of the CaM lobe, four helices of CaM can be flexibly fitted. Based on the flexibly fitted atomic model, it was observed that CaM not only binds to the tip of the CH1 helix (residues 691–703 aa) and the loop between the TRP helix and CH1 helix (residues 677–690 aa) but is also close to the linker helices loop (273–277 aa) [9] (Fig. 1.5E). The interaction peptide observed in the structure overlaps with the previously reported CIRB motif of TRPC4 (residues 695–724 aa) [58]. Compared to the apo structure of TRPC4 in the same condition, the CaM-bound state of TRPC4 exhibits a pronounced, stably longer stretch between the TRP helix and CH1 (residues 677–692 aa). As a result, the binding of CaM to TRPC4 reduces the overall flexibility of the channel and stabilizes it in the closed state [9]. The observed CaM binding domain is not conserved in TRPC3/6/7 (residues 784–853 in hTRPC6), which is located in an unstructured region with no obvious density in all resolved TRPC3/6 structures.

1.5.2 $G_{i/o}$ Proteins

$G_{i/o}$ proteins also play an important role in the activation of TRPC4 and TRPC5. Nearly 20 years ago, researchers conducted patch-clamp recordings on guinea-pig single ileal smooth muscle cells and discovered that the muscarinic agonist, carbachol, could activate endogenous cation currents, representing the opening of cationic channels, which were later identified to be mainly composed of TRPC4 [59]. Remarkably, the muscarinic cation currents were dependent on the co-activation of both M2 and M3 muscarinic receptors [60] and inhibited by pertussis toxin [61]. Subsequent studies also demonstrated the $G_{i/o}$ protein dependence of heterologously expressed TRPC4 and TRPC5 channels, especially the involvement of the Gα subunits [62–65]. In order to understand the molecular mechanism of direct regulation of the channel by Gα proteins, the human TRPC5-$G\alpha_{i3}$ protein complex was resolved in 2023 [14]. By improving the local resolution at the interface between TRPC5 and $G\alpha_{i3}$, which is affected by the inherent flexibilities and various binding stoichiometries of the bound $G\alpha_{i3}$ proteins, a resolution of 4.2 Å was achieved [14]. The $G\alpha_{i3}$-bound TRPC5 structure reveals that each protomer of TRPC5 can bind one $G\alpha_{i3}$ protein [14], meaning that a tetrameric TRPC5 can bind a maximum of four $G\alpha_{i3}$ subunits.

The $G\alpha_{i3}$ proteins are bound to the edge of the ARD, and the binding interface is formed by the Ank1-2 of TRPC5 and the α2 and α3 helices of $G\alpha_{i3}$ (Fig. 1.5F). The Arg205 and Arg208 on the α2 helix of $G\alpha_{i3}$ interact with Ile57 of TRPC5, while the Arg205 and Glu245 of $G\alpha_{i3}$ form hydrogen bonds with Tyr58 of TRPC5. Additionally, Trp211, Ile212, and Phe215 on the α2 helix of $G\alpha_{i3}$ participate in hydrophobic interactions with Tyr59 of TRPC5 [14]. The importance of these three residues (Ile57, Tyr58, and Tyr59) was confirmed by the loss of function of the triple alanine mutation in inside-out recording and live-cell imaging experiments. The triple alanine mutant was unable to be activated by the constitutively active form of $G\alpha_{i3}$; however, its open probability was still increased by diC_8-PIP_2 [14]. It is worth noting that these three residues (IYY) are conserved in the corresponding region of TRPC4 (as IYF), but not in any other TRPC subtypes, providing an explanation why $G\alpha_i$ specifically activates TRPC4/5 channels.

1.5.3 Zinc Ion

A conserved zinc finger motif was observed in the ICD of both TRPC5 and TRPC3/6 channels [11, 12]. This motif consists of three cysteines and one histidine located in the Ank4-linker helix (LH1) region (Fig. 1.5G). The presence of zinc ions was detected during the purification of TRPC channels, indicating a high affinity for zinc. Interestingly, mutations in the zinc finger motif do not affect TRPC5 activation by calcium or Englerin A [11], suggesting that this zinc-binding site might play a structural role rather than directly influencing ion channel gating.

1.6 CONCLUSION AND FURTHER PERSPECTIVE

Thanks to the recent breakthroughs in cryo-electron microscopy (cryo-EM), the field has experienced a resolution revolution [66], exemplified by the pioneering work on the structure determination of TRPV1 [24, 40]. As a result, there has been an explosive increase in the high-resolution structures of various TRPC channel members, some bound to different ligands. These structures have offered never-before-seen, highly detailed information about TRPC channels, revealing insights into how the channels are formed, how they recognize ligands, and how they interact with regulatory proteins. Additionally, these structures serve as crucial near-atomic models for investigating a wide range of chemical compounds in the pursuit of drug discovery.

Despite the significant progress made, the solved TRPC structures are all in a closed state, which limits our understanding of the structural gating mechanism of TRPC channels. Like other channels, TRPC channels undergo transitions between a resting closed state, an activated open state, and a desensitized closed state. What does the structure of a TRPC channel look like in a specific gating state? How does the channel transition between these states? Although seemingly separate, these questions are intricately connected to the structural pharmacology of TRPC channels, which primarily focuses on identifying and characterizing ligand binding sites as well as conducting mechanistic studies related to ligand–channel interactions. Understanding the structural basis of TRPC channel gating and the conformational changes involved in different states is essential for unraveling the mechanisms underlying ligand recognition and the subsequent functional modulation of these channels. By bridging these areas of research, we can gain a comprehensive understanding of TRPC channel function and dysregulation, paving the way for the development of targeted therapeutics.

For one example, both inhibitors BTDM and SAR7334 have been found to stabilize TRPC6 in distinct closed structures, as aforementioned. However, it is currently unknown which structure represents the resting closed state and which represents the desensitized closed state. Obtaining this information is crucial for the design and optimization of channel inhibitors. In another example, riluzole serves as a TRPC5 activator, theoretically increasing the open probability of the channel. However, the structure of TRPC5 in complex with riluzole closely resembles the apo resting state, leaving the mechanism by which riluzole activates TRPC5 largely unknown. As a result, there is limited knowledge on how to design new activators. In a third example, $G\alpha_i$ is known to activate TRPC5 channels. Interestingly, the structures of TRPC5 channels remain unchanged with or without $G\alpha_i$. The mechanism by which $G\alpha_i$ activates TRPC5 remains elusive and requires further investigation. These examples highlight the need for comprehensive studies on the gating mechanisms of TRPC channels in the context of structural pharmacology. Gaining insights into these mechanisms will not only enhance our understanding of TRPC channel function but also enable the development of more effective inhibitors, activators, and therapeutic strategies.

ACKNOWLEDGMENT

We thank Michael X. Zhu for his helpful suggestions. This work is supported by grants from the National Natural Science Foundation of China (32225027 and 31821091 to L.C.) and the Center for

Life Sciences (CLS to L.C.). W.G. is supported by the Peking-Tsinghua Center for Life Sciences Postdoctoral Fellowship of Peking University.

REFERENCES

1. Feng, S. TRPC channel structure and properties [J]. *Adv Exp Med Biol*, 2017, 976: 9–23.
2. Tang, Q., Guo, W., Zheng, L., Wu, J. X., Liu, M., Zhou, X., Zhang, X., Chen, L. Structure of the receptor-activated human TRPC6 and TRPC3 ion channels [J]. *Cell Res*, 2018, 28: 746–755.
3. Fan, C., Choi, W., Sun, W., Du, J., Lü, W. Structure of the human lipid-gated cation channel TRPC3 [J]. *Elife*, 2018, 7: e36852.
4. Duan, J., Li, J., Zeng, B., Chen, G. L., Peng, X., Zhang, Y., Wang, J., Clapham, D. E., Li, Z., Zhang, J. Structure of the mouse TRPC4 ion channel [J]. *Nat Commun*, 2018, 9: 3102.
5. Vinayagam, D., Mager, T., Apelbaum, A., Bothe, A., Merino, F., Hofnagel, O., Gatsogiannis, C., Raunser, S. Electron cryo-microscopy structure of the canonical TRPC4 ion channel [J]. *Elife*, 2018, 7: e36615.
6. Duan, J., Li, J., Chen, G. L., Ge, Y., Liu, J., Xie, K., Peng, X., Zhou, W., Zhong, J., Zhang, Y., Xu, J., Xue, C., Liang, B., Zhu, L., Liu, W., Zhang, C., Tian, X. L., Wang, J., Clapham, D. E., Zeng, B., Li, Z., Zhang, J. Cryo-EM structure of TRPC5 at 2.8-Å resolution reveals unique and conserved structural elements essential for channel function [J]. *Sci Adv*, 2019, 5: eaaw7935.
7. Sierra-Valdez, F., Azumaya, C. M., Romero, L. O., Nakagawa, T., Cordero-Morales, J. F. Structure-function analyses of the ion channel TRPC3 reveal that its cytoplasmic domain allosterically modulates channel gating [J]. *J Biol Chem*, 2018, 293: 16102–16114.
8. Bai, Y., Yu, X., Chen, H., Horne, D., White, R., Wu, X., Lee, P., Gu, Y., Ghimire-Rijal, S., Lin, D. C., Huang, X. Structural basis for pharmacological modulation of the TRPC6 channel [J]. *Elife*, 2020, 9: e53311.
9. Vinayagam, D., Quentin, D., Yu-Strzelczyk, J., Sitsel, O., Merino, F., Stabrin, M., Hofnagel, O., Yu, M., Ledeboer, M. W., Nagel, G., Malojcic, G., Raunser, S. Structural basis of TRPC4 regulation by calmodulin and pharmacological agents [J]. *Elife*, 2020, 9: e60603.
10. Wright, D. J., Simmons, K. J., Johnson, R. M., Beech, D. J., Muench, S. P., Bon, R. S. Human TRPC5 structures reveal interaction of a xanthine-based TRPC1/4/5 inhibitor with a conserved lipid binding site [J]. *Commun Biol*, 2020, 3: 704.
11. Song, K., Wei, M., Guo, W., Quan, L., Kang, Y., Wu, J. X., Chen, L. Structural basis for human TRPC5 channel inhibition by two distinct inhibitors [J]. *Elife*, 2021, 10: e63429.
12. Guo, W., Tang, Q., Wei, M., Kang, Y., Wu, J. X., Chen, L. Structural mechanism of human TRPC3 and TRPC6 channel regulation by their intracellular calcium-binding sites [J]. *Neuron*, 2022, 110: 1023–1035.e1025.
13. Yang, Y., Wei, M., Chen, L. Structural identification of riluzole-binding site on human TRPC5 [J]. *Cell Discov*, 2022, 8: 67.
14. Won, J., Kim, J., Jeong, H., Kim, J., Feng, S., Jeong, B., Kwak, M., Ko, J., Im, W., So, I., Lee, H. H. Molecular architecture of the Gα(i)-bound TRPC5 ion channel [J]. *Nat Commun*, 2023, 14: 2550.
15. Guo, W., Chen, L. Recent progress in structural studies on canonical TRP ion channels [J]. *Cell Calcium*, 2019, 83: 102075.
16. Jin, P., Bulkley, D., Guo, Y., Zhang, W., Guo, Z., Huynh, W., Wu, S., Meltzer, S., Cheng, T., Jan, L. Y., Jan, Y. N., Cheng, Y. Electron cryo-microscopy structure of the mechanotransduction channel NOMPC [J]. *Nature*, 2017, 547: 118–122.
17. Huang, Y., Fliegert, R., Guse, A. H., Lü, W., Du, J. A structural overview of the ion channels of the TRPM family [J]. *Cell Calcium*, 2020, 85: 102111.
18. Ruan, Z., Haley, E., Orozco, I. J., Sabat, M., Myers, R., Roth, R., Du, J., Lü, W. Structures of the TRPM5 channel elucidate mechanisms of activation and inhibition [J]. *Nat Struct Mol Biol*, 2021, 28: 604–613.
19. Yu, X., Xie, Y., Zhang, X., Ma, C., Liu, L., Zhen, W., Xu, L., Zhang, J., Liang, Y., Zhao, L., Gao, X., Yu, P., Luo, J., Jiang, L. H., Nie, Y., Yang, F., Guo, J., Yang, W. Structural and functional basis of the selectivity filter as a gate in human TRPM2 channel [J]. *Cell Rep*, 2021, 37: 110025.
20. Yin, Y., Zhang, F., Feng, S., Butay, K. J., Borgnia, M. J., Im, W., Lee, S. Y. Activation mechanism of the mouse cold-sensing TRPM8 channel by cooling agonist and PIP(2) [J]. *Science*, 2022, 378: eadd1268.
21. Zhao, C., Mackinnon, R. Structural and functional analyses of a GPCR-inhibited ion channel TRPM3 [J]. *Neuron*, 2023, 111: 81–91.e87.

22. Nadezhdin, K. D., Correia, L., Narangoda, C., Patel, D. S., Neuberger, A., Gudermann, T., Kurnikova, M. G., Chubanov, V., Sobolevsky, A. I. Structural mechanisms of TRPM7 activation and inhibition [J]. *Nat Commun*, 2023, 14: 2639.
23. Paulsen, C. E., Armache, J. P., Gao, Y., Cheng, Y., Julius, D. Structure of the TRPA1 ion channel suggests regulatory mechanisms [J]. *Nature*, 2015, 520: 511–517.
24. Liao, M., Cao, E., Julius, D., Cheng, Y. Structure of the TRPV1 ion channel determined by electron cryo-microscopy [J]. *Nature*, 2013, 504: 107–112.
25. Maier, T., Follmann, M., Hessler, G., Kleemann, H. W., Hachtel, S., Fuchs, B., Weissmann, N., Linz, W., Schmidt, T., Löhn, M., Schroeter, K., Wang, L., Rütten, H., Strübing, C. Discovery and pharmacological characterization of a novel potent inhibitor of diacylglycerol-sensitive TRPC cation channels [J]. *Br J Pharmacol*, 2015, 172: 3650–3660.
26. Yu, M., Ledeboer, M. W., Daniels, M., Malojcic, G., Tibbitts, T. T., Coeffet-Le Gal, M., Pan-Zhou, X. R., Westerling-Bui, A., Beconi, M., Reilly, J. F., Mundel, P., Harmange, J. C. Discovery of a potent and selective TRPC5 inhibitor, efficacious in a focal segmental glomerulosclerosis model [J]. *ACS Med Chem Lett*, 2019, 10: 1579–1585.
27. Richter, J. M., Schaefer, M., Hill, K. Clemizole hydrochloride is a novel and potent inhibitor of transient receptor potential channel TRPC5 [J]. *Mol Pharmacol*, 2014, 86: 514–521.
28. Zhu, Y., Lu, Y., Qu, C., Miller, M., Tian, J., Thakur, D. P., Zhu, J., Deng, Z., Hu, X., Wu, M., Mcmanus, O. B., Li, M., Hong, X., Zhu, M. X., Luo, H. R. Identification and optimization of 2-aminobenzimidazole derivatives as novel inhibitors of TRPC4 and TRPC5 channels [J]. *Br J Pharmacol*, 2015, 172: 3495–3509.
29. Zhou, Y., Castonguay, P., Sidhom, E. H., Clark, A. R., Dvela-Levitt, M., Kim, S., Sieber, J., Wieder, N., Jung, J. Y., Andreeva, S., Reichardt, J., Dubois, F., Hoffmann, S. C., Basgen, J. M., Montesinos, M. S., Weins, A., Johnson, A. C., Lander, E. S., Garrett, M. R., Hopkins, C. R., Greka, A. A small-molecule inhibitor of TRPC5 ion channels suppresses progressive kidney disease in animal models [J]. *Science*, 2017, 358: 1332–1336.
30. Miller, R. G., Mitchell, J. D., Moore, D. H. Riluzole for amyotrophic lateral sclerosis (ALS)/motor neuron disease (MND) [J]. *Cochrane Database Syst Rev*, 2012, 2012: Cd001447.
31. Zarate, C. A., Manji, H. K. Riluzole in psychiatry: A systematic review of the literature [J]. *Expert Opin Drug Metab Toxicol*, 2008, 4: 1223–1234.
32. Grant, P., Song, J. Y., Swedo, S. E. Review of the use of the glutamate antagonist riluzole in psychiatric disorders and a description of recent use in childhood obsessive-compulsive disorder [J]. *J Child Adolesc Psychopharmacol*, 2010, 20: 309–315.
33. Bellingham, M. C. A review of the neural mechanisms of action and clinical efficiency of riluzole in treating amyotrophic lateral sclerosis: What have we learned in the last decade? [J]. *CNS Neurosci Ther*, 2011, 17: 4–31.
34. Richter, J. M., Schaefer, M., Hill, K. Riluzole activates TRPC5 channels independently of PLC activity [J]. *Br J Pharmacol*, 2014, 171: 158–170.
35. Lichtenegger, M., Tiapko, O., Svobodova, B., Stockner, T., Glasnov, T. N., Schreibmayer, W., Platzer, D., De La Cruz, G. G., Krenn, S., Schober, R., Shrestha, N., Schindl, R., Romanin, C., Groschner, K. An optically controlled probe identifies lipid-gating fenestrations within the TRPC3 channel [J]. *Nat Chem Biol*, 2018, 14: 396–404.
36. Svobodova, B., Lichtenegger, M., Platzer, D., Di Giuro, C. M. L., De La Cruz, G. G., Glasnov, T., Schreibmayer, W., Groschner, K. A single point mutation in the TRPC3 lipid-recognition window generates supersensitivity to benzimidazole channel activators [J]. *Cell Calcium*, 2019, 79: 27–34.
37. Just, S., Chenard, B. L., Ceci, A., Strassmaier, T., Chong, J. A., Blair, N. T., Gallaschun, R. J., Del Camino, D., Cantin, S., D'amours, M., Eickmeier, C., Fanger, C. M., Hecker, C., Hessler, D. P., Hengerer, B., Kroker, K. S., Malekiani, S., Mihalek, R., Mclaughlin, J., Rast, G., Witek, J., Sauer, A., Pryce, C. R., Moran, M. M. Treatment with HC-070, a potent inhibitor of TRPC4 and TRPC5, leads to anxiolytic and antidepressant effects in mice [J]. *PLoS ONE*, 2018, 13: e0191225.
38. Rubaiy, H. N., Ludlow, M. J., Henrot, M., Gaunt, H. J., Miteva, K., Cheung, S. Y., Tanahashi, Y., Hamzah, N., Musialowski, K. E., Blythe, N. M., Appleby, H. L., Bailey, M. A., Mckeown, L., Taylor, R., Foster, R., Waldmann, H., Nussbaumer, P., Christmann, M., Bon, R. S., Muraki, K., Beech, D. J. Picomolar, selective, and subtype-specific small-molecule inhibition of TRPC1/4/5 channels [J]. *J Biol Chem*, 2017, 292: 8158–8173.
39. Minard, A., Bauer, C. C., Chuntharpursat-Bon, E., Pickles, I. B., Wright, D. J., Ludlow, M. J., Burnham, M. P., Warriner, S. L., Beech, D. J., Muraki, K., Bon, R. S. Potent, selective, and subunit-dependent activation of TRPC5 channels by a xanthine derivative [J]. *Br J Pharmacol*, 2019, 176: 3924–3938.

40. Cao, E., Liao, M., Cheng, Y., Julius, D. TRPV1 structures in distinct conformations reveal activation mechanisms [J]. *Nature*, 2013, 504: 113–118.
41. Gao, Y., Cao, E., Julius, D., Cheng, Y. TRPV1 structures in nanodiscs reveal mechanisms of ligand and lipid action [J]. *Nature*, 2016, 534: 347–351.
42. Beck, A., Speicher, T., Stoerger, C., Sell, T., Dettmer, V., Jusoh, S. A., Abdulmughni, A., Cavalié, A., Philipp, S. E., Zhu, M. X., Helms, V., Wissenbach, U., Flockerzi, V. Conserved gating elements in TRPC4 and TRPC5 channels [J]. *J Biol Chem*, 2013, 288: 19471–19483.
43. Zubcevic, L., Lee, S. Y. The role of π-helices in TRP channel gating [J]. *Curr Opin Struct Biol*, 2019, 58: 314–323.
44. Clapham, D. E. Calcium signaling [J]. *Cell*, 2007, 131: 1047–1058.
45. Hasan, R., Zhang, X. Ca(2+) regulation of TRP ion channels [J]. *Int J Mol Sci*, 2018, 19: 1256.
46. Kwon, Y., Hofmann, T., Montell, C. Integration of phosphoinositide- and calmodulin-mediated regulation of TRPC6 [J]. *Mol Cell*, 2007, 25: 491–503.
47. Zitt, C., Obukhov, A. G., Strübing, C., Zobel, A., Kalkbrenner, F., Lückhoff, A., Schultz, G. Expression of TRPC3 in Chinese hamster ovary cells results in calcium-activated cation currents not related to store depletion [J]. *J Cell Biol*, 1997, 138: 1333–1341.
48. Boulay, G., Brown, D. M., Qin, N., Jiang, M., Dietrich, A., Zhu, M. X., Chen, Z., Birnbaumer, M., Mikoshiba, K., Birnbaumer, L. Modulation of Ca(2+) entry by polypeptides of the inositol 1,4, 5-trisphosphate receptor (IP3R) that bind transient receptor potential (TRP): Evidence for roles of TRP and IP3R in store depletion-activated Ca(2+) entry [J]. *Proc Natl Acad Sci U S A*, 1999, 96: 14955–14960.
49. Kamouchi, M., Philipp, S., Flockerzi, V., Wissenbach, U., Mamin, A., Raeymaekers, L., Eggermont, J., Droogmans, G., Nilius, B. Properties of heterologously expressed hTRP3 channels in bovine pulmonary artery endothelial cells [J]. *J Physiol*, 1999, 518(Pt 2): 345–358.
50. Lintschinger, B., Balzer-Geldsetzer, M., Baskaran, T., Graier, W. F., Romanin, C., Zhu, M. X., Groschner, K. Coassembly of Trp1 and Trp3 proteins generates diacylglycerol- and Ca^{2+}-sensitive cation channels [J]. *J Biol Chem*, 2000, 275: 27799–27805.
51. Zhang, Z., Tang, J., Tikunova, S., Johnson, J. D., Chen, Z., Qin, N., Dietrich, A., Stefani, E., Birnbaumer, L., Zhu, M. X. Activation of Trp3 by inositol 1,4,5-trisphosphate receptors through displacement of inhibitory calmodulin from a common binding domain [J]. *Proc Natl Acad Sci U S A*, 2001, 98: 3168–3173.
52. Boulay, G. Ca^{2+}-calmodulin regulates receptor-operated Ca(2+) entry activity of TRPC6 in HEK-293 cells [J]. *Cell Calcium*, 2002, 32: 201–207.
53. Shi, J., Mori, E., Mori, Y., Mori, M., Li, J., Ito, Y., Inoue, R. Multiple regulation by calcium of murine homologues of transient receptor potential proteins TRPC6 and TRPC7 expressed in HEK293 cells [J]. *J Physiol*, 2004, 561: 415–432.
54. Polat, O. K., Uno, M., Maruyama, T., Tran, H. N., Imamura, K., Wong, C. F., Sakaguchi, R., Ariyoshi, M., Itsuki, K., Ichikawa, J., Morii, T., Shirakawa, M., Inoue, R., Asanuma, K., Reiser, J., Tochio, H., Mori, Y., Mori, M. X. Contribution of coiled-coil assembly to Ca(2+)/Calmodulin-dependent inactivation of TRPC6 channel and its impacts on FSGS-associated phenotypes [J]. *J Am Soc Nephrol*, 2019, 30: 1587–1603.
55. Friedlova, E., Grycova, L., Holakovska, B., Silhan, J., Janouskova, H., Sulc, M., Obsilova, V., Obsil, T., Teisinger, J. The interactions of the c-terminal region of the TRPC6 channel with calmodulin [J]. *Neurochem Int*, 2010, 56: 363–366.
56. Schaefer, M., Plant, T. D., Obukhov, A. G., Hofmann, T., Gudermann, T., Schultz, G. Receptor-mediated regulation of the nonselective cation channels TRPC4 and TRPC5 [J]. *J Biol Chem*, 2000, 275: 17517–17526.
57. Jung, S., Mühle, A., Schaefer, M., Strotmann, R., Schultz, G., Plant, T. D. Lanthanides potentiate TRPC5 currents by an action at extracellular sites close to the pore mouth [J]. *J Biol Chem*, 2003, 278: 3562–3571.
58. Tang, J., Lin, Y., Zhang, Z., Tikunova, S., Birnbaumer, L., Zhu, M. X. Identification of common binding sites for calmodulin and inositol 1,4,5-trisphosphate receptors on the carboxyl termini of TRP channels [J]. *J Biol Chem*, 2001, 276: 21303–21310.
59. Zholos, A. V., Zholos, A. A., Bolton, T. B. G-protein-gated TRP-like cationic channel activated by muscarinic receptors: Effect of potential on single-channel gating [J]. *J Gen Physiol*, 2004, 123: 581–598.
60. Bolton, T. B., Zholos, A. V. Activation of M2 muscarinic receptors in guinea-pig ileum opens cationic channels modulated by M3 muscarinic receptors [J]. *Life Sci*, 1997, 60: 1121–1128.
61. Pucovský, V., Zholos, A. V., Bolton, T. B. Muscarinic cation current and suppression of Ca^{2+} current in guinea pig ileal smooth muscle cells [J]. *Eur J Pharmacol*, 1998, 346: 323–330.

62. Tabata, H., Tanaka, S., Sugimoto, Y., Kanki, H., Kaneko, S., Ichikawa, A. Possible coupling of prostaglandin E receptor EP(1) to TRP5 expressed in Xenopus laevis oocytes [J]. *Biochem Biophys Res Commun*, 2002, 298: 398–402.
63. Jeon, J. P., Hong, C., Park, E. J., Jeon, J. H., Cho, N. H., Kim, I. G., Choe, H., Muallem, S., Kim, H. J., So, I. Selective Gαi subunits as novel direct activators of transient receptor potential canonical (TRPC)4 and TRPC5 channels [J]. *J Biol Chem*, 2012, 287: 17029–17039.
64. Thakur, D. P., Tian, J. B., Jeon, J., Xiong, J., Huang, Y., Flockerzi, V., Zhu, M. X. Critical roles of Gi/o proteins and phospholipase C-δ1 in the activation of receptor-operated TRPC4 channels [J]. *Proc Natl Acad Sci U S A*, 2016, 113: 1092–1097.
65. Jeon, J., Tian, J. B., Zhu, M. X. TRPC4 as a coincident detector of G(i/o) and G(q/11) signaling: Mechanisms and pathophysiological implications [J]. *Curr Opin Physiol*, 2020, 17: 34–41.
66. Kühlbrandt, W. Biochemistry. The resolution revolution [J]. *Science*, 2014, 343: 1443–1444.

2 Characterizing the Structures and Regulation of TRPV Channels Using Cryo-EM

Kaihua Zhang

2.1 INTRODUCTION

Transient receptor potential (TRP) channels play crucial roles in detecting noxious and innocuous stimuli as polymodal sensors and mediating the flux of monovalent, divalent and even large organic cations across biological membranes. These channels are tightly linked to a plethora of genetic and acquired diseases and appreciated as therapeutic targets (TRP channelopathies).[1–3] Of six mammalian TRP subfamilies, the vanilloid group (TRPV) is involved in numerous biological processes such as thermosensation, osmosensation and calcium homeostasis. The name "Vanilloid" originates from the first member of the subfamily Vanilloid Receptor 1.[4] The TRPV subfamily can be further subdivided into 6 isoforms, TRPV1-6. Within this subfamily, TRPV1-4 are categorized as thermoTRPV channels with varied temperature activation profiles, and their sequence identity is 40–50%.[5] TRPV5 and TRPV6 have evolved into highly selective ion channels for calcium and share approximately 75% sequence homology with each other but only ~20% identity with thermoTRPV channels.[6,7]

The founding member TRPV1 is highly expressed in small/medium-sized dorsal root ganglion (DRG) and trigeminal neurons and pertinent to thermal hyperalgesia during inflammation or after tissue injury.[8–10] TRPV1 channel senses changes in ambient temperature characterized by a thermal activation threshold of ~43°C and a steep temperature coefficient ($Q_{10} > 20$),[4] and also responds to a rich array of natural products such as capsaicin from hot chili pepper and tarantula peptide toxin DkTx, synthetic chemicals, extracellular protons, and endogenous molecules. Intriguingly, functional studies have indicated that heat- and capsaicin-dependent activation pathways are separate but allosterically linked, since mutations specifically silence one modality over the other.[11–13] DkTx and a vanilloid agonist (either resiniferatoxin (RTX) from *Euphorbia* cactus or capsaicin) can bind synergistically or independently to TRPV1 by targeting the outer pore region and vanilloid-binding pocket in the vicinity of the lower gate region, respectively.[14,15] Indeed, coapplication of the two agonists has been successfully conducted to capture the fully open state of the channel.[16–18]

TRPV2 is expressed in but not restricted to medium and large diameter Aδ DRG neurons. The TRPV2 channel is activated at temperatures greater than 52°C, and it exhibits use dependence and is insensitive to vanilloids.[19,20] Nevertheless, the study of TRPV2-deficient mice raised doubts about the involvement of TRPV2 in thermal nociception and heat hyperalgesia *in vivo*.[21] TRPV2 has also been implicated in maintaining physiological cardiomyocyte function, participating in early phagocytosis and influencing cancer progression.[22,23] However, the pharmacological profile of TRPV2 is less understood compared to other thermoTRP channels, albeit the presence of some non-selective modulators, exemplified by cannabinoids and 2-aminoethoxydiphenyl borate (2-APB) as exogenous activators.[24,25]

TRPV3 is mainly expressed in keratinocytes and has a lower temperature threshold (~33°C) than TRPV1 and TRPV2.[26] The TRPV3 channel is sensitized by repeated heat stimulation, mediates larger initial currents elicited by rapid heating than by slow heating, and displays steep temperature dependence above the activation threshold.[27,28] The physiologic functions of TRPV3 in keratinocytes

DOI: 10.1201/9781003402732-2

and its significance in thermosensation[29,30] are indicative of the important roles of TRPV3 as a peripheral sensor of both noxious and innocuous temperatures. In addition to maintenance of the skin barrier and thermal sensation, TRPV3 has also been implicated in itch, wound healing, and hair growth.[31–33] As a pharmacological target, TRPV3 is capsaicin-insensitive but responds to a large number of chemical compounds, such as camphor, carvacrol and 2-APB.

TRPV4 has a threshold temperature similar to TRPV3 (>27°C).[34] As opposed to TRPV3, TRPV4 desensitizes either at temperatures outside the physiological range (>40°C) or in response to repetitive heat stimulation.[35,36] Of note, heat-induced activation of TRPV4 may occur downstream of intracellular signaling cascades or require other molecular interactions.[36–40] Whereas TRPV4 does not seem to play a role in noxious heat sensation,[41,42] it has been implicated in the calcium-dependent osmotic and mechanical responses in various cell types. Moreover, warm temperatures can sensitize the activation of TRPV4 by other stimuli, such as hypotonic stress, 4α-PDD, and shear stress.[43–45] Due to its biological versatility, TRPV4 has been recognized as a valuable therapeutical target under numerous pathophysiological conditions, and its inhibitors have been advanced to clinical trials.[46–49]

TRPV5 and TRPV6 are found on the apical membranes of epithelial cells in the kidney (TRPV5) or that in the colon, prostate, pancreas and other epithelial tissues (TRPV6).[6,50] These channels are uniquely selective for Ca^{2+} (permeability ratio $P_{Ca}/P_{Na} > 100$) and indispensably associated with calcium homeostasis and calcium-related disease development. For instance, TRPV6 governs both intestinal absorption and renal reabsorption of Ca^{2+},[51] and its expression levels are elevated in noninvasive tumor areas.[52] TRPV5 and TRPV6 are constitutively active in the cellular membrane even in the absence of any modulators.[53,54] Studies using purified channels incorporated into lipid bilayers further revealed the important role of phosphatidylinositol 4,5-bisphosphate ($PI(4,5)P_2$) in maintaining the constitutive activity of TRPV5 and TRPV6.[55,56]

Prompted by the cryo-EM resolution revolution, an exponentially increasing number of high-solution structures, covering every TRPV subfamily member, have been solved following the first high-resolution structures of TRPV1 determined by the Cheng & Julius group in 2013.[16,57] The focus of this review will be on the structures of the TRPV subfamily rather than all members of TRP channels and the mechanistic insights they reveal, including ligand binding, ion selectivity, architectural tetramerization, the roles of channel termini and so forth.

2.2 THE NATURE OF THE SELECTIVITY FILTER

To date, as revealed by independent studies, there exit multiple structures from disparate species or solved in nonidentical hydrophobic environments (nanodiscs, amphipols or detergents) corresponding to a homogeneous conformation of a single member within the TRPV subfamily. Undoubtedly, these structural evidences emerge to advance our understanding of whether the selectivity filter universally serves as a restriction site across TRPV channels and how the configuration of the selectivity filter correlates with ion selectivity and channel gating.

Divergent from voltage-gated ion channels (VGICs), TRPV channels architecturally feature a short selectivity filter wherein backbone carbonyls of glycine or side chains of aspartate from each subunit point into the central pathway below which the pore widens into a large hydrophobic cavity (Fig. 2.1). The selectivity filter is consistently restricted throughout closed structures of TRPV1 deposited in the Protein Data Bank but remarkably wide in closed structures of both human and *Xenopus tropicalis* TRPV4.[58–60] Like TRPV1, closed structures of rabbit and mouse TRPV2 reveal the narrow configuration of the selectivity filter, whichever system (amphipols or nanodiscs) was employed to incorporate the channels.[61,62] Nevertheless, an enlarged selectivity filter in TRPV2 has been observed in other closed structures, including one in a two-fold symmetric conformation determined by crystallizing an engineered rabbit construct[63] and the other for a full-length rat channel in nanodiscs, wherein the residue methionine rotated away from the pore and the carbonyl of the glycine residue shifted away by 5 Å.[64] In contrast to the opened selectivity filter, the same dataset of rat TRPV2 yielded the second type of apo structure representing a restricted selectivity

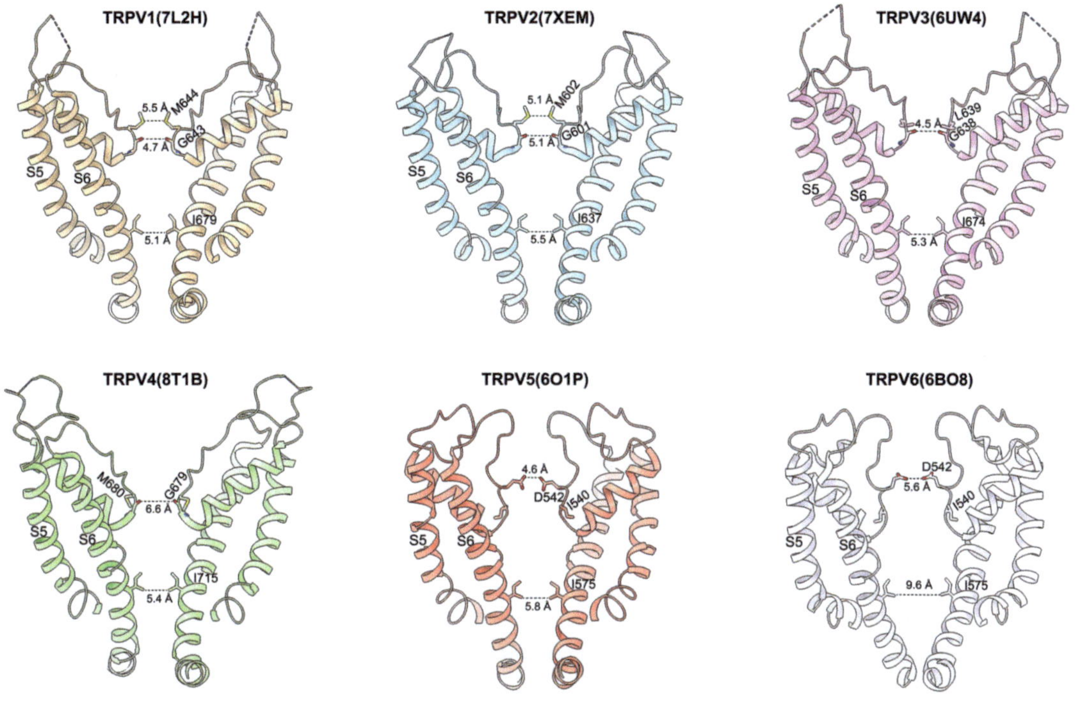

FIGURE 2.1 Representative structures of TRPV channels in apo state. With the residues lining the pore as indicated, the constriction sites are highlighted by measuring the distances of two diagonal residues at the selectivity filter and lower gate. Only two subunits are shown for clarity.

filter. These different observations highlight the conformational variability of the selectivity filter in TRPV2, which may be related to a certain level of constitutive activity of the channel or reflect the discrepancy in sample preparation or the existence of multiple apo states.

Narrow selectivity filter has also been seen in the closed structures of human and mouse TRPV3 in lipid nanodiscs,[65,66] as well as multiple sensitized states of mouse TRPV3 in detergents.[67] However, other apo structures of TRPV3 in amphipols or detergents indicate that the selectivity filter is adequately wide for ion conduction.[68,69] Given the roles of lipids as indispensable structural and functional components for numerous integral membrane proteins,[17,70] the selectivity filter is more likely to favor a narrow configuration when the channel is inactivated due to the tight association of lipids with TRPV3 *in vivo,* as revealed by the lipid-like densities surrounding the selectivity filter in the cryo-EM structures in lipid nanodiscs.[65]

To minimize the effect of separate systems used for sample preparation, the nature of the selectivity filter is analyzed by purposedly focusing on the structures of TRPV members determined in nanodiscs, which mimic the native membrane environment. By measuring the distance between centers of opposing oxygen atoms at the narrowest point along the permeation pathway, it was found that the diameters of the selectivity filters vary even when the closely related TRPV channels are in apo state (Fig. 2.1). The selectivity filters of TRPV1-3 have the diameter in the range of 4.5–5.1 Å at the narrowest point defined by the glycine residue, while that in TRPV4 is exceptionally expanded by ~2 Å. Meanwhile, in both TRPV1 and TRPV2, side chains of the methionine residues at the selectivity filter protrude into the central axis, forming a narrow and hydrophobic seal. In contrast to the configuration of the selectivity filter in TRPV1 and TRPV2, the leucine residues in TRPV3 which define the selectivity filter point away from the central pathway in closed structures, suggesting that the selectivity filter of TRPV3 may allow the passage of dehydrated or partially dehydrated ions in the apo state.

Albeit these structural observations, thiol-reactive ions are shown to pass through the selectivity filters of TRPV1-3 channels in the absence of activating stimuli,[71] reminiscent of the flickery behaviors observed during activation of these channels.[72,73] Though all apo structures of TRPV1 reported so far reveal the restricted configuration of the selectivity filter, the transition between closed and open states has been seen at the selectivity filter according to the structures of TRPV1 in complex with a vanilloid agonist RTX or peptide toxin DkTx.[18,74] The structurally dynamic selectivity filters in TRPV1-3 may not function as activation gates, but they enable external stimuli to allosterically cause the opening of the lower gate and allow large organic cations, such as the large monovalent cation NMDG (195 Da) and the sodium channel blocker QX-314 (263 Da), to move through the channel pores of at least TRPV1 and TRPV2.[75,76] Indeed, a host of mutagenesis studies suggest that the outer pore and pore helix domains of TRPV1 are critical for gating, wherein extracellular protons and toxin target to exert allosteric regulatory effects.[11,77,78] Moreover, a couple of mutations selectively abolishing thermal activation and single residues undergoing temperature-dependent conformational change map to the outer pore regions of TRPV1 and TRPV3.[79,80] In agreement with these functional results, cryo-EM structures reveal the rearrangement of the interactions between the turret (the loop between S5 and the pore helix) and the pore loop (the loop between the selectivity filter and S6) required for channel opening. Conversely, no physiological stimuli have been reported to activate the TRPV4 channel by targeting the selectivity filter.

By comparison, the selectivity filters of TRPV5 and TRPV6 bear different sequences (IDA/GP versus GM/LGD/E of TRPV1-4), whose conformation is relatively stationary between the inactivated state and open state. Different from that of TRPV1-4, the narrowest point at the selectivity filter of TRPV5/6 diverges to aspartate, whose side chains directly coordinate cations in a tight fashion given the typical Ca^{2+}-oxygen distance of ~2.4 Å (Fig. 2.1), conferring a much higher selectivity for calcium over monovalent cations ($P_{Ca}/P_{Na} > 100$) than other TRPV subtypes ($P_{Ca}/P_{Na} \approx 3\text{–}10$).[81–83]

Aside from the pore diameter correlating with the flux of ions, the knowledge about the detailed coordination of cations in the selectivity filter is indispensable for understanding the channel permeability. Anomalous X-ray diffraction signals are commonly used to identify the binding of the permeant ions in the pore. Until recently, the cryo-EM density maps played an increasingly important role in this regard. Intriguingly, high cryo-EM densities are present in the pore in apo structures of TRPV1, presumably representing ions stabilized in the selectivity filter. With the dilated diameter of the selectivity filter induced by DkTx binding, the densities appear to progressively elongate along the pore axis and eventually split into two main peaks,[18] reflecting the ion movement along the dynamic selectivity filter. According to the crystal structure of rabbit TRPV2, one putative Ca^{2+} is present in the selectivity filter coordinated by the residues located at C termini of pore helices, below which the other Ca^{2+} is observed in the cavity.[63] Similarly, the pore at the selectivity filter harbors a cation density in the apo state of TRPV3.[65] In the crystal and cryo-EM structures of *Xenopus tropicalis* TRPV4, the cation-binding site is located on the pore axis near to the carbonyls of G675, and the diameter of the selectivity filter is large enough to accommodate hydrated ions at this site.[58] In the case of TRPV6 with exceptional Ca^{2+} selectivity, the anomalous difference peaks in its crystal structures are indicative of the presence of different cation-binding sites along the central pore axis, including the position near the same plane as D541 side chains and another two sites below, supporting a "knock-off" mechanism proposed for calcium permeation by TRPV6.[83,84] The homologous site for Ca^{2+} binding in the position at D541 has also been observed in the cryo-EM structure of TRPV6 in open state,[82] supporting the important role of this site in ion permeation.

2.3 THE ROLE OF THE S1–S4 DOMAIN

TRP channels are structurally homologous to VGICs in regard to their overall transmembrane topology and tetrameric subunit composition, including the S1–S4 domain functioning as a voltage sensor in VGICs versus the voltage-sensing-like domain in the TRP channels, a short helical linker between S4 and S5 termed S4–S5 linker, and TRP domain parallel to the membrane and wedging into the intracellular bottom of S1–S4 bundle. Intriguingly, a more recent study revealed a pentameric state of TRPV3 by combing high-speed atomic force microscopy and cryo-EM, which highlights the membrane diffusion-dependent equilibrium between a rare pentamer and the canonical tetramer.[85]

As revealed in the cryo-EM reconstructions of TRPV channels, the S1–S4 domain of one subunit is packed against the pore domain of the adjacent subunit, creating a characteristic domain-swapped organization and highly electronegative extracellular vestibule. Notably, an extended extracellular linker between S1 and S2 is uniquely present in TRPV5 and TRPV6, which interacts with the pore region of the adjacent subunit and contains an N-glycosylation site. Such interactions may constrain the flexibility of the selectivity filter, thereby enabling the visualization of the turret in cryo-EM density maps of TRPV5 and TRPV6. Though a non-swapped architecture was initially observed in a model of TRPV6 construct resolved by crystallization ($TRPV6_{cryst}$),[83] the follow-up study by the same group showed that the non-swapped arrangement of $TRPV6_{cryst}$ resulted from the mutation of residue L495 to Q, which projects its side chain into a hydrophobic pocket formed by residues of S5 and S6 of one subunit and S4–S5 elbow of the adjacent subunit.[84] In the structures of the truncated *Xenopus tropicalis* TRPV4 channel carrying the mutation N647Q, the architectural divergences from other TRPV channels and even human TRPV4 include non-swapped pore domains, the movement of S2 towards the vanilloid binding pocket, and an ordered loop structure of the S4–S5 linker rather than a more common α-helix,[58–60] which may reflect the inter-species differences.

In VGICs, the extensive side-chain interactions between S3 and S4 helices contribute to a charge-transfer center that facilitates the movement of S4 in response to voltage changes of the cell membrane, and the S4–S5 linker plays a critical role in coupling the movement of the S1–S4 domain to gating of the pore. Instead of positive charges in S4 of VGICs, TRPV channels are characterized by hydrophobic interactions formed by aromatic side chains in S1–S4, which appear to rigidify the helical bundle conformation. In TRPV1, the superimposition between apo and activated structures by capsaicin and DkTx/RTX, respectively, reveals that the S1-S4 domain remains relatively static as a rigid bundle,[16,17] which is commonly seen in other TRPV channels and indicative of more active roles of the S4–S5 linker, S5-P-S6 pore and TRP domain during channel gating, which is consistent with the presence of gain-of-function mutations within these regions rather than within the S1–S4 domain.[77,86,87] Nevertheless, the binding of DkTx alone to the outer pore region progressively induces a rigid body-like tilting and twisting of the S1–S4 domain in TRPV1, which allosterically leads to the movement of S6 and the opening of the lower gate.[18]

It has also been reported that the binding by an agonist 2-APB to the top of the S1-S4 domain led to the opening of TRPV3 by squeezing out the lipid located at the interface between the S1–S4 domain and pore domain.[69] In addition, agonist-dependent TRPV4 gating took place after the binding by a ligand to the cavity between the S1–S4 domain and the TRP domain, which was followed by a rearrangement of the interaction network among the S1–S4 domain, TRP domain, and coupling domain. These observations highlight not only the existence of distinct gating mechanisms mediated by the S1–S4 domain between TRP channels and VGICs but also the involvement of the S1–S4 domain in channel gating in diverse manners.

2.4 THE MODULATORS

In the realm of ion channel research, a wide range of modulators, including agonists and antagonists, appear to influence channel functions by interacting with specific binding sites that are finely tuned for modulation of a single modality. These interactions induce conformational changes that are linked to processes such as sensitization, activation, or desensitization. The ligand-binding sites, including the vanilloid-binding pocket in TRPV channels, represent important loci for modulation by direct-acting and allosteric molecules, and thereby are of great clinical and therapeutic relevance. An increasing number of TRP structures obtained in complex with modulators provide valuable insights into the molecular mechanism underlying the ligand-mediated modulation of channel gating. Additionally, these structures serve as templates for the rational design of next-generation drugs (Table 2.1).

TABLE 2.1
List for the Available Structures of TRPV Channels in Complex with Modulators

	Name of Modulators	Type of Modulators	Binding Sites	References
TRPV1	Capsaicin, RTX, DkTx, Proton, Capsazepine, SB-366791	Agonism or antagonism	Vanilloid-binding pocket; Outer pore region	16–18,74,88–90
TRPV2	RTX, Cholesterol, 2-APB, Piperlongumine, Cannabidiol, Δ^9-tetrahydrocannabiorcol (C16), Probenecid	Agonism or antagonism	Vanilloid-binding pocket; Pocket formed by S4 and the S5 and S6 helices of an adjacent protomer; Hydrophobic pocket located between S5 and S6 of an adjacent subunit; The interface of S5 and the S4-S5 linker of the adjacent protomer; Putative/ambiguous binding site	62–64,91–94
TRPV3	2-APB, Dyclonine, Osthole	Agonism or antagonism	Intracellular half of the S1–S4 domain and the C-terminal portion of the TRP domain; The nexus of the linker domain, pre-S1 helix, and TRP domain; Extracellular portions of helices S1-S4; Side portals connecting the membrane environment to the pore's central cavity	65,68,69,95–97
TRPV4	GSK2798745, GSK1016790A, 4α-PDD, HC-067047	Agonism or antagonism	The cavity between the S1-S4 domain and the TRP domain	59,60
TRPV5	Econazole, diC$_8$ PI(4,5)P$_2$, ZINC9155420, ZINC17988990, Calmodulin (CaM)	Agonism or antagonism	Vanilloid-binding pocket; Binding site formed by N-linker, S4-S5 linker, and S6; The interface between the S4-S5 linker and S6 of an adjacent protomer; The intracellular half of the S1-S4 bundle; Intracellular entrance	98–102
TRPV6	Activating lipid, cis-22a, Genistein, Econazole, Tetrahydrocannabivarin (THCV), 2-APB, Ruthenium red (RR), CaM	Agonism or antagonism	Vanilloid-binding pocket; The interface between S5 and S4 of an adjacent subunit; Side portals connecting the membrane environment to the pore's central cavity; Intracellular half of the S1-S4 domain and the C-terminal portion of the TRP domain; Extracellular pore entrance; Intracellular entrance	82,103–109

As far as TRPV subtypes are concerned, it is well known that TRPV1 accommodates vanilloid agonists or chemical antagonists in the vanilloid-binding pocket, such as RTX, capsaicin and capsazepine. However, it remains to be determined how the less-studied modulators act on TRPV1, such as the small synthetic compound 2-APB and the lipid PI(4,5)P$_2$. Except for one member of this subfamily, TRPV1, exhibiting vanilloid-dependent modulation, the other TRPV channels fail to respond to vanilloids. Nevertheless, a TRPV2 mutant becomes responsive to vanilloid compounds after modification of four critical residues located in the vanilloid-binding pocket.[63,91,110,111] Additionally, TRPV2 is activated by cannabinoids and inhibited by PI(4,5)P_2. TRPV3 is activated by camphor, carvacrol, eugenol, and thymol, as well as by 2-APB, while inhibited by PI(4,5)P_2, dyclonine, FTP-THQ and GRC 15300. On the contrary, PI(4,5)P_2 facilitates the activation of TRPV4, and it effectively activates TRPV5 and TRPV6.[112–114] Of note, PI(4,5)P_2 has also been reported to inhibit TRPV4.[115] Aside from small molecules econazole and miconazole inhibiting both TRPV5 and TRPV6, as well as 2-APB inhibiting TRPV6, novel TRPV5 inhibitors have been identified through structure-based virtual screening, such as the compound ZINC 1788990 targeting the intracellular half of the S1–S4 bundle.[101] Different from econazole and 2-APB displacing the resident lipids, other known inhibitors of TRPV6 directly block the channel pore, such as calmodulin (CaM), (4-phenylcyclohexyl)piperazine derivatives and ruthenium red (RR). Furthermore, an activating lipid is present in TRPV6, whose location coincides with the vanilloid-binding site of TRPV1. The functional mutant R470E targeting this site displayed calcium uptake about ten times slower than that of wild-type channels,[82] suggesting the important role of lipids at this site. Similarly, an analogous mutation has been found to eliminate capsaicin-evoked currents in TRPV1.[116] The detailed information about ligand-binding sites in TRPV channels can also be found in another review.[117] This section will focus on a few shared modulators within the members of the TRPV subfamily, represented by PI(4,5)P_2, 2-APB and cannabinoid.

PI(4,5)P_2 is a general modulator of TRP channels. It is a well-established activator of both TRPV5 and TRPV6, permitting these two channels to be constitutively active. In contrast, there is a controversy surrounding the regulation of TRPV1 by PI(4,5)P_2.[118,119] Whereas dual regulation has been hypothesized in the previous studies,[120,121] introduction of phosphoinositides into a chemically defined system enabled the characterization of PI(4,5)P_2 as a negative regulator of TRPV1.[122] In the case of other thermoTRPVs, PI(4,5)P_2 appears to play either an inhibitory (TRPV2, TRPV3 and TRPV4) or activating (TRPV4) role. Albeit a dearth of their corresponding structures, early examination combining mutagenesis and molecular modeling indicated that a positively charged residue R575 located within the S4-S5 linker, which is absent in TRPV5 and TRPV6, is responsible for high-affinity binding of PI(4,5)P_2 to the TRPV1 channel.[123] Introducing an arginine residue in place of an equivalent residue in TRPV6 results in increased binding to PI(4,5)P_2 and abolishment of Ca^{2+}-dependent inactivation of TRPV6.[124] Subsequently, in light of the cryo-EM structures of TRPV5 and its phosphomimetic mutant T709D determined at pH 8 or 6 in the presence of a soluble form of PI(4,5)P_2, one gains additional insights into the binding of PI(4,5)P_2 in a position located between the N-linker, S4–S5 linker and S6 helix as well as channel opening through PI(4,5)P_2-induced conformational changes in S6 and the S4–S5 linker.[99,102] In addition to phosphoinositides, cholesterol has also been reported to be important for TRPV1, TRPV2 and TRPV3 activity. For instance, cholesterol was shown to directly bind to and modulate the TRPV1's responses to drugs and temperature,[73,125,126] and to increase the sensitivity of TRPV3 to agonists and temperature.[127] Given the limited structural evidence, more structural work is needed to probe the regulation of the TRPV subfamily by phosphoinositides, cholesterol and other lipids.

2-APB is able to activate TRPV1-3,[25] inhibit TRPV6, but has no effect on the functions of TRPV4 and TRPV5.[103,128] 2-APB fails to modulate the functions of TRPV4 and TRPV5, possibly by virtue of the inability of 2-APB to displace the tightly bound lipids. The Sobolevsky group structurally visualized three sites for 2-APB association with TRPV3,[69] including one site located at the intracellular half of the S1-S4 domain and the C-terminal portion of the TRP domain, the second site formed by extracellular portions of helices S1-S4, and the third site located at the nexus of the linker domain, pre-S1 helix, and TRP domain (Table 2.1). However, two of them were not separately confirmed by the Lee group and Yuan group.[65,95] Intriguingly, the critical histidine residue of TRPV3 at the functionally determined 2-APB binding site is not conserved in TRPV2,[129] suggesting that different binding sites (or modes) may be used for the activation of TRPV2 and TRPV3 by 2-APB. Consistent with this view, the Moiseenkova-Bell group revealed a novel site of 2-APB positioned between S5 of one protomer and the S4-S5 linker of an adjacent protomer in TRPV2, which rotates the channel as a wedge followed by the disruption of the endogenous lipid binding in the vanilloid pocket.[93] Despite its inhibitory role in TRPV6 activity, 2-APB was identified to bind to the same site as one of three binding sites in TRPV3, as discussed above.[103] These observations argue for the complicated regulation of TRPV channels by 2-APB.

Cannabinoid-based bioactive compounds have been used in medicine and have the potential to treat a broad spectrum of diseases, such as neurodegenerative disorders, cardiovascular disorders, cancer and so forth.[130] Aside from G protein-coupled receptors CB1 and CB2, cannabinoids have also been shown to act on multiple TRP channels, including TRPA1, TRPM8 and thermoTRPV channels. Cannabidiol (CBD) is one of several cannabinoids that activate TRPV1-4.[24,131–133] Another phytocannabinoid, Δ^9-tetrahydrocannabivarin (THCV), has been shown to inhibit TRPV5 and TRPV6 channels.[134] However, only a few of the cannabinoids have been structurally characterized to interact with TRPV channels to date, including Δ^9-tetrahydrocannabiorcol (C16),[94] an analog of Δ^9-tetrahydrocannabinol (Δ^9-THC) lacking CB1 and CB2 receptor activity, and CBD that activates TRPV2,[64,93] and THCV that inhibits TRPV6.[107] CBD occupies the hydrophobic pocket located between S5 of one protomer and the S6 helix of the adjacent subunit, but the pore of TRPV2 stays non-conducting in the presence of CBD alone, presumably representing a pre-open state or desensitized state.[64] In another study, a putative density in the cryo-EM map suggests that C16 binds at the interface between two protomers, where it destabilizes the interaction of the lipid within the vanilloid binding pocket of TRPV2, thereby promoting channel opening. This underpins a lipid-displacement-activation-mechanism.[94] Based on the structural reconstruction, two densities per subunit corresponding to the shape and size of THCV molecules were identified in TRPV6. The structure supports their binding at a side portal of TRPV6, which connects the surrounding membrane environment with the central cavity of the ion channel pore. Additionally, another binding site was observed next to the first one, but it is more peripheral with respect to the pore. Analyzing their localized interactions led the authors to propose that the portal site is the main inhibitory site of THCV,[107] reminiscent of CBD binding in TRPV2[64, 93] and the binding of local anesthetic dyclonine in TRPV3.[97]

2.5 THE NARROWEST PORE CONSTRICTION OF LOWER GATE

In spite of the low sequence identity throughout the TRPV subtypes, the S6 helix, forming the lower gate, exhibits high sequence conservation (Fig. 2.2). Strikingly, all members harbor an isoleucine and methionine at positions equivalent to I679 and M682 in rat TRPV1, respectively (Fig. 2.2). The transitions in S6 from low-energy α-helices to high-energy and bulkier π-helices have been proposed to play important roles in channel gating.[135] Of note, the π-helix features a bulge in S6 by introducing a helical-register shift of one residue, resulting in a rotation of the C-terminal part of S6 and a switch of the pore-forming residues between methionine (Met) of the α-helical S6 and isoleucine (Ile) of the π-helical S6. To analyze and understand the correlation of S6 with channel gating, the structural information is herein summarized for TRPV channels (Table 2.2).

To date, all apo structures of TRPV1 reveal that the pore-forming residues Ile in S6 constitute the most constricted point in the lower gate. In the absence of ligands, both rabbit and rat TRPV2 adopt a canonical α helix conformation and the Met residue from each subunit forms a hydrophobic seal near the S6 helical bundle-crossing region,[61,64] which positions the lower gate closer to the cytosol by one helical turn than that seen in the apo structures of TRPV1. In contrast, residues Ile constrict the lower gate of mouse TRPV2 according to its apo structure embedded into nanodiscs,[62] indicating the critical roles of endogenous molecules in TRPV2 conformation. Resembling TRPV2, there are multiple apo structures of TRPV3 exhibiting either α- or π-helix conformation in S6. Given the different systems used for structural studies, the visualization of π-helical S6 in lipid nanodiscs highlights the crosstalk between lipids and the TRPV3 channel.[66]

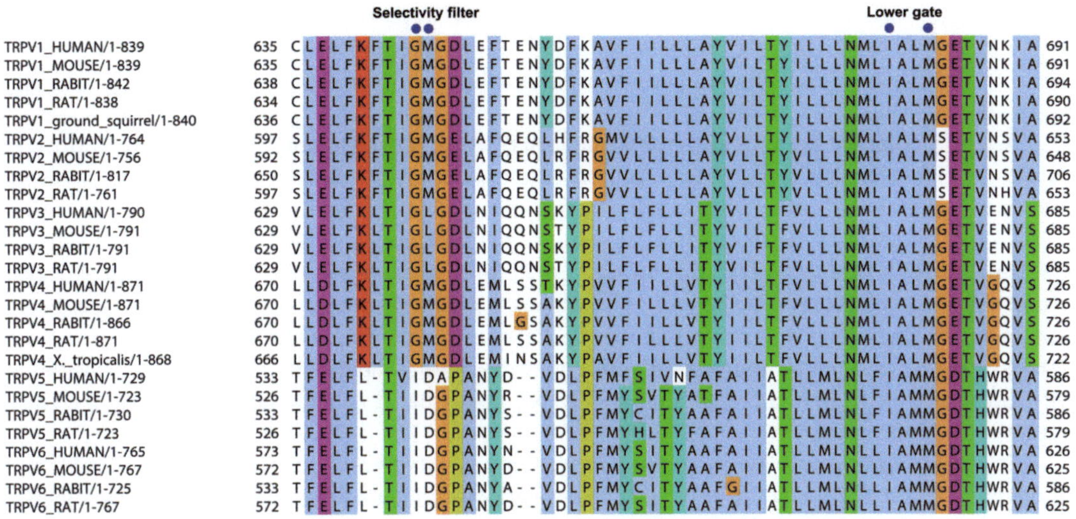

FIGURE 2.2 Sequence alignment covering the localized regions of the selectivity filter and lower gate for members within TRPV subfamily from the species used in structural studies.

TABLE 2.2
Biochemical Systems Used for Cryo-EM Studies and the Structural Characterization of S6 Helices in TRPV Channels

	Systems Used for Structural Determination	The State of Channels	The Narrowest Point at Lower Gate	π-helical or α-helical S6
TRPV1	In amphipols or nanodiscs	Apo	I679 (Rat) or I681 (Ground squirrel) or I680 (Human)	π-helix
	In nanodiscs	Closed	M682 (Rat) or I681 (Ground squirrel) or I679 (Rat) or I680 (Human)	α-helix, π-helix
		Intermediate-open or open	I679 (Rat) or M682 (Rat)	π-helix, α-helix
TRPV2	In amphipols or nanodiscs	Apo	M643 (Rabbit) or M645 (Rat) or I635 (Mouse)	α-helix, π-helix
	In nanodiscs	Closed	M645 (Rat)	α-helix
		Open or activated	M645 (Rat)	α-helix
TRPV3	In detergents or amphipols or nanodiscs	Apo	M677 (Mouse) or M677 (Human) or I674 (Human)	α-helix, π-helix
	In detergents or nanodiscs	Sensitized or inactivated	I674 (Human) or I674 (Mouse) or M677 (Human)	π-helix, α-helix
		Open	I674 (Mouse) or I674 (Human)	π-helix
TRPV4	In detergents	Apo	M714 (*X. tropicalis*) or M718 (human) or I715 (human)	α-helix, π-helix
		Closed	M718 (human)	α-helix
		Open	I715 (human)	π-helix
TRPV5	In nanodiscs	Apo or open	I575 (Rabbit)	π-helix
	In detergents or nanodiscs	Closed or inactivated	I575 (Rabbit)	π-helix
TRPV6	In nanodiscs	Apo or open	I575 (human)	π-helix
	In amphipols or detergents	Closed or inactivated	M578 (human) or I575 (human)	α-helix, π-helix

Though the currently available structures of TRPV4 were solved in detergents, both α- and π-helices have been observed in S6 in apo state. As two members with the constitutive activity *in vivo*, TRPV5 and TRPV6 are characteristic of π-helical S6 as observed in ligand-free or open structures.

When in complex with agonists or antagonists, which shift the equilibrium between different states, TRPV channels have been captured in closed, sensitized, intermediate-open, open, and desensitized states. Both α- and π-helical S6 conformations occur in the closed or open state of TRPV1, wherein the flexible conformation of the pore-forming residue M682 was observed.[18] For TRPV2, the α-helix is a preferred conformation of S6 as observed in the opening and closed structures. TRPV3 is structurally shown to possess a π-helical S6 in an open state and α- or π-helical S6 in a non-conducting state, respectively. In TRPV4, S6 adopts a α-helix conformation in a closed state versus π-helix conformation in an open state. Excluding the density map at limited resolution, π-helical S6 is present in a closed state of TRPV5 and either α- or π-helical S6 occurs in a closed state of TRPV6.

These structural observations provide important insights into the conformation of the S6 helix and the pore constriction at the lower gate in unliganded and ligand-bound states of TRPV channels. For instance, the α- or π-helical conformation of S6 is not necessarily required for the pore opening or closure of TRPV channels and may be associated with the presence of lipids surrounding the channels. It is also likely that a certain member of the TRPV subfamily favors α- or π-helical S6 in a certain state during the gating cycle. It is worth noting that the current comparison is founded on the existing pool of available structures. As new structures continue to emerge, the conclusive analysis will require updates to remain accurate and comprehensive.

2.6 THE CYTOPLASMIC REGIONS

TRPV channels have large cytoplasmic regions flanking the transmembrane domain, including an N-terminal ankyrin repeat domain (ARD), featuring six ankyrin-like repeats and a C-terminal domain (CTD). The intracellular N- and C-termini are involved in canonical tetramerization, membrane trafficking, modulatory protein–protein interactions and thermosensitivity.[136–146] For instance, modifications of amino acid residues in the third ankyrin-like repeat of TRPV6 abolished the tetramerization and produced a nonfunctional channel.[137] A large number of mutations that alter the activities of TRPV1 and other TRPV channels lie in the vicinity of the putative ATP-binding site,[77,147] indicating that the ARD serves as an important locus for channel modulation. ARD exerts the impact on channel behavior via a rigid-body movement originating from ATP, calmodulin and RhoA binding, suggesting the intrinsic stability of the overall structure of ARD. The recent structures demonstrate that RhoA interacts with ARD of TRPV4,[59,60] which serves as a basis to understand the crosstalk between ion channels and small GTPases and channelopathies caused by their altered interactions.[148]

In the cryo-EM structures of TRPV channels, a β-sheet from the linker domain and CTD associates with finger 3 and the fourth ankyrin-like repeat of an adjacent ARD. Additionally, the distal region of CTD appears to wrap around the β-sheet of the coupling domain at the interface between two adjacent protomers.[68] This structural feature is proposed to play a role in the channel inactivation, whereas in the open state, it unwraps and projects toward the cytoplasm.[67,95] The acidic residues E751 and D752 in the distal CTD of human TRPV3 have been visualized to form salt-bridge interactions with K169 of the adjacent ARD.[95] Breaking the salt bridges disrupted the interactions between CTD and ARD, leading to altered sensitivities of TRPV1, TRPV3 and TRPV4 to modulators and a drastic secondary structure change at the distal CTD of human TRPV3 from an extended loop to an α-helix.[65,95] In TRPV5 and TRPV6, a helix preceding the first ankyrin-like repeat forms a major interface between adjacent subunits at each corner, functioning as an intersubunit "glue",[83,89,98] and the C-terminus does not wrap around the three-stranded β-sheet.

Pacsin3 is proposed to not only regulate TRPV4 endocytosis but also directly modulate channel activity by interacting with the unstructured N-terminus.[40,149,150] By comparison, CaM binds to the usually unstructured C-termini of TRPV channels,[151–155] and the C-terminal helix is only observed in the CaM-bound structures of TRPV5 and TRPV6. The N- and C-terminal lobes of CaM, displaying differential affinity to Ca^{2+}, bind to the proximal and distal regions of their C termini, respectively. Studies have revealed a stoichiometry of one CaM molecule per TRPV6 tetramer [109,154,156] and either one or two CaM molecule(s) per TRPV5 tetramer.[100,152] The structural observations also showed that lysine 115 (K115), located in the C-terminal lobe of CaM, plugs into a tight cage formed by four tryptophan residues from each subunit of the channel at the S6 bundle crossing. This interaction is stabilized by the cation-π interaction between the indole rings of tryptophan and the positively charged side chain of lysine.[99,100,109] This scheme thereby illustrates a plausible mechanism for the Ca^{2+}-dependent inactivation of the channels by CaM, which adopts the head-to-tail orientation right underneath the intracellular entrance of the pore. The published structures so far have resolved the stabilized regions of the N- and C-termini of the channels, but more examinations need to be conducted to interpret the roles of channel termini in channel modulation. For instance, CaM has also been found to regulate the functions of thermoTRPV channels, but it remains to be explored whether alternative binding modes of CaM exist since thermoTRPVs lack the tryptophan ring as seen in TRPV5 and TRPV6.

2.7 DISCUSSION

The truncated functional constructs of TRPV channels have served as valid model systems to decipher the gating mechanisms. In more recent years, further efforts have also been made to structurally examine the functionalities of the wild-type channels. Presumably attributed to the conformational heterogeneity, the pore turrets are not fully resolved in the majority of cryo-EM structures of TRPVs so far. Nonetheless, the extended polypeptide densities between S5 and the pore helix indicate that the pore turrets from individual subunits of the tetrameric TRPV1 project towards the central axis and appear as an unstructured or cap-like density atop the extracellular entrance of the pore,[18,74,88,89] which may provide an additional way to regulate the passage of cations through the channels. The pore turret of TRPV2 has been proposed to lay parallel to the lipid membrane, but the quality of the published cryo-EM map is insufficient to allow precise structural assignments of channel elements.[157] On the contrary, the findings concerning the pore turrets of TRPV2 by other groups clearly show that they are more likely to orient perpendicular to the membrane.[64,91] Intriguingly, TRPV5 and TRPV6 display the structured pore turret, likely owing to the stabilization by the contacting interactions between the S1–S2 extracellular loops and the S5-P and P-S6 loops of the adjacent subunit. Moreover, the S1–S2 loop harbors a conserved N-linked glycosylation site, whose modification is a likely reason for activation of TRPV5 and TRPV6 by the beta glucuronidase klotho.[158]

Symmetry reduction has been observed in the cryo-EM and X-ray structures of RTX-bound TRPV2[63,91] and cryo-EM structures of TRPV3 in the presence of 2-APB and its mutant Y564A at 37°C,[67,68] which hypothetically represents an intermediate arrangement during the gating cycle of these channels before returning to the four-fold symmetry upon final opening. Notably, the overall structure of genistein-bound TRPV6 is almost C4 symmetric, while the localized regions at the S4–S5 linker and S6-TRP helix adopt a two-fold symmetrical conformation due to genistein binding,[108] reminiscent of sub-stoichiometric binding of RTX or DkTx to TRPV1.[18] In stark contrast, symmetry reduction is not typically observed in all TRPV channels during gating, based on their available structures, and it is not seen in the datasets of TRPV2 in complex with CBD either.[64] It remains to be further elucidated to what extent symmetry reduction is involved in the gating of TRPV channels.

The unsuccessful attempts to pertinently visualize the fully heat-dependent opening of TRPV1 in nanodiscs or detergents suggest that the *in vitro* systems may be inadequate to capture such a state, given the lack of membrane voltage and lateral pressure as physiologically present in the cell membrane.[88] Another potential explanation could be that the heated channel had a low open probability, requiring intense thermal stimulation to capture the open state. Nevertheless, the pre-treatment with capsaicin enabled the observation of the heat-activated open state of TRPV1, revealing the global conformational transition followed by outer pore rearrangements.[88] It is still worth noting that the contribution of capsaicin cannot be ruled out due to the necessity of TRPV1 being sensitized by this ligand during the heat-induced gating. In the case of TRPV3 sensing warm temperatures, it has also proven to be challenging to structurally capture the channel opening without using the temperature-sensitizing mutations.[67] In the follow-up study,[159] the Sobolevsky group proceeded with the wild-type TRPV3 in MSP2N2 nanodiscs for heat-evoked activation instead of the previously used mutant. Neither of the classified reconstructions represented the open state. Interestingly, the cNW11-reconstituted TRPV3 yielded a conducting structure with a wide pore. However, it remains enigmatic in regard to the mechanism underlying the divergent structural observations of TRPV3 under thermal stimulation in MSP2N2 versus cNW11 nanodiscs. The current findings lead the authors to hypothesize that the conformational wave propagating among different domains may account for the molecular mechanism of heat-induced activation in TRPV channels.[74,159] It is anticipated that significant advancements will continue to be made in the field of TRP channels as the power of cryo-EM and other advanced techniques further boost research endeavors in this area.

ACKNOWLEDGEMENTS

The author would like to thank Drs. Yifan Cheng and Michael X. Zhu for their comments and suggestions. The work in the author's laboratory is supported by a grant from the National Natural Science Foundation of China (32301013), the Shanghai Pujiang Program (22PJ1402100), and a start-up fund from Fudan University. The author also apologizes to those whose important work could not be cited due to space limitations.

REFERENCES

1. Clapham, D. E. TRP channels as cellular sensors. *Nature*, 2003, 426: 517–524.
2. Julius, D. TRP channels and pain. *Annu Rev Cell Dev Biol*, 2013, 29: 355–384.
3. Nilius, B., Szallasi, A. Transient receptor potential channels as drug targets: From the science of basic research to the art of medicine. *Pharmacol Rev*, 2014, 66: 676–814.
4. Caterina, M. J., Schumacher, M. A., Tominaga, M., et al. The capsaicin receptor: A heat-activated ion channel in the pain pathway. *Nature*, 1997, 389: 816–824.
5. Patapoutian, A., Peier, A. M., Story, G. M., Viswanath, V. Thermotrp channels and beyond: Mechanisms of temperature sensation. *Nat Rev Neurosci*, 2003, 4: 529–539.
6. Peng, J. B. TRPV5 and TRPV6 in transcellular Ca2+ transport: Regulation, Gene duplication, and polymorphisms in African populations. *Adv Exp Med Biol*, 2011, 704: 239–275.
7. Peralvarez-Marin, A., Donate-Macian, P., Gaudet, R. What do we know about the transient receptor potential vanilloid 2 (TRPV2) ion channel? *FEBS J*, 2013, 280: 5471–5487.
8. Caterina, M. J., Leffler, A., Malmberg, A. B., et al. Impaired nociception and pain sensation in mice lacking the capsaicin receptor. *Science*, 2000, 288: 306–313.
9. Davis, J. B., Gray, J., Gunthorpe, M. J., et al. Vanilloid receptor-1 is essential for inflammatory thermal hyperalgesia. *Nature*, 2000, 405: 183–187.
10. Vilceanu, D., Honore, P., Hogan, Q. H., Stucky, C. L. Spinal nerve ligation in mouse upregulates TRPV1 heat function in injured IB4-positive nociceptors. *J Pain*, 2010, 11: 588–599.
11. Grandl, J., Kim, S. E., Uzzell, V., et al. Temperature-induced opening of TRPV1 ion channel is stabilized by the pore domain. *Nat Neurosci*, 2010, 13: 708–714.
12. Jara-Oseguera, A., Bae, C., Swartz, K. J. An external sodium ion binding site controls allosteric gating in TRPV1 channels. *Elife*, 2016, 5: e13356.
13. Jordt, S. E., Julius, D. Molecular basis for species-specific sensitivity to "hot" chili peppers. *Cell*, 2002, 108: 421–430.
14. Geron, M., Kumar, R., Zhou, W. C., et al. 2018. TRPV1 pore turret dictates distinct DkTx and capsaicin gating. *Proc Natl Acad Sci U S A*, 2018, 115: E11837–E11846.

15. Bohlen, C. J., Priel, A., Zhou, S., et al. A bivalent tarantula toxin activates the capsaicin receptor, TRPV1, by targeting the outer Pore domain. *Cell*, 2010, 141: 834–845.
16. Cao, E., Liao, M., Cheng, Y., Julius, D. TRPV1 structures in distinct conformations reveal activation mechanisms. *Nature*, 2013, 504: 113–118.
17. Gao, Y., Cao, E., Julius, D., Cheng, Y. TRPV1 structures in nanodiscs reveal mechanisms of ligand and lipid action. *Nature*, 2016, 534: 347–351.
18. Zhang, K., Julius, D., Cheng, Y. Structural snapshots of TRPV1 reveal mechanism of polymodal functionality. *Cell*, 2021, 184: 5138–5150. e5112.
19. Caterina, M. J., Rosen, T. A., Tominaga, M., Brake, A. J., Julius, D. A capsaicin-receptor homologue with a high threshold for noxious heat. *Nature*, 1999, 398: 436–441.
20. Liu, B. Y., Qin, F. Use dependence of heat sensitivity of vanilloid receptor TRPV2. *Biophys J*, 2016, 110: 1523–1537.
21. Park, U., Vastani, N., Guan, Y., et al. TRP vanilloid 2 knock-out mice are susceptible to perinatal lethality but display normal thermal and mechanical nociception. *J Neurosci*, 2011, 31: 11425–11436.
22. Katanosaka, Y., Iwasaki, K., Ujihara, Y., et al. TRPV2 is critical for the maintenance of cardiac structure and function in mice. *Nat Commun*, 2014, 5:3932.
23. Link, T. M., Park, U., Vonakis, B. M., et al. TRPV2 has a pivotal role in macrophage particle binding and phagocytosis. *Nat Immunol*, 2010, 11: 232–U237.
24. Muller, C., Morales, P., Reggio, P. H. Cannabinoid ligands targeting TRP channels. *Front Mol Neurosci*, 2019, 11:487.
25. Hu, H. Z., Gu, Q. H., Wang, C. B., et al. 2-aminoethoxydiphenyl borate is a common activator of TRPV1, TRPV2, and TRPV3. *J Biol Chem*, 2004, 279: 35741–35748.
26. Peier, A. M., Reeve, A. J., Andersson, D. A., et al. A heat-sensitive TRP channel expressed in keratinocytes. *Science*, 2002, 296: 2046–2049.
27. Xu, H. X., Ramsey, I. S., Kotecha, S. A., et al. TRPV3 is a calcium-permeable temperature-sensitive cation channel. *Nature*, 2002, 418: 181–186.
28. Smith, G. D., Gunthorpe, J., Kelsell, R. E., et al. TRPV3 is a temperature-sensitive vanilloid receptor-like protein. *Nature*, 2002, 418: 186–190.
29. Moqrich, A., Hwang, S. W., Earley, T. J., et al. Impaired thermosensation in mice lacking TRPV3, a heat and camphor sensor in the skin. *Science*, 2005, 307: 1468–1472.
30. Chung, M. K., Lee, H., Caterina, M. J. Warm temperatures activate TRPV4 in mouse 308 keratinocytes. *J Biol Chem*, 2003, 278: 32037–32046.
31. Cheng, X. P., Jin, J., Hu, L., et al. TRP channel regulates EGFR signaling in hair morphogenesis and skin barrier formation. *Cell*, 2010, 141: 331–343.
32. Aijima, R., Wang, B., Takao, T., et al. The thermosensitive TRPV3 channel contributes to rapid wound healing in oral epithelia. *FASEB J*, 2015, 29: 182–192.
33. Imura, K., Yoshioka, T., Hikita, I., et al. Influence of TRPV3 mutation on hair growth cycle in mice. *Biochem Biophys Res Commun*, 2007, 363: 479–483.
34. Guler, A. D., Lee, H. S., Iida, T., et al. Heat-evoked activation of the ion channel, TRPV4. *J Neurosci*, 2002, 22: 6408–6414.
35. Chung, M. K., Lee, H., Mizuno, A., Suzuki, M., Caterina, M. J. TRPV3 and TRPV4 mediate warmth-evoked currents in primary mouse keratinocytes. *J Biol Chem*, 2004, 279: 21569–21575.
36. Watanabe, H., Vriens, J., Suh, S. H., et al. Heat-evoked activation of TRPV4 channels in a HEK293 cell expression system and in native mouse aorta endothelial cells. *J Biol Chem*, 2002, 277: 47044–47051.
37. Kotttgen, M., Buchholz, B., Garcia-Gonzalez, M. A., et al. TRPP2 and TRPV4 form a polymodal sensory channel complex. *J Cell Biol*, 2008, 182: 437–447.
38. Vriens, J., Watanabe, H., Janssens, A., et al. Cell swelling, heat, and chemical agonists use distinct pathways for the activation of the cation channel TRPV4. *Proc Natl Acad Sci U S A*, 2004, 101: 396–401.
39. Wegierski, T., Lewandrowski, U., Muller, B., Sickmann, A., Walz, G. Tyrosine phosphorylation modulates the activity of TRPV4 in response to defined stimuli. *J Biol Chem*, 2009, 284: 2923–2933.
40. D'hoedt, D., Owsianik, G., Prenen, J., et al. Stimulus-specific modulation of the cation channel TRPV4 by PACSIN 3. *J Biol Chem*, 2008, 283: 6272–6280.
41. Todaka, H., Taniguchi, J., Satoh, J., Mizuno, A., Suzuki, M. Warm temperature-sensitive transient receptor potential vanilloid 4 (TRPV4) plays an essential role in thermal hyperalgesia. *J Biol Chem*, 2004, 279: 35133–35138.
42. Suzuki, M., Mizuno, A., Kodaira, K., Imai, M. Impaired pressure sensation in mice lacking TRPV4. *J Biol Chem*, 2003, 278: 22664–22668.

43. Gao, X., Wu, L., O'Neil, R. G. Temperature-modulated diversity of TRPV4 channel gating - activation by physical stresses and phorbol ester derivatives through protein kinase C-dependent and -independent pathways. *J Biol Chem*, 2003, 278: 27129–27137.
44. Liedtke, W., Choe, Y., Marti-Renom, M. A., et al. Vanilloid receptor-related osmotically activated channel (VR-OAC), a candidate vertebrate osmoreceptor. *Cell*, 2000, 103: 525–535.
45. Liedtke, W., Tobin, D. M., Bargmann, C. I., Friedman, J. M. Mammalian TRPV4 (VR-OAC) directs behavioral responses to osmotic and mechanical stimuli in Caenorhabditis elegans. *Proc Natl Acad Sci U S A*, 2003, 100: 14531–14536.
46. Rosenbaum, T., Benitez-Angeles, M., Sanchez-Hernandez, R., et al. TRPV4: A physio and pathophysiologically significant ion channel. *Int J Mol Sci*, 2020, 21:3837.
47. Goyal, N., Skrdla, P., Schroyer, R., et al. Clinical pharmacokinetics, safety, and tolerability of a novel, first-in-class TRPV4 ion channel inhibitor, GSK2798745, in healthy and heart failure subjects. *Am J Cardiovasc Drug*, 2019, 19: 335–342.
48. Thorneloe, K. S., Cheung, M., Bao, W. K., et al. An orally active TRPV4 channel blocker prevents and resolves pulmonary edema induced by heart failure. *Sci Transl Med* 2012, 4:159ra148.
49. Ludbrook, V. J., Hanrott, K. E., Kreindler, J. L., et al. Adaptive study design to assess effect of TRPV4 inhibition in patients with chronic cough. *ERJ Open Res*, 2021, 7:00269-2021.
50. van Goor, M. K. C., Hoenderop, J. G. J., van der Wijst, J. TRP channels in calcium homeostasis: From hormonal control to structure-function relationship of TRPV5 and TRPV6. *Biochim Biophys Acta Mol Cell Res*, 2017, 1864: 883–893.
51. Khattar, V., Wang, L., Peng, J. B. Calcium selective channel TRPV6: Structure, function, and implications in health and disease. *Gene*, 2022, 817: 146192.
52. Dhennin-Duthille, I., Gautier, M., Faouzi, M., et al. High expression of transient receptor potential channels in human breast cancer epithelial cells and tissues: Correlation with pathological parameters. *Cell Physiol Biochem*, 2011, 28: 813–822.
53. Na, T., Peng, J.-B. (2014). TRPV5: A Ca2+ Channel for the Fine-Tuning of Ca2+ Reabsorption. In *Mammalian Transient Receptor Potential (TRP) Cation Channels*: Volume I, B. Nilius, and V. Flockerzi, eds. (Springer Berlin Heidelberg), pp. 321–357. https://doi.org/10.1007/978-3-642-54215-2_13
54. Peng, J. B., Chen, X. Z., Berger, U. V., et al. Molecular cloning and characterization of a channel-like transporter mediating intestinal calcium absorption. *J Biol Chem*, 1999, 274: 22739–22746.
55. Zakharian, E., Cao, C., Rohacs, T. Intracellular ATP supports TRPV6 activity via lipid kinases and the generation of PtdIns(4,5)p-2. *FASEB J*, 2011, 25: 3915–3928.
56. Rohacs, T. Phosphoinositide regulation of TRP channels. *Handb Exp Pharmacol*, 2014, 223: 1143–1176.
57. Liao, M., Cao, E., Julius, D., Cheng, Y. Structure of the TRPV1 ion channel determined by electron cryo-microscopy. *Nature*, 2013, 504: 107–112.
58. Deng, Z., Paknejad, N., Maksaev, G., et al. Cryo-EM and X-ray structures of TRPV4 reveal insight into ion permeation and gating mechanisms. *Nat Struct Mol Biol*, 2018, 25: 252–260.
59. Kwon, D. H., Zhang, F., McCray, B. A., et al. TRPV4-rho GTPase complex structures reveal mechanisms of gating and disease. *Nat Commun*, 2023, 14: 3732.
60. Nadezhdin, K. D., Talyzina, I. A., Parthasarathy, A., et al. Structure of human TRPV4 in complex with GTPase RhoA. *Nat Commun*, 2023. 14: 3733.
61. Zubcevic, L., Herzik, M. A. Jr., Chung, B. C., et al. Cryo-electron microscopy structure of the TRPV2 ion channel. *Nat Struct Mol Biol*, 2016, 23: 180–186.
62. Su, N., Zhen, W., Zhang, H., et al. Structural mechanisms of TRPV2 modulation by endogenous and exogenous ligands. *Nat Chem Biol*, 2023, 19: 72–80.
63. Zubcevic, L., Le, S., Yang, H., Lee, S. Y. Conformational plasticity in the selectivity filter of the TRPV2 ion channel. *Nat Struct Mol Biol*, 2018, 25: 405–415.
64. Pumroy, R. A., Samanta, A., Liu, Y., et al. Molecular mechanism of TRPV2 channel modulation by cannabidiol. *Elife*, 2019, 8:e48792.
65. Deng, Z., Maksaev, G., Rau, M., et al. Gating of human TRPV3 in a lipid bilayer. *Nat Struct Mol Biol*, 2020, 27: 635–644.
66. Shimada, H., Kusakizako, T., Dung Nguyen, T. H., et al. The structure of lipid nanodisc-reconstituted TRPV3 reveals the gating mechanism. *Nat Struct Mol Biol*, 2020, 27: 645–652.
67. Singh, A. K., McGoldrick, L. L., Demirkhanyan, L., et al. Structural basis of temperature sensation by the TRP channel TRPV3. *Nat Struct Mol Biol*, 2019, 26: 994–998.
68. Zubcevic, L., Herzik, M. A. Jr., Wu, M., et al. Conformational ensemble of the human TRPV3 ion channel. *Nat Commun*, 2018, 9: 4773.
69. Singh, A. K., McGoldrick, L. L., Sobolevsky, A. I. Structure and gating mechanism of the transient receptor potential channel TRPV3. *Nat Struct Mol Biol*, 2018, 25: 805–813.

70. Hille, B., Dickson, E. J., Kruse, M., Vivas, O., Suh, B. C. Phosphoinositides regulate ion channels. *Bba-Mol Cell Biol L*, 2015, 1851: 844–856.
71. Jara-Oseguera, A., Huffer, K. E., Swartz, K. J. The ion selectivity filter is not an activation gate in TRPV1-3 channels. *Elife*, 2019, 8: e51212.
72. Hui, K. K., Liu, B. Y., Qin, F. Capsaicin activation of the pain receptor, VR1: Multiple open states from both partial and full binding. *Biophys J*, 2003, 84: 2957–2968.
73. Liu, B. Y., Hui, K. Y., Qin, F. Thermodynamics of heat activation of single capsaicin ion channels VR1. *Biophys J*, 2003, 85: 2988–3006.
74. Kwon, D. H., Zhang, F., Fedor, J. G., Suo, Y., Lee, S. Y. Vanilloid-dependent TRPV1 opening trajectory from cryoEM ensemble analysis. *Nat Commun*, 2022, 13: 2874.
75. Chung, M. K., Guler, A. D., Caterina, M. J. TRPV1 shows dynamic ionic selectivity during agonist stimulation. *Nat Neurosci*, 2008, 11: 555–564.
76. Nabissi, M., Morelli, M. B., Santoni, M., Santoni, G. Triggering of the TRPV2 channel by cannabidiol sensitizes glioblastoma cells to cytotoxic chemotherapeutic agents. *Carcinogenesis*, 2013, 34: 48–57.
77. Myers, B. R., Bohlen, C. J., Julius, D. A yeast genetic screen reveals a critical role for the pore helix domain in TRP channel gating. *Neuron*, 2008, 58: 362–373.
78. Ryu, S. J., Liu, B. Y., Yao, J., Fu, Q., Qin, F. Uncoupling proton activation of vanilloid receptor TRPV1. *J Neurosci*, 2007, 27: 12797–12807.
79. Grandl, J., Hu, H. Z., Bandell, M., et al. Pore region of TRPV3 ion channel is specifically required for heat activation. *Nat Neurosci*, 2008, 11: 1007–1013.
80. Kim, S. E., Patapoutian, A., Grandl, J. Single residues in the outer Pore of TRPV1 and TRPV3 have temperature-dependent conformations. *PLoS One*, 2013, 8:e59593.
81. Owsianik, G., Talavera, K., Voets, T., Nilius, B. Permeation and selectivity of TRP channels. *Annu Rev Physiol*, 2006, 68: 685–717.
82. McGoldrick, L. L., Singh, A. K., Saotome, K., et al. Opening of the human epithelial calcium channel TRPV6. *Nature*, 2018, 553: 233–237.
83. Saotome, K., Singh, A. K., Yelshanskaya, M. V., Sobolevsky, A. I. Crystal structure of the epithelial calcium channel TRPV6. *Nature*, 2016, 534: 506–511.
84. Singh, A. K., Saotome, K., Sobolevsky, A. I. Swapping of transmembrane domains in the epithelial calcium channel TRPV6. *Sci Rep*, 2017, 7: 10669.
85. Lansky, S., Betancourt, J. M., Zhang, J., et al. A pentameric TRPV3 channel with a dilated pore. *Nature*, 2023, 621: 206–214.
86. Dai, J., Cho, T. J., Unger, S., et al. TRPV4-pathy, a novel channelopathy affecting diverse systems. *J Hum Genet*, 2010, 55: 400–402.
87. Lin, Z. M., Chen, Q., Lee, M. Y., et al. Exome sequencing reveals mutations in TRPV3 as a cause of Olmsted syndrome. *Am J Hum Genet*, 2012, 90: 558–564.
88. Kwon, D. H., Zhang, F., Suo, Y., et al. Heat-dependent opening of TRPV1 in the presence of capsaicin. *Nat Struct Mol Biol*, 2021, 28: 554–563.
89. Nadezhdin, K. D., Neuberger, A., Nikolaev, Y. A., et al. Extracellular cap domain is an essential component of the TRPV1 gating mechanism. *Nat Commun*, 2021, 12: 2154.
90. Neuberger, A., Oda, M., Nikolaev, Y. A., et al. Human TRPV1 structure and inhibition by the analgesic SB-366791. *Nat Commun*, 2023, 14: 2451.
91. Zubcevic, L., Hsu, A. L., Borgnia, M. J., Lee, S. Y. Symmetry transitions during gating of the TRPV2 ion channel in lipid membranes. *Elife* 2019, 8:e45779.
92. Conde, J., Pumroy, R. A., Baker, C., et al. Allosteric antagonist modulation of TRPV2 by piperlongumine impairs glioblastoma progression. *ACS Cent Sci*, 2021, 7: 868–881.
93. Pumroy, R. A., Protopopova, A. D., Fricke, T. C., et al. Structural insights into TRPV2 activation by small molecules. *Nat Commun*, 2022, 13: 2334.
94. Zhang, L., Simonsen, C., Zimova, L., et al. Cannabinoid non-cannabidiol site modulation of TRPV2 structure and function. *Nat Commun*, 2022, 13: 7483.
95. Zubcevic, L., Borschel, W. F., Hsu, A. L., Borgnia, M. J., Lee, S. Y. Regulatory switch at the cytoplasmic interface controls TRPV channel gating. *Elife*, 2019, 8:e47746.
96. Neuberger, A., Nadezhdin, K. D., Zakharian, E., Sobolevsky, A. I. Structural mechanism of TRPV3 channel inhibition by the plant-derived coumarin osthole. *EMBO Rep*, 2021, 22: e53233.
97. Neuberger, A., Nadezhdin, K. D., Sobolevsky, A. I. Structural mechanism of TRPV3 channel inhibition by the anesthetic dyclonine. *Nat Commun*, 2022, 13: 2795.
98. Hughes, T. E. T., Lodowski, D. T., Huynh, K. W., et al. Structural basis of TRPV5 channel inhibition by econazole revealed by cryo-EM. *Nat Struct Mol Biol*, 2018, 25: 53–60.

99. Hughes, T. E. T., Pumroy, R. A., Yazici, A. T., et al. Structural insights on TRPV5 gating by endogenous modulators. *Nat Commun*, 2018, 9: 4198.
100. Dang, S., van Goor, M. K., Asarnow, D., et al. Structural insight into TRPV5 channel function and modulation. *Proc Natl Acad Sci U S A*, 2019, 116: 8869–8878.
101. Hughes, T. E., Del Rosario, J. S., Kapoor, A., et al. Structure-based characterization of novel TRPV5 inhibitors. *Elife*, 2019, 8:e49572.
102. Fluck, E. C., Yazici, A. T., Rohacs, T., Moiseenkova-Bell, V. Y. Structural basis of TRPV5 regulation by physiological and pathophysiological modulators. *Cell Rep*, 2022, 39: 110737.
103. Singh, A. K., Saotome, K., McGoldrick, L. L., Sobolevsky, A. I. Structural bases of TRP channel TRPV6 allosteric modulation by 2-APB. *Nat Commun*, 2018, 9: 2465.
104. Bhardwaj, R., Lindinger, S., Neuberger, A., et al. Inactivation-mimicking block of the epithelial calcium channel TRPV6. *Sci Adv*, 2020, 6:eabe1508.
105. Neuberger, A., Nadezhdin, K. D., Sobolevsky, A. I. Structural mechanisms of TRPV6 inhibition by ruthenium red and econazole. *Nat Commun*, 2021, 12: 6284.
106. Yelshanskaya, M. V., Nadezhdin, K. D., Kurnikova, M. G., Sobolevsky, A. I. Structure and function of the calcium-selective TRP channel TRPV6. *J Physiol*, 2021, 599: 2673–2697.
107. Neuberger, A., Trofimov, Y. A., Yelshanskaya, M. V., et al. Molecular pathway and structural mechanism of human oncochannel TRPV6 inhibition by the phytocannabinoid tetrahydrocannabivarin. *Nat Commun*, 2023, 14: 4630.
108. Neuberger, A., Trofimov, Y. A., Yelshanskaya, M. V., et al. Structural mechanism of human oncochannel TRPV6 inhibition by the natural phytoestrogen genistein. *Nat Commun*, 2023, 14: 2659.
109. Singh, A. K., McGoldrick, L. L., Twomey, E. C., Sobolevsky, A. I. Mechanism of calmodulin inactivation of the calcium-selective TRP channel TRPV6. *Sci Adv*, 2018, 4: eaau6088.
110. Zhang, F., Hanson, S. M., Jara-Oseguera, A., et al. Engineering vanilloid-sensitivity into the rat TRPV2 channel. *Elife*, 2016, 5:e16409.
111. Yang, F., Vu, S., Yarov-Yarovoy, V., Zheng, J. Rational design and validation of a vanilloid-sensitive TRPV2 ion channel. *Proc Natl Acad Sci U S A*, 2016, 113: E3657–E3666.
112. Garcia-Elias, A., Mrkonjic, S., Pardo-Pastor, C., et al. Phosphatidylinositol-4,5-biphosphate-dependent rearrangement of TRPV4 cytosolic tails enables channel activation by physiological stimuli. *Proc Natl Acad Sci U S A*, 2013, 110: 9553–9558.
113. Lee, J., Cha, S. K., Sun, T. J., Huang, C. L. PIP2 activates TRPV5 and releases its inhibition by intracellular Mg2+. *J Gen Physiol*, 2005, 126: 439–451.
114. Thyagarajan, B., Lukacs, V., Rohacs, T. Hydrolysis of phosphatidylinositol 4,5-bisphosphate mediates calcium-induced inactivation of TRPV6 channels. *J Biol Chem*, 2008, 283: 14980–14987.
115. Harraz, O. F., Longden, T. A., Hill-Eubanks, D., Nelson, M. T. PIP(2) depletion promotes TRPV4 channel activity in mouse brain capillary endothelial cells. *Elife*, 2018, 7:e38689.
116. Boukalova, S., Marsakova, L., Teisinger, J., Vlachova, V. Conserved residues within the putative S4-S5 region serve distinct functions among thermosensitive vanilloid transient receptor potential (TRPV) channels. *J Biol Chem*, 2010, 285: 41455–41462.
117. Yelshanskaya, M. V., Sobolevsky, A. I. Ligand-binding sites in vanilloid-subtype TRP channels. *Front Pharmacol*, 2022, 13:900623.
118. Chuang, H. H., Prescott, E. D., Kong, H., et al. Bradykinin and nerve growth factor release the capsaicin receptor from PtdIns(4,5)P2-mediated inhibition. *Nature*, 2001, 411: 957–962.
119. Klein, R. M., Ufret-Vincenty, C. A., Hua, L., Gordon, S. E. Determinants of molecular specificity in phosphoinositide regulation: Phosphatidylinositol (4,5)-bisphosphate (PI(4,5)P2) is the endogenous lipid regulating TRPV1. *J Biol Chem*, 2008, 283: 26208–26216.
120. Lukacs, V., Thyagarajan, B., Varnai, P., et al. Dual regulation of TRPV1 by phosphoinositides. *J Neurosci*, 2007, 27: 7070–7080.
121. Yazici, A. T., Gianti, E., Kasimova, M. A., et al. Dual regulation of TRPV1 channels by phosphatidylinositol via functionally distinct binding sites. *J Biol Chem*, 2021, 296.
122. Cao, E., Cordero-Morales, J. F., Liu, B. Y., Qin, F., Julius, D. TRPV1 channels are intrinsically heat sensitive and negatively regulated by phosphoinositide lipids. *Neuron*, 2013, 77: 667–679.
123. Poblete, H., Oyarzun, I., Olivero, P., et al. Molecular determinants of phosphatidylinositol 4,5-bisphosphate (PI(4,5)P2) binding to transient receptor potential V1 (TRPV1) channels. *J Biol Chem*, 2015, 290: 2086–2098.
124. Velisetty, P., Borbiro, I., Kasimova, M. A., et al. A molecular determinant of phosphoinositide affinity in mammalian TRPV channels. *Sci Rep*, 2016, 6: 27652.
125. Jansson, E. T., Trkulja, C. L., Ahemaiti, A., et al. Effect of cholesterol depletion on the pore dilation of TRPV1. *Molecular Pain* 2013, 9:1.

126. Picazo-Juarez, G., Romero-Suarez, S., Nieto-Posadas, A., et al. Identification of a binding motif in the S5 helix that confers cholesterol sensitivity to the TRPV1 ion channel. *J Biol Chem*, 2011, 286: 24966–24976.
127. Klein, A. S., Tannert, A., Schaefer, M. Cholesterol sensitises the transient receptor potential channel TRPV3 to lower temperatures and activator concentrations. *Cell Calcium*, 2014, 55: 59–68.
128. Gao, L., Yang, P., Qin, P., et al. Selective potentiation of 2-APB-induced activation of TRPV1-3 channels by acid. *Sci Rep*, 2016, 6: 20791.
129. Hu, H. Z., Grandl, J., Bandell, M., Petrus, M., Patapoutian, A. Two amino acid residues determine 2-APB sensitivity of the ion channels TRPV3 and TRPV4. *Proc Natl Acad Sci U S A*, 2009, 106: 1626–1631.
130. Kogan, N. M., Mechoulam, R. Cannabinoids in health and disease. *Dialogues Clin Neurosci*, 2007, 9: 413–430.
131. De Petrocellis, L., Orlando, P., Moriello, A. S., et al. Cannabinoid actions at TRPV channels: Effects on TRPV3 and TRPV4 and their potential relevance to gastrointestinal inflammation. *Acta Physiol*, 2012, 204: 255–266.
132. Qin, N., Neeper, M. P., Liu, Y., et al. TRPV2 is activated by cannabidiol and mediates CGRP release in cultured rat dorsal root ganglion neurons. *J Neurosci*, 2008, 28: 6231–6238.
133. Smart, D., Gunthorpe, M. J., Jerman, J. C., et al. The endogenous lipid anandamide is a full agonist at the human vanilloid receptor (hVR1). *Br J Pharmacol*, 2000, 129: 227–230.
134. Janssens, A., Silvestri, C., Martella, A., et al. Δ9-tetrahydrocannabivarin impairs epithelial calcium transport through inhibition of TRPV5 and TRPV6. *Pharmacol Res*, 2018, 136: 83–89.
135. Zubcevic, L., Lee, S. Y. The role of pi-helices in TRP channel gating. *Curr OpinStruct Biol*, 2019, 58: 314–323.
136. Chang, Q., Gyftogianni, E., van de Graaf, S. F., et al. Molecular determinants in TRPV5 channel assembly. *J Biol Chem*, 2004, 279: 54304–54311.
137. Erler, I., Hirnet, D., Wissenbach, U., Flockerzi, V., Niemeyer, B. A. Ca2+-selective transient receptor potential v channel architecture and function require a specific ankyrin repeat. *J Biol Chem*, 2004, 279: 34456–34463.
138. Arniges, M., Fernandez-Fernandez, J. M., Albrecht, N., Schaefer, M., Valverde, M. A. Human TRPV4 channel splice variants revealed a key role of ankyrin domains in multimerization and trafficking. *J Biol Chem*, 2006, 281: 1580–1586.
139. Al-Ansary, D., Bogeski, I., Disteldorf, B. M. J., Becherer, U., Niemeyer, B. A. ATP modulates Ca2+ uptake by TRPV6 and is counteracted by isoform-specific phosphorylation. *FASEB J*, 2010, 24: 425–435.
140. Lishko, P. V., Procko, E., Jin, X., Phelps, C. B., Gaudet, R. The ankyrin repeats of TRPV1 bind multiple ligands and modulate channel sensitivity. *Neuron*, 2007, 54: 905–918.
141. Phelps, C. B., Procko, E., Lishko, P. V., Wang, R. R., Gaudet, R. Insights into the roles of conserved and divergent residues in the ankyrin repeats of TRPV ion channels. *Channels*, 2007, 1: 148–151.
142. Stokes, A. J., Wakano, C., del Carmen, K. A., Koblan-Huberson, M., Turner, H. Formation of a physiological complex between TRPV2 and RGA protein promotes cell surface expression of TRPV2. *J Cell Biochem*, 2005, 94: 669–683.
143. Garcia-Sanz, N., Fernandez-Carvajal, A., Morenilla-Palao, C., et al. Identification of a tetramerization domain in the C terminus of the vanilloid receptor. *J Neurosci*, 2004, 24: 5307–5314.
144. Becker, D., Muller, M., Leuner, K., Jendrach, M. The c-terminal domain of TRPV4 is essential for plasma membrane localization. *Mol Membr Biol*, 2008, 25: 139–U131.
145. Liu, B. Y., Ma, W. J., Ryu, S. J., Qin, F. Inhibitory modulation of distal C-terminal on protein kinase C-dependent phospho-regulation of rat TRPV1 receptors. *J Physiol-London*, 2004, 560: 627–638.
146. Brauchi, S., Orta, G., Salazar, M., Rosenmann, E., Latorre, R. A hot-sensing cold receptor: C-terminal domain determines thermosensation in transient receptor potential channels. *J Neurosci*, 2006, 26: 4835–4840.
147. Inada, H., Procko, E., Sotomayor, M., Gaudet, R. Structural and biochemical consequences of disease-causing mutations in the ankyrin repeat domain of the human TRPV4 channel. *Biochemistry*, 2012, 51: 6195–6206.
148. McCray, B. A., Diehl, E., Sullivan, J. M., et al. Neuropathy-causing TRPV4 mutations disrupt TRPV4-RhoA interactions and impair neurite extension. *Nat Commun*, 2021, 12:1444.
149. Cuajungco, M. P., Grimm, C., Oshima, K., et al. PACSINs bind to the TRPV4 cation channel - PACSIN 3 modulates the subcellular localization of TRPV4. *J Biol Chem*, 2006. 281: 18753–18762.
150. Goretzki, B., Glogowski, N. A., Diehl, E., et al. Structural basis of TRPV4 N terminus interaction with Syndapin/PACSIN1-3 and PIP2. *Structure*, 2018, 26: 1583.
151. Lau, S. Y., Procko, E., Gaudet, R. Distinct properties of Ca2+-calmodulin binding to N- and C-terminal regulatory regions of the TRPV1 channel. *J Gen Physiol*, 2012, 140: 541–555.

152. de Groot, T., Kovalevskaya, N. V., Verkaart, S., et al. Molecular mechanisms of calmodulin action on TRPV5 and modulation by parathyroid hormone. *Mol Cell Biol*, 2011, 31: 2845–2853.
153. Bokhovchuk, F. M., Bate, N., Kovalevskaya, N. V., et al. The structural basis of calcium-dependent inactivation of the transient receptor potential vanilloid 5 channel. *Biochemistry*, 2018, 57: 2623–2635.
154. Bate, N., Caves, R. E., Skinner, S. P., et al. A novel mechanism for calmodulin-dependent inactivation of transient receptor potential vanilloid 6. *Biochemistry*, 2018, 57: 2611–2622.
155. Kovalevskaya, N. V., Bokhovchuk, F. M., Vuister, G. W. The TRPV5/6 calcium channels contain multiple calmodulin binding sites with differential binding properties. *J Struct Funct Genomics*, 2012, 13: 91–100.
156. Hirnet, D., Olausson, J., Fecher-Trost, C., et al. The TRPV6 gene, cDNA and protein. *Cell Calcium*, 2003, 33: 509–518.
157. Dosey, T. L., Wang, Z., Fan, G., et al. Structures of TRPV2 in distinct conformations provide insight into role of the pore turret. *Nat Struct Mol Biol*, 2019, 26: 40–49.
158. Lu, P., Boros, S., Chang, Q., Bindels, R. J., Hoenderop, J. G. The beta-glucuronidase klotho exclusively activates the epithelial Ca2+ channels TRPV5 and TRPV6. *Nephrol Dial Transplant*, 2008, 23: 3397–3402.
159. Nadezhdin, K. D., Neuberger, A., Trofimov, Y. A., et al. Structural mechanism of heat-induced opening of a temperature-sensitive TRP channel. *Nat Struct Mol Biol*, 2021, 28: 564–572.

3 Role of the TRPM2 Channel as an ROS Sensing and Signaling Mechanism in Ischemia-Reperfusion Brain Damage

Huifang Hou, Yanping Luo, Jing Yao, Lin Lu, Luyu Ye, Ning Hua, Li Meng, Lu Wang, Yaling Yin, Sebastian Roger, Yang Wei and Lin-Hua Jiang

3.1 INTRODUCTION

Reactive oxygen species (ROS) refer to a collection of chemically reactive, oxygen-derived radical or non-radical molecules [1]. Under physiological conditions, mammalian cells produce ROS in a modest amount during cellular metabolism, and ROS as redox signaling molecules plays a crucial role in maintaining normal cellular and tissue homeostasis [1]. However, ROS can alter via oxidization the structure and function of nucleic acids, proteins and other biologically important macromolecules, and, as such, phagocytes generate a large quantity of ROS as a weaponry to kill invading pathogens [1–3]. Conceivably, accumulation of high concentrations of ROS, due to excessive generation of ROS and/or reduced antioxidant capacity of the cell, and development of oxidative stress can cause oxidative damage to healthy cells and tissues. Oxidative stress has been well documented as a prominent and common feature of numerous conditions, ranging from ischemic damage, neurodegenerative diseases, cardiovascular disorders, chronic kidney disease to liver injury. More importantly, as supported by increasing evidence, oxidative stress is a critical factor or mediator in the pathogenesis and progression of these pathological conditions [4–14].

The brain is high in energy consumption to maintain its normal function but displays substantial vulnerability to oxidative damage. Stroke occurring to the brain represents the second leading cause of mortality, the third leading cause of disability and the leading cause of dementia worldwide [15]. As recently reported by the World Stroke Organization [16], more than 63% of all stroke cases are ischemic stroke, which mainly arises from cerebral ischemia. Ischemia results from restricted blood circulation and thus the supply of oxygen and glucose to brain tissues and, if extended, can cause severe or even fatal brain damage. Advances in introducing preventive measurements, developing pharmacological and surgical interventions of ischemia and making them available to the public have been made, resulting in a significant reduction in the incidence and damage of ischemic stroke, most noticed in developed countries [15]. Reperfusion following transient ischemia can cause further brain damage, commonly referred to as ischemia-reperfusion brain damage, which is an important contributor to substantial neurological and cognitive deficits experienced by ischemic stroke survivors. In addition to cerebral ischemic stroke, ischemia-reperfusion brain damage can happen, albeit at low occurrence, following transient global brain ischemia, for example, as a result of cardiac arrest or drowsiness. Currently, treatment to alleviate ischemia-reperfusion brain damage is limited but highly desirable, considering the incidence and prevalence of ischemic stroke remaining alarmingly high [15, 16]. Decades of studies indicate that ischemia-reperfusion brain damage results from a complicated and highly intertwined cascade of pathophysiological events, including neuronal death, neuroinflammation, impairment in neurovascular unit and blood-brain

DOI: 10.1201/9781003402732-3

barrier (BBB) function and increased infiltration of peripheral immune cells into the brain [17–20]. Oxidative stress is known to be causatively associated with these pathophysiological events. Therefore, it is vital to understand the molecular mechanisms that sense and signal ROS, or oxidative stress, in order to identify novel targets for the development of therapeutics and rehabilitation strategies to lessen or prevent ischemia-reperfusion brain damage.

TRPM2 is the second melastatin-related member of the transient receptor potential (TRP) ion channel superfamily [21] and functions as a Ca^{2+}-permeable cation channel that is specifically gated by intracellular ADP-ribose (ADPR) [22]. ADPR is long known as a by-product of NAD metabolism in response to oxidative stress, making the TRPM2 channel highly sensitive to oxidative stress or pathologically relevant concentrations of ROS [23–25]. Thus, the TRPM2 channel has been suggested to be critically involved in ROS sensing and signaling [26–28]. The TRPM2 channel is widely expressed and found to be located on the cell surface in most cell types and, additionally or exclusively, in the membranes of intracellular organelles, such as lysosomes and mitochondria, in some cell types [29]. Studies have disclosed an important role of the TRPM2 channel in multiple physiological functions, including glucose-induced insulin secretion by pancreatic β-cells [30–33], regulation of immune cell function [34–36], warmth sensing by sensory neurons [37, 38], embryonic neurogenesis in hyperthermia [39] and protection against cisplatin-induced acute kidney injury [40]. More impressively, studies have accumulated a large body of evidence that supports the TRPM2 channel as a critical mechanism linking a diversity of oxidative stress-inducing pathological factors to associated pathologies, including ischemia-reperfusion brain damage [20, 29, 41–47].

In this chapter, we provide a concise summary of the structural features of the TRPM2 channel, particularly the recent progress in elucidating the structural basis for ADPR binding and gating, and discuss the current understanding of the role and mechanisms of the TRPM2 channel in ischemia-reperfusion brain damage.

3.2 STRUCTURAL FEATURES AND ACTIVATION OF THE TRPM2 CHANNEL

TRPM2 channels from multiple species, including human (hsTRPM2), mouse (mTRPM2), rat (rTRPM2), nematostella vectensis (nvTRPM2) and zebrafish (drTRPM2), have been molecularly cloned and examined, using electrophysiology and mutagenesis and, more recently, with structural determinations. These studies have significantly advanced our understanding of the structural features and activation of the TRPM2 channels.

3.2.1 Structural Features

The mammalian TRPM2 proteins consist of 1503 (hsTRPM2) or 1507 (mTRPM2 and rTRPM2) amino acid residues and have a size of approximately 170 kDa [26]. TRPM2 channels are permeable to cations including Ca^{2+}, Na^+ and K^+, with the ionic currents characterized by a linear current/voltage relationship [22, 48]. TRP channels, including TRPM2, have a membrane topology or architecture that resembles that of the voltage-gated K^+ channels and, by extension, the voltage-gated ion channels [49]. Namely, the TRPM2 channel is formed by four subunits, and each subunit is composed of an intracellular N-terminus, a transmembrane domain with six membrane-spanning segments (S1-S6), and an intracellular C-terminus (Fig. 3.1). As revealed in the structures [50], the transmembrane domain assumes a canonical domain-swapping conformation, in which the peripheral S1-S4 domain in one subunit is positioned adjacent to the central S5-S6 domain from the neighboring subunit. The S5-S6 domain, including S5, S6 and the re-entrant loop between them, from each subunit constitute the ion-permeating pathway in the center of the tetrameric complex. The N-terminus is mainly comprised of four TRPM homology regions (MHR1-4), with MHR1/2 adopting the Rossmann-like fold and MHR3/4 being α-helical. The C-terminus contains a TRP signature domain, a coiled-coil tetramerization domain, and the NUDT9-H domain that was named

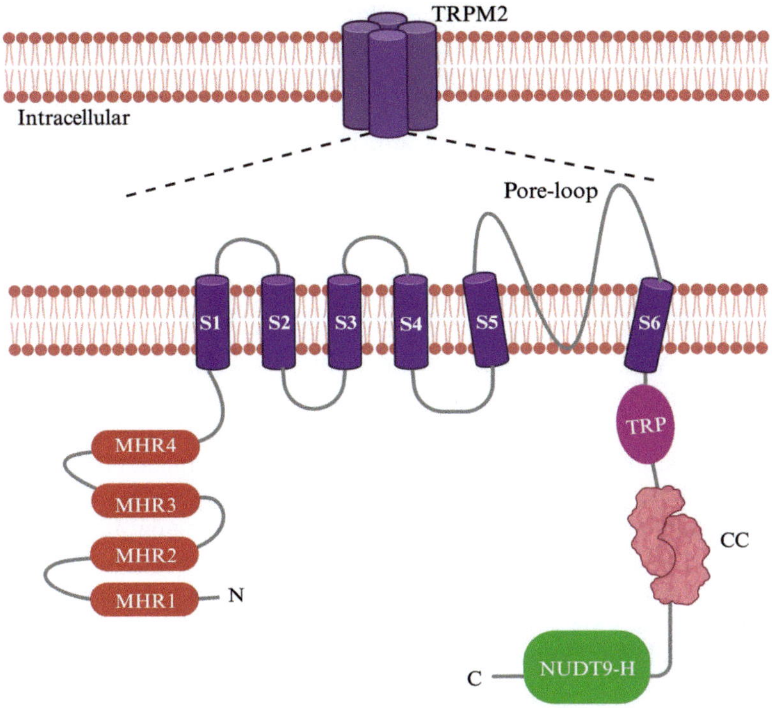

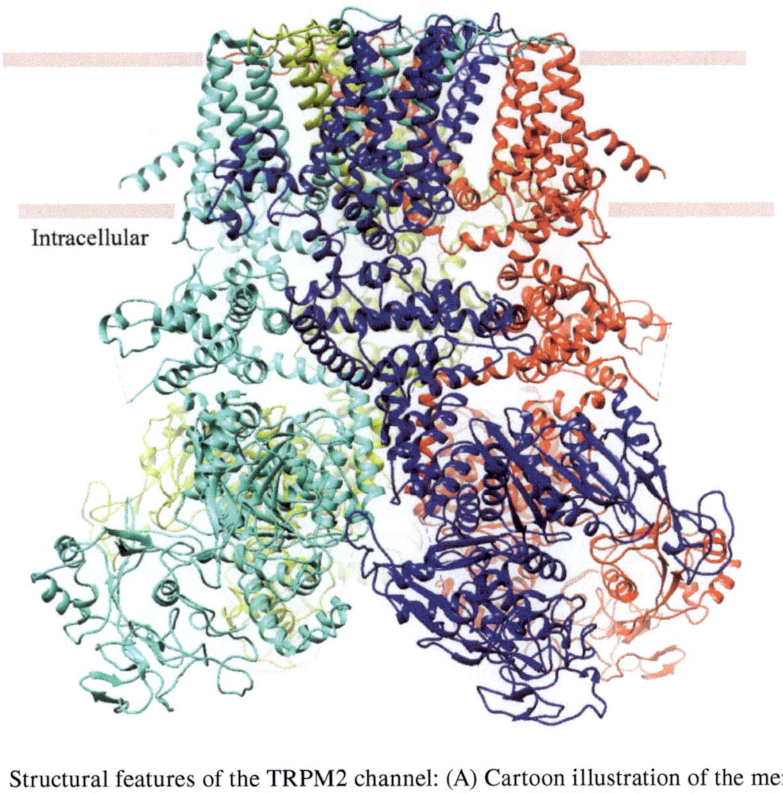

FIGURE 3.1 Structural features of the TRPM2 channel: (A) Cartoon illustration of the membrane topology of the tetrameric TRPM2 channel and its single subunit. MHR, TRPM homology region; S1–S6, six membrane-spanning segments; CC, coiled-coil domain; NUDT9-H, NUDT9 homology domain. (B) The atomic structure of the tetrameric hsTRPM2 channel [62].

for its sequence homology to the cytosolic ADPR pyrophosphate hydrolase [22] (Fig. 3.1). The NUDT9-H domain in the distant C-terminus is unique to the TRPM2 channels and, as elaborated next, is important for ADPR to interact with the TRPM2 channels [22]. It is worth mentioning that in addition to the full-length isoform (TRPM2-L), several short alternative splicing isoforms have been identified, including the TRPM2-S [24, 25]. The TRPM2-S isoform only has the N-terminus and the first two membrane-spanning segments and thus does not form an ion channel on its own, but can interact with the TRPM2 channel formed by the TRPM2-L isoform and prevent its activation [51].

3.2.2 Channel Activation

TRPM2 channels are activated by intracellular ADPR, more effectively in the presence of or in synergy with intracellular Ca^{2+} [52]. The EC_{50} value for ADPR to activate the TRPM2 channels varies in the range of micromolar concentrations, depending upon the experimental conditions used [26]. ADPR was initially thought to bind to the aforementioned NUDT9-H domain in the C-terminus [22]. Recent studies using cyro-electron microscopy (cyro-EM) in conjunction with patch-clamp recording provide substantial insights into the molecular basis for ADPR and Ca^{2+} to bind to the TRPM2 channels [50, 53–60]. In the nvTRPM2 channel, the C-terminal NUDT9-H domain exhibits the activity of catalyzing ADPR hydrolysis, indicating it is an ADPR-binding domain, but studies have shown that the nvTRPM2 channel activation by ADPR is independent of the C-terminal NUDT9-H domain [57–59]. Instead, ADPR has been proposed to gate the nvTRPM2 channel by interacting with site(s) other than the NUDT9-H domain, even though such ADPR-binding sites remain unidentified. The cryo-EM structures of the drTRPM2 channel have disclosed direct ADPR binding to the N-terminus with the site mainly formed by the MHR1/2 domain but no ADPR binding to the C-terminal NUDT9-H domain, and, nonetheless, as demonstrated by functional studies, the C-terminal NUDT9-H domain is indispensable for ADPR-induced activation of the drTRPM2 channel [53]. The ADPR-bound structures of the hsTRPM2 channel have revealed two ADPR-binding sites, located in the N-terminal MHR1/2 domain and in the C-terminal NUDT9-H domain, respectively. The ADPR-binding pocket in the MHR1/2 domain is small, shallow and oblate and has been proposed to serve the orthostatic ADPR-binding site, whereas the ADPR-binding pocket formed by the NUDT9-H domain is comparatively wide and long and acts as a second ADPR-binding site only in vertebrates to facilitate channel activation [54]. Intracellular Ca^{2+}, as suggested by an earlier study combining mutagenesis and patch-clamp recording, activates the hsTRPM2 channel through the calmodulin-binding IQ-like motif in the N-terminus [61]. However, as identified in the recently reported structures of the nvTRPM2, drTRPM2 and hsTRPM2 channels, intracellular Ca^{2+} binds to the membrane-cytosol interface and the binding site is formed by the intracellular ends of S2 and S3 and the S2-S3 linker [50, 53, 54]. It is worth pointing out that considerable structural differences exist in the reported Ca^{2+}-bound structures and Ca^{2+}-binding configurations. For example, in the cADPR-free closed state, the nvTRPM2 channel adopts a fourfold symmetry, but the drTRPM2 channel shows a twofold symmetry in the N-terminal MHR1/2 and MHR3 domains. In the ADPR-bound open-state structures of both hsTRPM2 and drTRPM2 channels, the Ca^{2+}-binding site is closer to and interacts with the TRP domain. Examinations of the hsTRPM2 and nvTRPM2 channels by combining cryo-EM, computational modeling, mutagenesis and patch-clamp recording also suggest significant structural differences in the ion selectivity filter located within the re-entrant loop domain between the S5 and S6 [62]. In the hsTRPM2 channel, the ion selectivity filter around the Phe^{979}-Gly-Gln-Ile^{982} microdomain assumes a narrow configuration restricting ion access to the transmembrane ion-conducting pathway in the closed state and widens in the open state, thus acting as an outer gate, and, by contrast, the ion selectivity filter in the nvTRPM2 channel adopts a much wider configuration even in the closed state, imposing no restriction to ion movements [62]. In summary, studies have shed light on the common and different structural features of the TRPM2 channels across species, particularly the structural basis for ADPR binding, and further efforts are required to develop a clear understanding of the channel gating mechanism [60].

In addition to ADPR, several other NAD metabolites, including cyclic ADPR (cADPR), 2′-O-acetyl-ADPR, ADPR-2'-phosphate, 2'-deoxy-ADPR and nicotinic acid-adenine dinucleotide phosphate, have been reported to activate TRPM2 channels [30, 63–66]. cADPR is a well-known intracellular signaling molecule participating in the regulation of muscle contraction, insulin release, neurotransmitter secretion and oocyte fertilization, among other physiological processes [67, 68]. As reported in early studies, cADPR can activate the TRPM2 channels [63, 64, 69] or in synergy with warm temperature [30]. Such channel activation by cADPR was however attributed by subsequent studies to contamination of commercially sourced cADPR with ADPR [70, 71]. Further efforts have not resolved such inconsistence. In one study, cADPR, synthesized in-house and purified with high-pressure liquid chromatography, was shown for its ability to bind to the C-terminal NUDT9-H domain derived from the hsTRPM2 channel using a surface plasmon resonance assay and induce hsTRPM2 channel activation using patch-clamp recording [72]. Moreover, mutations of critical residues in the ADPR-binding site in the C-terminal NUDT9-H domain resulted in differential effects on the ability of ADPR and cADPR to activate the hsTRPM2 channel. These results support the notion that cADPR is a direct agonist at the hsTRPM2 channel by interacting with an overlapping but not the same subset of residues in the C-terminal ADPR-binding pocket [72, 73]. However, in other independent studies [74, 75], cADPR, pretreated with the purified phosphohydrolase domain of the nvTRPM2 channel that selectively degrades ADPR but not cADPR, failed to activate the TRPM2 channels. In short, there is no consensus with respect to whether cADPR is a TRPM2 channel agonist.

Bilirubin, the end product of heme metabolism, and structurally related compounds, including biliverdin and xanthobilirubic acid methyl ester, have been recently reported as agonists of the TRPM2 channels with a micromolar or sub-micromolar potency [45]. Bilirubin and its derivatives differ from ADPR in chemical structures and are proposed to activate the TRPM2 channels from the extracellular side, and, moreover, the channel activation by bilirubin is independent of intracellular Ca^{2+} [45]. As already alluded to above, the TRPM2 channel is thermosensitive, with a temperature threshold of 42°C under *in vivo* conditions and 48°C in *in vitro* experimental settings, which is significantly higher, possibly due to loss of unknown endogenous regulation [76, 77]. Indeed, as shown for the mTRPM2 channel, the temperature threshold was elevated upon phosphorylation of Thr^{782} in the N-terminus by protein kinase C [78] or lowered to body temperature (37°C) by oxidization of Met^{214} in the N-terminus by H_2O_2 [79]. The temperature threshold shows strong dependence on intracellular ADPR and Ca^{2+} concentrations, and the sensitivity of the hsTRPM2 channel to body temperature is thought to be governed by the concerted actions of ADPR and Ca^{2+} restricted to the narrow range of their physiological concentrations [74]. Introduction of mutations in the C-terminal NUDT9-H domain was shown to compromise temperature-dependent activation of the hsTRPM2 channel by ADPR [80]. Nonetheless, the structural mechanism for temperature-induced activation of the TRPM2 channel remains to be delineated. In summary, studies indicate that the TRPM2 channel is gated by ADPR or other NAD metabolites together with Ca^{2+} from the inside and bilirubin and derivatives from outside (Fig. 3.2).

The TRPM2 channel, as introduced above, is potently activated by ROS at high and pathologically relevant concentrations. As has been widely documented in a variety of cell types [81–84], such channel activation is indirect, depending on the mechanism that is mediated by poly(ADPR) polymerase (PARP), particularly PARP1, and poly(ADPR)glycohydrolase (PARG) in the nucleus to promote the production of ADPR from NAD (Fig. 3.3). ROS has also been reported to activate the TRPM2 channel by stimulating NADase-mediated generation of ADPR from NAD in the mitochondria [85] (Fig. 3.3).

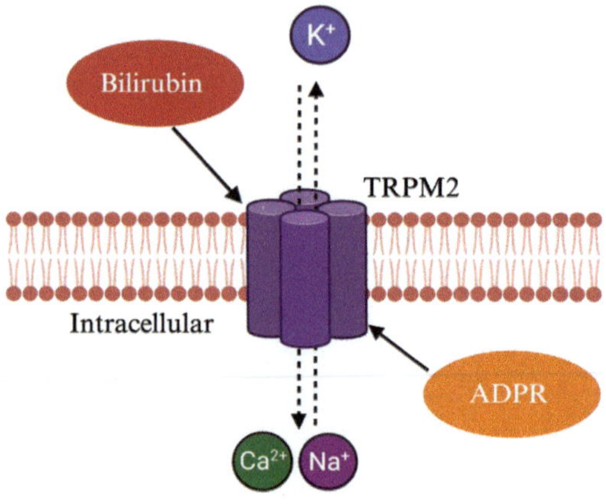

FIGURE 3.2 Activation of the TRPM2 channel by intracellular ADPR and extracellular bilirubin.

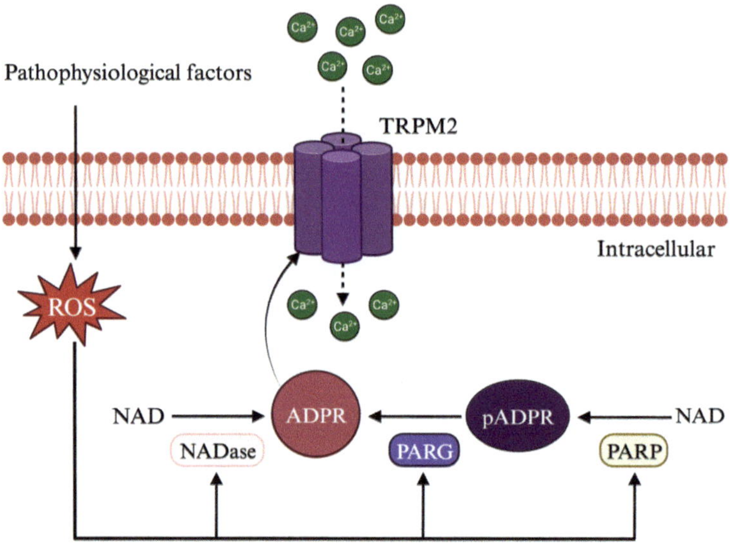

FIGURE 3.3 Molecular mechanisms of TRPM2 channel activation by reactive oxygen species. Accumulation of ROS inside the cell in response to a variety of pathophysiological factors activates the mechanism mediated by poly(ADPR)polymerase (PARP) and poly(ADPR)glycohydrolase (PARG) in the nucleus or NADase in the mitochondria to produce ADPR from NAD. ADPR binds to and opens the Ca^{2+}-permeable TRPM2 channel.

3.3 ROLE OF THE TRPM2 CHANNEL IN ISCHEMIA AND REPERFUSION BRAIN DAMAGE

Ischemia and reperfusion, particularly during reperfusion following transient ischemia, is known to induce excessive production of ROS and oxidative stress. Thus, since the recognition of the TRPM2 channel as a ROS sensing mechanism, its role in ischemia and reperfusion brain damage has been examined using *in vitro* and *in vivo* models in conjunction with pharmacological and genetic interventions of TRPM2 channel function or expression.

3.3.1 Ischemia-Reperfusion Using OGD/R In Vitro Model

Transient oxygen and glucose deprivation (OGD) followed by reoxygenation of extracellular solutions and supplementation with glucose (OGD/R) is a widely used *in vitro* model to examine ischemia-reperfusion brain damage, particularly neuronal death in cultured neurons or brain slices. OGD/R-induced neuronal death was significantly reduced in cultured cortical and hippocampal neurons by treatment, prior to OGD, with 2-aminoethoxy diphenyl borate (2-APB), clotrimazole (CTZ), flufenamic acid or N-(p-amylcinnamoyl) anthranilic acid (ACA) [86, 87], a group of structurally different chemicals that are known to inhibit the TRPM2 channel [88–91]. Similarly, in human neuroblastoma SH-SY5Y cells, a commonly used human neuronal cell model, OGD/R-induced cell death was attenuated by treatment, starting before reperfusion, with A23, an ACA derivative with a better selectivity toward the TRPM2 channel as well as with an antioxidant [97]. Consistent with these results from pharmacological studies, OGD/R induced significantly less neuronal death in cultured cortical and hippocampal neurons with the TRPM2 expression reduced, prior to OGD, by infection with lentivirus expressing TRPM2-specific short hairpin RNA (TRPM2-shRNA) [86, 87]. OGD/R-induced neuronal death was also significantly lower in cultured cortical neurons from the TRPM2-KO mice than from the WT mice [92]. Similarly, OGD/R-induced neuronal death was substantially mitigated in CA1 pyramidal neurons in hippocampal slice preparations from the TRPM2-KO mice compared to that from the WT mice [93]. Of notice, neuronal death in cultured hippocampal neurons was also inhibited by treatment with CTZ, post re-oxygenation, and the inhibition was almost as effective as that by treatment with CTZ starting prior to OGD, suggesting that post-ischemia TRPM2 channel activation is critical in driving neuronal death. Nonetheless, as shown in a recent study, neuronal death in cultured cortical neurons induced by OGD without introducing reperfusion was also attenuated by TRPM2-KO [43].

3.3.2 Ischemic Stroke Using MCAO/R In Vivo Model

Transient middle cerebral artery occlusion (MCAO) and reperfusion (MCAO/R) is an *in vivo* model to examine ischemic stroke-related brain damage in rodents, with the damage level determined by the infarction size or volume, as well as neuronal death and neurological deficits. MCAO/R-induced infarction in the cortex and striatum or in the whole ischemic hemisphere was significantly mitigated in the mice by treatment with CTZ via subcutaneous injections twice, one immediately after MCAO and the other at the beginning of reperfusion [86]. Similarly, MCAO/R-induced infarction and neurological deficits were attenuated by treatment with A23 [45, 97]. MCAO/R-induced infarction was significantly lessened in the TRPM2-KO mice than in the WT mice [43, 45, 92, 94, 95] and, more specifically, in neuron-selective TRPM2-KO mice [43]. Of notice, MCAO/R-induced brain damage in the TRPM2-KO mice was not reduced further by treatment with CTZ, suggesting that CTZ targets the TRPM2 channel to protect brain damage [92]. In the WT mice, MCAO/R-induced infarction was reduced within the striatal region that was injected, prior to MCAO, with lentivirus expressing TRPM2-specific shRNA [86]. As described above for the protective effects of treatment with CTZ, MCAO/R-induced brain damage in the WT mice was alleviated by treatment, prior to MCAO, with tat-M2NX, a cell-permeable peptide inhibitor of the TRPM2 channel, and such a protective effect was not observed in the TRPM2-KO mice [96]. The inhibition of MCAO/R-induced brain damage in the WT mice resulting from treatment with tat-M2NX lasted for several days after ischemia. MCAO/R-induced brain damage in aged mice was also effectively alleviated by treatment with tat-M2NX starting after reperfusion, a finding that is therapeutically important as most ischemic stroke cases occur to adults, particularly the elderly [15, 16]. Moreover, MCAO/R-induced neuronal death in the neocortex and neurological deficits in mice were mitigated by TRPM2-KO [95]. Similarly, MCAO/R-induced neuronal death was mitigated by administration of CTZ after reperfusion [96]. These findings support a critical role for the TRPM2 channel in brain damage due to ischemia-reperfusion. In contrast, brain damage caused by MCAO-induced permanent ischemia without the reperfusion phase in mice was not significantly affected by TRPM2-KO [94]. These studies consistently support the critical role of the TRPM2 channel in ischemic stroke brain damage and, more importantly, the TRPM2 channel as a promising target to treat ischemic stroke brain damage and associated neurological dysfunction.

3.3.3 Global Ischemia-Reperfusion In Vivo Models

In addition to brain damage due to cerebral ischemia, attention has been devoted to the role of the TRPM2 channel in mediating ischemia-reperfusion brain damage caused by global brain ischemia. CA/R-induced neuronal cell death in the hippocampal CA1 region in the WT mice was reduced by subcutaneous injection of CTZ after resuscitation [98, 99]. Transient bilateral common carotid artery occlusion and reperfusion (BCCAO/R) is another *in vivo* model used to examine global ischemia-reperfusion brain damage. BCCAO/R-induced neuronal death in the hippocampal CA1 region was lessened in the TRPM2-KO mice compared to that in the WT mice [93]. Moreover, BCCAO/R-induced impairments in learning and memory in mice were mitigated by TRPM2-KO [93].

In summary, studies using *in vitro* and *in vivo* rodent models have gathered compelling evidence to support an important role of the TRPM2 channel in mediating ischemic stroke and other ischemia-reperfusion brain damage and complications to neurological and cognitive functions.

3.4 TRPM2-MEDIATED MECHANISMS OF ISCHEMIA-REPERFUSION BRAIN DAMAGE

The TRPM2 channel is expressed in neuronal and non-neuronal cell types present in the brain. Increasing evidence indicates that activation of the TRPM2 channel in different cell types during ischemia and reperfusion contributes to ischemia-reperfusion brain damage (Fig. 3.4).

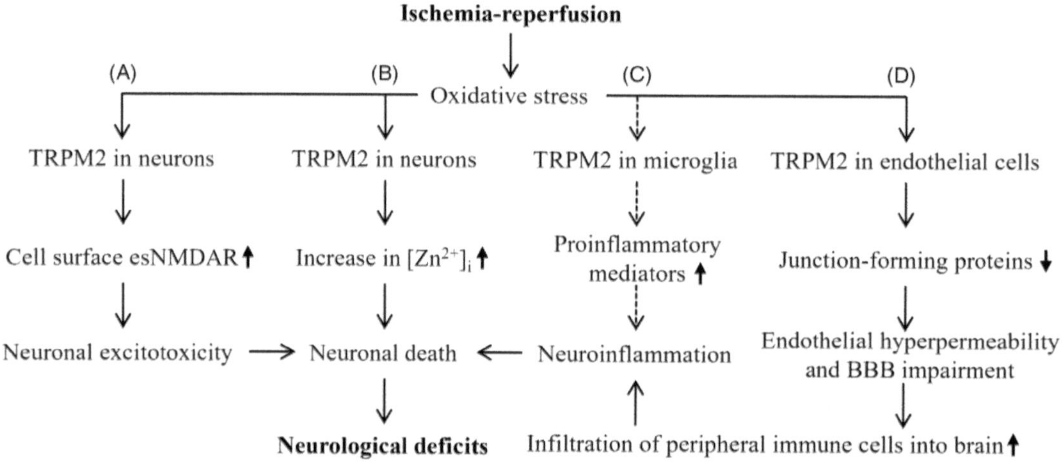

FIGURE 3.4 Summary of multiple TRPM2-mediated cellular mechanisms of ischemia-reperfusion brain damage. (A) The TRPM2 channel in neuronal cells mediates ischemia-induced cell surface expression of pro-death extra-synaptic NMDA receptors (esNMDAR) and NMDAR-mediated neuronal excitotoxicity. (B) Activation of the TRPM2 channel in neuronal cells during reperfusion results in elevation in the $[Zn^{2+}]_i$, leading to mitochondrial disruption and neuronal death. (C) The TRPM2 channel expressed in microglial cells is known to promote microglial activation and production of proinflammatory mediators induced by neurodegenerative disease-associated oxidative stress but, as highlighted by the dashed lines, such a role in ischemia-reperfusion brain damage remains to be established. (D) Activation of the TRPM2 channel in neurovascular endothelial cells promotes degradation of endothelial junction-forming proteins, leading to endothelial hyperpermeability and impairment in blood-brain barrier (BBB) function that facilitates infiltration of peripheral immune cells into the brain to exacerbate neuroinflammation.

3.4.1 Neuronal TRPM2 Channel-Mediated Neuronal Death

Neurons are the major executors of brain functions; for example, neurons in the hippocampus, cortex and striatum are critical for mediating neurological and cognitive functions. Studies in conjunction with pharmacological and genetic interventions have gathered strong evidence to support a vital role of the neuronal TRPM2 channel in mediating neuronal death induced by exposure to pathophysiological concentrations of ROS or pathological conditions known to promote ROS generation, including ischemia-reperfusion [29, 100]. Distinctive molecular and signaling mechanisms have been proposed for the neuronal TRPM2 channel to mediate neuronal death associated with ischemia-reperfusion brain damage.

It is well established that ischemia results in excessive release of glutamate, the major excitatory neurotransmitter in the brain, which induces neuronal excitotoxicity via the glutamate receptors, particularly the Ca^{2+}-permeable N-methyl-D-aspartate receptors (NMDARs) [117]. Studies have revealed a critical role of the TRPM2 channel in mediating neuronal death through regulating the NMDAR-dependent signaling pathways [43, 94]. As found in the aforementioned study examining the role of the TRPM2 channel in ischemic stroke brain damage, the GluN2A expression was upregulated and the GluN2B expression downregulated in hippocampal neurons as a result of TRPM2-KO. In addition, the synaptic excitability was increased in the TRPM2-KO mice, which was reversed by inhibiting the GluN2A-containing NMDARs [94]. Therefore, as has been hypothesized based on detailed analysis of the expression levels of the downstream signaling pathways, the TRPM2 channel on one hand enhances cell surface expression of the extra-synaptic GluN2B-containing NMDARs, resulting in extra-synaptic Ca^{2+} influx to induce neuronal death. On the other hand, the TRPM2 channel suppresses the GluN2A expression and the cell surface expression of the synaptic GluN2A-containing NMDARs, thereby reducing synaptic Ca^{2+} influx and subsequent activation of PI3 kinase and Akt and inhibition of GSK3β, the proapoptotic factor, to protect neurons [94]. In other words, the TRPM2 channel is thought to stimulate the extra-synaptic GluN2B-containing NMDAR-mediated pro-death signaling pathway and attenuate the synaptic GluN2A-containing NMDAR-mediated neuroprotective signaling pathway. However, as shown in a more recent study using OGD *in vitro* applied to cultured mouse cortical neurons or using MCAO *in vivo* in mice [43], ischemia, by promoting the TRPM2 channel to interact with GluN2A and GluN2B, augments the surface expression of extra-synaptic NMDARs and thereby enhances pro-death NMDAR-mediated neuronal excitotoxicity. Interestingly, administration via intraperitoneal injection of a membrane-permeable peptide to disrupt the TRPM2-NMDAR interactions, before or after MCAO, strongly protected MCAO/R-induced infarction and neurological deficits in the WT mice, but such a neuroprotective effect was absent in the mice with neuron-selective TRPM2-KO [43]. In short, these findings have led to the notion that ischemia promotes the TRPM2-NMDAR interaction and upregulates the cell surface expression of pro-death extra-synaptic NMDAR and thereby exacerbates neuronal excitotoxicity.

The homeostasis of intracellular ions, particularly intracellular Ca^{2+}, is vital for a plethora of cell functions, including neuronal death. It is known that aberrant intracellular Ca^{2+} homeostasis in neurons occurs during ischemia and reperfusion and induces neuronal death [100–102]. For the TRPM2 channel with a substantial permeability to Ca^{2+}, accumulating evidence from studies using cultured hippocampal, striatal and cortical neurons suggests that neuronal TRPM2-mediated Ca^{2+} signaling can induce neuronal death [43, 103, 104] and, nonetheless, the role of neuronal TRPM2-mediated Ca^{2+} signaling in ischemia-reperfusion-induced neuronal death remains to be defined. Zn^{2+}, one of the rich trace metal ions in the human body, is particularly known for its potent ability to induce neurotoxicity in the brain [102, 105–107]. While whether or not the TRPM2 channel is permeable to Zn^{2+} remains unclear, both OGD/R-induced increase in the $[Zn^{2+}]_i$ *in vitro* in hippocampal neurons in mouse brain slice preparations and BCCAO/R-induced increase in the $[Zn^{2+}]_i$ *in vivo* in hippocampal pyramidal neurons in the mice were reduced by TRPM2-KO [93]. In addition, hippocampal neuronal death induced by OGD/R or BCCAO/R was reduced by treatment with Zn^{2+} chelators. As further shown in hippocampal slice preparations using single cell imaging, OGD induced similar increases in the $[Zn^{2+}]_i$ in WT and TRPM2-KO hippocampal neurons, and the $[Zn^{2+}]_i$ remained persistently high in WT neurons but was rapidly lost in TRPM2-KO neurons during reperfusion, supporting a vital role for the TRPM2 channel in mediating post-ischemia increase in the $[Zn^{2+}]_i$ [93]. Accumulation of Zn^{2+} as well as Ca^{2+} in the mitochondria can disrupt mitochondrial function and induce mitochondrial ROS generation [106]. BCCAO/R-induced ROS generation in the hippocampus was strongly attenuated by TRPM2-KO [93]. As shown in human neuroblastoma SH-SY5Y cells and cultured mouse hippocampal neurons, exposure to high concentrations of H_2O_2 induced neuronal death, an increase in the $[Zn^{2+}]_i$, mitochondrial Zn^{2+} accumulation, mitochondrial dysfunction and ROS generation, all of which were inhibited by treatment with 2-APB as well as with a Zn^{2+} chelator [108, 109]. Collectively, these studies suggest that activation of the neuronal TRPM2 channel forms a vicious positive feedback cycle to drive neuronal death induced by ischemia-reperfusion by triggering an increase in the $[Zn^{2+}]_i$ (Fig. 3.3B) and possibly $[Ca^{2+}]_i$ that subsequently disrupts mitochondrial function and induces mitochondrial ROS generation, as has been proposed for amyloid beta-induced neuronal death and cognitive dysfunction associated with AD [100]. It is worth pointing out that multiple death signaling mechanisms, including apoptosis, necrosis, ferroptosis and autophagy, are known or proposed to contribute to neuronal death resulting from ischemia-reperfusion [110–115]. There is evidence to support the role of the TRPM2 channel in mediating neuronal death during ischemia and reperfusion via apoptosis and necrosis [43, 93]. Evidence also exists to suggest that TRPM2-mediated inhibition of autophagy by downregulating the AMPK/mTOR pathway is involved in MCAO/R-induced neuronal death [116]. Further investigations are required to dissect the role and contributions of different death mechanisms in TRPM2-mediated neuronal death associated with ischemia-reperfusion brain damage.

Bilirubin, as introduced above, is the product of heme metabolism and neurotoxic. It is known that the post-stroke serum bilirubin level is elevated among ischemic stroke patients and, importantly, as reported in a recent study, exhibits a positive correlation with the infarction volume in both ischemic stroke patients and in mice subjected to MCAO/R [45]. Bilirubin was found to act as a direct agonist at the TRPM2 channel. Furthermore, in the WT mice, intraperitoneal injection of bilirubin resulted in increases in the bilirubin levels in both serum and cerebrospinal fluid and exacerbated MCAO/R-induced infarction, both of which were attenuated in the TRPM2-KO mice. Exposure of cortical neurons in brain slice preparations from the WT mice to bilirubin increased neuronal excitability, which was also reduced by TRPM2-KO. Exposure of brain slice preparations from the WT mice to OGD triggered release of bilirubin, which was attenuated by treatment with A23 to inhibit the TRPM2 channel as well as by TRPM2-KO. Application of bilirubin to brain slice preparations from the WT mice exacerbated cortical neuronal death, which was also reduced by treatment with A23. In contrast, exposure to bilirubin of brain slice preparations from the knock-in mice expressing the bilirubin-insensitive TRPM2-D1066A mutant channel induced no increase in neuronal excitability [45]. Injection of bilirubin in the knock-in mice also failed to elevate the bilirubin levels in serum and cerebrospinal fluid and to exacerbate MCAO/R-induced infarction. Taken together, these findings support an important role of bilirubin-induced TRPM2 channel activation in exacerbating the vulnerability of the brain to ischemic stroke, adding another positive feedback loop in the vicious cycle of ischemic stroke brain damage, namely, the initial ischemic insult triggers release of bilirubin; bilirubin in turn activates the TRPM2 channel to induce neuronal excitability and excitotoxity, and damaged neurons release more bilirubin.

In summary, the neuronal TRPM2 channel participates in distinct molecular and signaling mechanisms mediating neuronal death associated with ischemia-reperfusion brain damage (Fig. 3.4A and B). It should be pointed out that these mechanisms are not necessarily mutually exclusive.

3.4.2 Non-neuronal TRPM2 Channel Mechanisms

In addition to neurons, the TRPM2 channel is expressed in multiple non-neuronal cells in the brain, such as microglial cells, BBB-forming neurovascular endothelial cells and pericytes as well as peripheral immune cells that infiltrate into the brain during ischemia-reperfusion. There is increasing evidence that the TRPM2 channel in these cells plays a critical role in ischemia-reperfusion brain damage by mediating microglial activation and neuroinflammation, disruption of endothelial cells and pericytes to impair BBB function, and infiltration of multiple peripheral immune cells into the brain to intensify neuroinflammation (Fig. 3.4C and D).

Microglial cells are the major immunocompetent cells in the brain, and, in response to brain tissue infection or damage, microglial cells become activated and initiate an immune response by generating proinflammatory mediators, including ROS, to restore brain tissue homeostasis. Such inflammatory mediators are also detrimental to neurons, and thus aberrant activation of microglial cells can provoke neuroinflammation, which represents a key factor in multiple brain pathologies, including ischemia-reperfusion brain damage [118]. The TRPM2 channel is highly expressed in microglial cells. There is compelling evidence to support an important role for the TRPM2 channel in mediating microglial activation and neuroinflammation in response to ROS or stimuli known to induce oxidative stress [119–121]. It has been reported that TRPM2-mediated activation of microglial cells contributes to chronic cerebral hypo-perfusion brain damage and neonatal hypoxic-ischemic brain damage [122, 123], but whether the TRPM2 channel plays a similar role in mediating microglial activation and neuroinflammation induced by ischemia-reperfusion remains unknown (Fig. 3.3C) and further investigations are required. The TRPM2 expression has been documented in peripheral immune cells. As shown in the mice, depletion of the TRPM2 expression in peripheral immune cells attenuated MCAO/R-induced infarction and neurological deficits by reducing activation of peripheral immune cells and infiltration into the brain. Therefore, the TRPM2 channel in peripheral immune cells contributes to ischemic stroke brain damage by promoting activation and infiltration of these cells into the brain to exacerbate neuroinflammation [95], which is consistent with a critical role of peripheral inflammation and brain-peripheral interactions in aggravating neuroinflammation and ischemia-reperfusion brain damage [124, 125].

BBB plays a crucial role in maintaining brain homeostasis and preventing the entry of harmful substances and peripheral immune cells into the brain. Impaired BBB function is a salient pathological feature of ischemic stroke, particularly in the early stages. Endothelial cells are a key cell component forming the BBB, and an increase in the endothelial permeability triggers the initial pathological changes, including infiltration of peripheral immune cells into the brain and intensification of neuroinflammation [126, 127]. The TRPM2 channel is highly expressed in neurovascular endothelial cells [128]. Activation of the TRPM2 channel in neurovascular endothelial cells in response to exposure to ROS or amyloid-beta results in an increase in the permeability of BBB, which contributes to AD pathogenesis [129]. As reported in a recent study examining the role of the TRPM2 channel in mediating BBB dysfunction associated with ischemic stroke [42], MCAO/R-induced infarction, plasma extravasation and infiltration of peripheral immune cells into the brain, as well as oxidative stress in endothelial cells in the mice, were alleviated by TRPM2-KO or endothelial cell-selective TRPM2-KO. Moreover, oxidative stress, an increase in the $[Ca^{2+}]_i$ and endothelial hyperpermeability in cultured cerebral endothelial cells induced by OGD/R or treatment with CD36 ligand thrombospondin-1 were suppressed by TRPM2-KO or inhibition of the TRPM2 channel [42]. Mechanistically, it is proposed that the TRPM2 channel and CD36 work together to support their activation and TRPM2-mediated Ca^{2+} influx into endothelial cells, and such Ca^{2+} signaling in turn promotes degradation of tight junction-forming proteins and endothelial hyperpermeability [42] (Fig. 3.4D). Pericytes are another cell type participating in BBB formation [130]. The TRPM2 channel is expressed in neurovascular pericytes. It was reported that exposure to zinc oxide nanoparticles (ZnO-NP) *in vitro* induced pericyte death and autophagy, which was reduced by pharmacological and genetic inhibition of the TRPM2 channel [131]. Autophagy was attenuated, and pericytes were better preserved in intact brain microvessels in the TRPM2-KO mice following treatment with ZnO-NP. Mechanically, ZnO-NP induces nitrosative stress that increases nitration of Tyr^{1485} in the C-terminus of the hsTRPM2 channel and, as a result, triggers autophagy of pericytes [131]. Further investigations are required to determine the role of the TRPM2 channel in pericytes in ischemia-reperfusion-induced impairment of BBB function and brain damage.

In summary, studies so far provide evidence to suggest that the TRPM2 channel in non-neuronal cells plays a significant role in mediating ischemia-reperfusion brain damage by promoting infiltration of peripheral immune cells into the brain to exacerbate neuroinflammation and disrupting neurovascular endothelial cells to impair the BBB function.

3.5 CONCLUDING REMARKS

Over the past more than twenty years since molecular cloning and functional demonstration of the TRPM2 channel as a Ca^{2+}-permeable channel gated by intracellular ADPR and Ca^{2+} and activated by ROS, substantial progress has been made in understanding the operational function of the TRPM2 channel, in particular, the structural basis for binding by ADPR/Ca^{2+}, thanks to recent structural studies. As highlighted above, more efforts are required to better understand the gating mechanism and the striking structural differences in terms of ADPR binding and other aspects of the TRPM2 channels across species. Studies have considerably advanced our knowledge regarding the cellular expression and subcellular location of the TRPM2 channels and their roles in physiological processes. The facts that oxidative stress plays significant roles in a variety of brain conditions and the TRPM2 channel shows a high sensitivity to oxidative stress have created continuous attention to the role of the TRPM2 channel and corresponding molecular and signaling mechanisms in the pathogenesis and progression of oxidative stress-mediated pathological conditions and diseases. Accumulating evidence supports an important role of the TRPM2 channel as a ROS sensing and signaling mechanism in coupling a number of oxidative stress-inducing pathological factors to associated conditions, including ischemia-reperfusion brain damage. Nonetheless, it is evident from the discussion above that more investigations are needed to develop a full comprehension of the role and mechanisms of the TRPM2 channels in different cell types and their contribution in ischemia-reperfusion brain. Finally, more efforts are required to develop selective and potent TRPM2-specific inhibitors [97, 132], particularly ones that can penetrate the BBB, and such inhibitors are important in providing the proof of concept that the TRPM2 channel is a therapeutic target for the development of pharmacological interventions to effectively alleviate brain damage and loss of neurological and cognitive functions due to ischemia-reperfusion.

ACKNOWLEDGEMENTS

The research from authors described in this chapter was supported by Wellcome Trust (072275/Z/03/Z). Alzheimer Research Trust (ART/PPG2009A/2), Natural Science Foundation of China (3147118, 82371421), Ministry of Science and Technology of China (G2022026006L), Henan Bureau of Education (16IRSTHN020) and Henan Bureau of Science and Technology (HNGD2022067).

REFERENCES

1. Sies, H., Jones, D. P. Reactive oxygen species (ROS) as pleiotropic physiological signalling agents. *Nat Rev Mol Cell Biol*, 2020, 21(7): 363–383. https://doi.org/10.1038/s41580-020-0230-3
2. Bedard, K., Krause, K. H. The NOX family of ROS-generating NADPH oxidases: Physiology and pathophysiology. *Physiol Rev*, 2007, 87(1): 245–313. https://doi.org/10.1152/physrev.00044.2005
3. Forrester, S. J., Kikuchi, D. S., Hernandes, M. S., Xu, Q., Griendling, K. K. Reactive oxygen species in metabolic and inflammatory signaling. *Circ Res*, 2018, 122(6): 877–902. https://doi.org/10.1161/CIRCRESAHA.117.311401
4. Chen, H., Yoshioka, H., Kim, G. S., et al. Oxidative stress in ischemic brain damage: Mechanisms of cell death and potential molecular targets for neuroprotection. *Antioxid Redox Signal*, 2011, 14(8): 1505–1517. https://doi.org/10.1089/ars.2010.3576

5. Jiang, T., Sun, Q., Chen, S. Oxidative stress: A major pathogenesis and potential therapeutic target of antioxidative agents in Parkinson's disease and Alzheimer's disease. *Prog Neurobiol*, 2016, 147: 1–19. https://doi.org/10.1016/j.pneurobio.2016.07.005
6. Förstermann, U., Xia, N., Li, H. Roles of vascular oxidative stress and nitric oxide in the pathogenesis of atherosclerosis. *Circ Res*, 2017, 120(4): 713–735. https://doi.org/10.1161/CIRCRESAHA.116.309326
7. Cadenas, S. ROS and redox signaling in myocardial ischemia-reperfusion injury and cardioprotection. *Free Radic Biol Med*, 2018, 117: 76–89. https://doi.org/10.1016/j.freeradbiomed.2018.01.024
8. Cheignon, C., Tomas, M., Bonnefont-Rousselot, D., Faller, P., Hureau, C., Collin, F. Oxidative stress and the amyloid beta peptide in Alzheimer's disease. *Redox Biol*, 2018, 14: 450–464. https://doi.org/10.1016/j.redox.2017.10.014
9. Orellana-Urzúa, S., Rojas, I., Líbano, L., Rodrigo, R. Pathophysiology of ischemic stroke: Role of oxidative stress. *Curr Pharm Des*, 2020, 26(34): 4246–4260. https://doi.org/10.2174/138161282666620070813 3912
10. Allameh, A., Niayesh-Mehr, R., Aliarab, A., Sebastiani, G., Pantopoulos, K. Oxidative stress in liver pathophysiology and disease. *Antioxidants (Basel)*, 2023, 12(9): 1653. https://doi.org/10.3390/antiox12091653
11. Amponsah-Offeh, M., Diaba-Nuhoho, P., Speier, S., Morawietz, H. Oxidative stress, antioxidants and hypertension. *Antioxidants (Basel)*, 2023, 12(2): 281. https://doi.org/10.3390/antiox12020281
12. Kishi, S., Nagasu, H., Kidokoro, K., Kashihara, N. Oxidative stress and the role of redox signalling in chronic kidney disease. *Nat Rev Nephrol*, 2024, 20(2): 101–119. https://doi.org/10.1038/s41581-023-00775-0
13. Zilberter, Y., Tabuena, D. R., Zilberter, M. NOX-induced oxidative stress is a primary trigger of major neurodegenerative disorders. *Prog Neurobiol*, 2023, 231: 102539. https://doi.org/10.1016/j.pneurobio.2023.102539
14. Perluigi, M., Di Domenico, F., Butterfield, D. A. Oxidative damage in neurodegeneration: Roles in the pathogenesis and progression of Alzheimer disease. *Physiol Rev*, 2024, 104(1): 103–197. https://doi.org/10.1152/physrev.00030.2022
15. Feigin, V. L., Owolabi, M. O. World Stroke Organization–Lancet Neurology Commission Stroke Collaboration Group Pragmatic solutions to reduce the global burden of stroke: A world stroke organization-lancet neurology commission. *Lancet Neurol*, 2023, 22(12): 1160–1206. https://doi.org/10.1016/S1474-4422(23)00277-6
16. Feigin, V. L., Brainin, M., Norrving, B., et al. World stroke organization (WSO): Global stroke fact sheet 2022. *Int J Stroke*, 2022, 17(1): 18–29. https://doi.org/10.1177/17474930211065917
17. Eltzschig, H. K., Eckle, T. Ischemia and reperfusion–from mechanism to translation. *Nat Med*, 2011, 17(11): 1391–1401. https://doi.org/10.1038/nm.2507
18. Iadecola, C., Anrather, J. The immunology of stroke: From mechanisms to translation. *Nat Med*, 2011, 17(7): 796–808. https://doi.org/10.1038/nm.2399
19. Sanderson, T. H., Reynolds, C. A., Kumar, R., Przyklenk, K., Hüttemann, M. Molecular mechanisms of ischemia-reperfusion injury in brain: Pivotal role of the mitochondrial membrane potential in reactive oxygen species generation. *Mol Neurobiol*, 2013, 47(1): 9–23. https://doi.org/10.1007/s12035-012-8344-z
20. Zhang, Y., Ying, F., Tian, X., et al. TRPM2 promotes atherosclerotic progression in a mouse model of atherosclerosis. *Cells*, 2022, 11(9):1423. https://doi.org/10.3390/cells11091423
21. Venkatachalam, K., Montell, C. TRP channels. *Annu Rev Biochem*, 2007, 76: 387–417. https://doi.org/10.1146/annurev.biochem.75.103004.142819
22. Perraud, A. L., Fleig, A., Dunn, C. A., et al. ADP-ribose gating of the calcium-permeable LTRPC2 channel revealed by Nudix motif homology. *Nature*, 2001, 411(6837): 595–599. https://doi.org/10.1038/35079100
23. Hara, Y., Wakamori, M., Ishii, M., et al. LTRPC2 Ca^{2+}-permeable channel activated by changes in redox status confers susceptibility to cell death. *Mol Cell*, 2002, 9(1): 163–173. https://doi.org/10.1016/s1097-2765(01)00438-5
24. Wehage, E., Eisfeld, J., Heiner, I., Jüngling, E., Zitt, C., Lückhoff, A. Activation of the cation channel long transient receptor potential channel 2 (LTRPC2) by hydrogen peroxide. A splice variant reveals a mode of activation independent of ADP-ribose. *J Biol Chem*, 2002, 277(26): 23150–23156. https://doi.org/10.1074/jbc.M112096200
25. Zhang, W., Chu, X., Tong, Q., et al. A novel TRPM2 isoform inhibits calcium influx and susceptibility to cell death. *J Biol Chem*, 2003, 278(18): 16222–16229. https://doi.org/10.1074/jbc.M300298200
26. Jiang, L. H., Yang, W., Zou, J., Beech, D. J. TRPM2 channel properties, functions and therapeutic potentials. *Expert Opin Ther Targets*, 2010, 14(9): 973–988. https://doi.org/10.1517/14728222.2010.510135

27. Sumoza-Toledo, A., Penner, R. TRPM2: A multifunctional ion channel for calcium signalling. *J Physiol*, 2011, 589(Pt 7): 1515–1525. https://doi.org/10.1113/jphysiol.2010.201855
28. Ru, X., Yao, X. TRPM2: A multifunctional ion channel for oxidative stress sensing. *Sheng Li Xue Bao*, 2014, 66(1): 7–15.
29. Malko, P., Jiang, L. H. TRPM2 channel-mediated cell death: An important mechanism linking oxidative stress-inducing pathological factors to associated pathological conditions. *Redox Biol*, 2020, 37: 101755. https://doi.org/10.1016/j.redox.2020.101755
30. Togashi, K., Hara, Y., Tominaga, T., et al. TRPM2 activation by cyclic ADP-ribose at body temperature is involved in insulin secretion. *EMBO J*, 2006, 25(9): 1804–1815. https://doi.org/10.1038/sj.emboj.7601083
31. Uchida, K., Dezaki, K., Damdindorj, B., et al. Lack of TRPM2 impaired insulin secretion and glucose metabolisms in mice. *Diabetes*, 2011, 60(1): 119–126. https://doi.org/10.2337/db10-0276
32. Yosida, M., Dezaki, K., Uchida, K., et al. Involvement of cAMP/EPAC/TRPM2 activation in glucose- and incretin-induced insulin secretion. *Diabetes*, 2014, 63(10): 3394–3403. https://doi.org/10.2337/db13-1868
33. Di, A., Gao, X. P., Qian, F., et al. The redox-sensitive cation channel TRPM2 modulates phagocyte ROS production and inflammation. *Nat Immunol*, 2011, 13(1): 29–34. https://doi.org/10.1038/ni.2171
34. Tripathi, J. K., Sharma, A., Sukumaran, P., et al. Oxidant sensor cation channel TRPM2 regulates neutrophil extracellular trap formation and protects against pneumoseptic bacterial infection. *FASEB J*, 2018, 32(12): fj201800605. https://doi.org/10.1096/fj.201800605
35. Gershkovitz, M., Caspi, Y., Fainsod-Levi, T., et al. TRPM2 mediates neutrophil killing of disseminated tumor cells. *Cancer Res*, 2018, 78(10): 2680–2690. https://doi.org/10.1158/0008-5472.CAN-17-3614
36. Wang, G., Cao, L., Liu, X., et al. Oxidant sensing by TRPM2 inhibits neutrophil migration and mitigates inflammation. *Dev Cell*, 2016, 38(5): 453–462. https://doi.org/10.1016/j.devcel.2016.07.014
37. Song, K., Wang, H., Kamm, G. B., et al. The TRPM2 channel is a hypothalamic heat sensor that limits fever and can drive hypothermia. *Science*, 2016, 353(6306): 1393–1398. https://doi.org/10.1126/science.aaf7537
38. Tan, C. H., McNaughton, P. A. The TRPM2 ion channel is required for sensitivity to warmth. *Nature*, 2016, 536(7617): 460–463. https://doi.org/10.1038/nature19074
39. Li, Y., Jiao, J. Deficiency of TRPM2 leads to embryonic neurogenesis defects in hyperthermia. *Sci Adv*, 2020, 6(1): eaay6350. https://doi.org/10.1126/sciadv.aay6350
40. Yu, B., Jin, L., Yao, X., et al. TRPM2 protects against cisplatin-induced acute kidney injury and mitochondrial dysfunction via modulating autophagy. *Theranostics*, 2023, 13(13): 4356–4375. https://doi.org/10.7150/thno.84655
41. Ali, E. S., Rychkov, G. Y., Barritt, G. J. TRPM2 non-selective cation channels in liver injury mediated by reactive oxygen species. *Antioxidants (Basel)*, 2021, 10(8): 1243. https://doi.org/10.3390/antiox10081243
42. Zong, P., Feng, J., Yue, Z., et al. TRPM2 deficiency in mice protects against atherosclerosis by inhibiting TRPM2-CD36 inflammatory axis in macrophages. *Nat Cardiovasc Res*, 2022, 1(4): 344–360. https://doi.org/10.1038/s44161-022-00027-7
43. Zong, P., Feng, J., Yue, Z., et al. Functional coupling of TRPM2 and extrasynaptic NMDARs exacerbates excitotoxicity in ischemic brain injury. *Neuron*, 2022, 110(12): 1944–1958.e8. https://doi.org/10.1016/j.neuron.2022.03.021
44. Zong, P., Feng, J., Li, C. X., et al. Activation of endothelial TRPM2 exacerbates blood-brain barrier degradation in ischemic stroke. *Cardiovasc Res*, 2024, 120(2): 188–202. https://doi.org/10.1093/cvr/cvad126
45. Liu, H. W., Gong, L. N., Lai, K., et al. Bilirubin gates the TRPM2 channel as a direct agonist to exacerbate ischemic brain damage. *Neuron*, 2023, 111(10): 1609–1625.e6. https://doi.org/10.1016/j.neuron.2023.02.022
46. Zhong, C., Yang, J., Zhang, Y., et al. TRPM2 mediates hepatic ischemia-reperfusion injury via Ca^{2+}-induced mitochondrial lipid peroxidation through increasing ALOX12 expression. *Research (Wash D C)*, 2023. 6: 0159.
47. Ye, P., Fang, Q., Hu, X., et al. TRPM2 as a conserved gatekeeper determines the vulnerability of DA neurons by mediating ROS sensing and calcium dyshomeostasis. *Prog Neurobiol*, 2023, 231: 102530. https://doi.org/10.1016/j.pneurobio.2023.102530
48. Sano, Y., Inamura, K., Miyake, A., et al. Immunocyte Ca^{2+} influx system mediated by LTRPC2. *Science*, 2001, 293(5533): 1327–1330. https://doi.org/10.1126/science.1062473

49. Yu, F. H., Yarov-Yarovoy, V., Gutman, G. A., Catterall, W. A. Overview of molecular relationships in the voltage-gated ion channel superfamily. *Pharmacol Rev*, 2005, 57(4): 387–395. https://doi.org/10.1124/pr.57.4.13
50. Zhang, Z., Tóth, B., Szollosi, A., Chen, J., Csanády, L. Structure of a TRPM2 channel in complex with Ca^{2+} explains unique gating regulation. *Elife*, 2018, 7: e36409. https://doi.org/10.7554/eLife.36409
51. Hecquet, C. M., Zhang, M., Mittal, M., et al. Cooperative interaction of trp melastatin channel transient receptor potential (TRPM2) with its splice variant TRPM2 short variant is essential for endothelial cell apoptosis. *Circ Res*, 2014, 114(3): 469–479. https://doi.org/10.1161/CIRCRESAHA.114.302414
52. McHugh, D., Flemming, R., Xu, S. Z., Perraud, A. L., Beech, D. J. Critical intracellular Ca^{2+} dependence of transient receptor potential melastatin 2 (TRPM2) cation channel activation. *J Biol Chem*, 2003, 278(13): 11002–11006. https://doi.org/10.1074/jbc.M210810200
53. Huang, Y., Winkler, P. A., Sun, W., Lü, W., Du, J. Architecture of the TRPM2 channel and its activation mechanism by ADP-ribose and calcium. *Nature*, 2018, 562(7725): 145–149. https://doi.org/10.1038/s41586-018-0558-4
54. Huang, Y., Roth, B., Lü, W., Du, J. Ligand recognition and gating mechanism through three ligand-binding sites of human TRPM2 channel. *Elife*, 2019, 8: e50175. https://doi.org/10.7554/eLife.50175
55. Wang, L., Fu, T. M., Zhou, Y., Xia, S., Greka, A., Wu, H. Structures and gating mechanism of human TRPM2. *Science*, 2018, 362(6421): eaav4809. https://doi.org/10.1126/science.aav4809
56. Yin, Y., Wu, M., Hsu, A. L., et al. Visualizing structural transitions of ligand-dependent gating of the TRPM2 channel. *Nat Commun*, 2019, 10(1): 3740. https://doi.org/10.1038/s41467-019-11733-5
57. Iordanov, I., Tóth, B., Szollosi, A., Csanády, L. Enzyme activity and selectivity filter stability of ancient TRPM2 channels were simultaneously lost in early vertebrates. *Elife*, 2019, 8: e44556. https://doi.org/10.7554/eLife.44556
58. Kühn, F. J., Kühn, C., Winking, M., Hoffmann, D. C., Lückhoff, A. ADP-ribose activates the TRPM2 channel from the sea anemone nematostella vectensis independently of the NUDT9H domain. *PLoS One*, 2016, 11(6): e0158060. https://doi.org/10.1371/journal.pone.0158060
59. Kühn, F. J. P., Watt, J. M., Potter, B. V. L., Lückhoff, A. Different substrate specificities of the two ADPR binding sites in TRPM2 channels of *Nematostella vectensis* and the role of IDPR. *Sci Rep*, 2019, 9(1): 4985. https://doi.org/10.1038/s41598-019-41531-4
60. Xia, S., Wang, L., Fu, T. M., Wu, H. Mechanism of TRPM2 channel gating revealed by cryo-EM. *FEBS J*, 2019, 286(17): 3333–3339. https://doi.org/10.1111/febs.14939
61. Du, J., Xie, J., Yue, L. Intracellular calcium activates TRPM2 and its alternative spliced isoforms. *Proc Natl Acad Sci U S A*, 2009, 106(17): 7239–7244. https://doi.org/10.1073/pnas.0811725106
62. Yu, X., Xie, Y., Zhang, X., et al. Structural and functional basis of the selectivity filter as a gate in human TRPM2 channel. *Cell Rep*, 2021, 37(7): 110025. https://doi.org/10.1016/j.celrep.2021.110025
63. Kolisek, M., Beck, A., Fleig, A., Penner, R. Cyclic ADP-ribose and hydrogen peroxide synergize with ADP-ribose in the activation of TRPM2 channels. *Mol Cell*, 2005, 18(1): 61–69. https://doi.org/10.1016/j.molcel.2005.02.033
64. Beck, A., Kolisek, M., Bagley, L. A., Fleig, A., Penner, R. Nicotinic acid adenine dinucleotide phosphate and cyclic ADP-ribose regulate TRPM2 channels in T lymphocytes. *FASEB J*, 2006, 20(7): 962–964. https://doi.org/10.1096/fj.05-5538fje
65. Grubisha, O., Rafty, L. A., Takanishi, C. L., et al. Metabolite of SIR2 reaction modulates TRPM2 ion channel. *J Biol Chem*. 2006; 281(20):14057–14065. https://doi.org/10.1074/jbc.M513741200
66. Fliegert, R., Bauche, A., Wolf Pérez, A. M., et al. 2'-deoxyadenosine 5'-diphosphoribose is an endogenous TRPM2 superagonist. *Nat Chem Biol*, 2017, 13(9): 1036–1044. https://doi.org/10.1038/nchembio.2415
67. Koch-Nolte, F., Haag, F., Guse, A. H., Lund, F., Ziegler, M. Emerging roles of NAD^+ and its metabolites in cell signaling. *Sci Signal*, 2009, 2(57): mr1. https://doi.org/10.1126/scisignal.257mr1
68. Lee, H. C. Cyclic ADP-ribose and nicotinic acid adenine dinucleotide phosphate (NAADP) as messengers for calcium mobilization. *J Biol Chem*, 2012, 287(38): 31633–31640. https://doi.org/10.1074/jbc.R112.349464
69. Lange, I., Penner, R., Fleig, A., Beck, A. Synergistic regulation of endogenous TRPM2 channels by adenine dinucleotides in primary human neutrophils. *Cell Calcium*, 2008, 44(6): 604–615. https://doi.org/10.1016/j.ceca.2008.05.001
70. Tóth, B., Csanády, L. Identification of direct and indirect effectors of the transient receptor potential melastatin 2 (TRPM2) cation channel. *J Biol Chem*, 2010, 285(39): 30091–30102. https://doi.org/10.1074/jbc.M109.066464

71. Tóth, B., Iordanov, I., Csanády, L. Ruling out pyridine dinucleotides as true TRPM2 channel activators reveals novel direct agonist ADP-ribose-2'-phosphate. *J Gen Physiol*, 2015, 145(5): 419–430. https://doi.org/10.1085/jgp.201511377
72. Yu, P., Liu, Z., Yu, X., et al. Direct gating of the TRPM2 channel by cADPR via specific interactions with the ADPR binding pocket. *Cell Rep*, 2019, 27(12): 3684–3695.e4. https://doi.org/10.1016/j.celrep.2019.05.067
73. Yu, P., Xue, X., Zhang, J., et al. Identification of the ADPR binding pocket in the NUDT9 homology domain of TRPM2. *J Gen Physiol*, 2017, 149(2): 219–235. https://doi.org/10.1085/jgp.201611675
74. Bartók, Á, Csanády, L. Dual amplification strategy turns TRPM2 channels into supersensitive central heat detectors. *Proc Natl Acad Sci U S A*, 2022, 119(48): e2212378119. https://doi.org/10.1073/pnas.2212378119
75. Riekehr, W. M., Sander, S., Pick, J., et al. cADPR does not activate TRPM2. *Int J Mol Sci*, 2022, 23(6): 3163. https://doi.org/10.3390/ijms23063163
76. Vilar, B., Tan, C. H., McNaughton, P. A. Heat detection by the TRPM2 ion channel. *Nature*, 2020, 584(7820): E5–E12. https://doi.org/10.1038/s41586-020-2510-7
77. Mulier, M., Vandewauw, I., Vriens, J., Voets, T. Reply to: Heat detection by the TRPM2 ion channel. *Nature*, 2020, 584(7820): E13–E15. https://doi.org/10.1038/s41586-020-2511-6
78. Kashio, M., Masubuchi, S., Tominaga, M. Protein kinase c-mediated phosphorylation of transient receptor potential melastatin type 2 Thr738 counteracts the effect of cytosolic Ca^{2+} and elevates the temperature threshold. *J Physiol*, 2022, 600(19): 4287–4302. https://doi.org/10.1113/JP283350
79. Kashio, M., Sokabe, T., Shintaku, K., et al. Redox signal-mediated sensitization of transient receptor potential melastatin 2 (TRPM2) to temperature affects macrophage functions. *Proc Natl Acad Sci U S A*, 2012, 109(17): 6745–6750. https://doi.org/10.1073/pnas.1114193109
80. Gattkowski, E., Johnsen, A., Bauche, A., et al. Novel CaM-binding motif in its NudT9H domain contributes to temperature sensitivity of TRPM2. *Biochim Biophys Acta Mol Cell Res*, 2019, 1866(7): 1162–1170. https://doi.org/10.1016/j.bbamcr.2018.12.010
81. Fonfria, E., Marshall, I. C., Benham, C. D., et al. TRPM2 channel opening in response to oxidative stress is dependent on activation of poly(ADP-ribose) polymerase. *Br J Pharmacol*, 2004, 143(1): 186–192. https://doi.org/10.1038/sj.bjp.0705914
82. Miller, B. A. Inhibition of TRPM2 function by PARP inhibitors protects cells from oxidative stress-induced death. *Br J Pharmacol*, 2004, 143(5): 515–516. https://doi.org/10.1038/sj.bjp.0705923
83. Buelow, B., Song, Y., Scharenberg, A. M. The poly(ADP-ribose) polymerase PARP-1 is required for oxidative stress-induced TRPM2 activation in lymphocytes. *J Biol Chem*, 2008, 283(36): 24571–24583. https://doi.org/10.1074/jbc.M802673200
84. Blenn, C., Wyrsch, P., Bader, J., Bollhalder, M., Althaus, F. R. Poly(ADP-ribose)glycohydrolase is an upstream regulator of Ca^{2+} fluxes in oxidative cell death. *Cell Mol Life Sci*, 2011, 68(8): 1455–1466. https://doi.org/10.1007/s00018-010-0533-1
85. Perraud, A. L., Takanishi, C. L., Shen, B., et al. Accumulation of free ADP-ribose from mitochondria mediates oxidative stress-induced gating of TRPM2 cation channels. *J Biol Chem*, 2005, 280(7): 6138–6148. https://doi.org/10.1074/jbc.M411446200
86. Jia, J., Verma, S., Nakayama, S., et al. Sex differences in neuroprotection provided by inhibition of TRPM2 channels following experimental stroke. *J Cereb Blood Flow Metab*, 2011, 31(11): 2160–2168. https://doi.org/10.1038/jcbfm.2011.77
87. Verma, S., Quillinan, N., Yang, Y. F., et al. TRPM2 channel activation following in vitro ischemia contributes to male hippocampal cell death. *Neurosci Lett*, 2012, 530(1): 41–46. https://doi.org/10.1016/j.neulet.2012.09.044
88. Togashi, K., Inada, H., Tominaga, M. Inhibition of the transient receptor potential cation channel TRPM2 by 2-aminoethoxydiphenyl borate (2-APB). *Br J Pharmacol*, 2008, 153(6): 1324–1330. https://doi.org/10.1038/sj.bjp.0707675
89. Hill, K., McNulty, S., Randall, A. D. Inhibition of TRPM2 channels by the antifungal agents clotrimazole and econazole. *Naunyn Schmiedebergs Arch Pharmacol*, 2004, 370(4): 227–237. https://doi.org/10.1007/s00210-004-0981-y
90. Hill, K., Benham, C. D., McNulty, S., Randall, A. D. Flufenamic acid is a pH-dependent antagonist of TRPM2 channels. *Neuropharmacology*, 2004, 47(3): 450–460. https://doi.org/10.1016/j.neuropharm.2004.04.014
91. Kraft, R., Grimm, C., Frenzel, H., Harteneck, C. Inhibition of TRPM2 cation channels by n-(p-amylcinnamoyl)anthranilic acid. *Br J Pharmacol*, 2006, 148(3): 264–273. https://doi.org/10.1038/sj.bjp.0706739

92. Shimizu, T., Macey, T. A., Quillinan, N., et al. Androgen and PARP-1 regulation of TRPM2 channels after ischemic injury. *J Cereb Blood Flow Metab*, 2013, 33(10): 1549–1555. https://doi.org/10.1038/jcbfm.2013.105
93. Ye, M., Yang, W., Ainscough, J. F., et al. TRPM2 channel deficiency prevents delayed cytosolic Zn^{2+} accumulation and CA1 pyramidal neuronal death after transient global ischemia. *Cell Death Dis*, 2014, 5(11): e1541. https://doi.org/10.1038/cddis.2014.494
94. Alim, I., Teves, L., Li, R., Mori, Y., Tymianski, M. Modulation of NMDAR subunit expression by TRPM2 channels regulates neuronal vulnerability to ischemic cell death. *J Neurosci*, 2013, 33(44): 17264–17277. https://doi.org/10.1523/JNEUROSCI.1729-13.2013
95. Gelderblom, M., Melzer, N., Schattling, B., et al. Transient receptor potential melastatin subfamily member 2 cation channel regulates detrimental immune cell invasion in ischemic stroke. *Stroke*, 2014, 45(11): 3395–3402. https://doi.org/10.1161/STROKEAHA.114.005836
96. Shimizu, T., Dietz, R. M., Cruz-Torres, I., et al. Extended therapeutic window of a novel peptide inhibitor of TRPM2 channels following focal cerebral ischemia. *Exp Neurol*, 2016, 283(Pt A): 151–156. https://doi.org/10.1016/j.expneurol.2016.06.015
97. Zhang, H., Yu, P., Lin, H., et al. The discovery of novel ACA derivatives as specific TRPM2 inhibitors that reduce ischemic injury both in vitro and in vivo. *J Med Chem*, 2021, 64(7): 3976–3996. https://doi.org/10.1021/acs.jmedchem.0c02129
98. Nakayama, S., Vest, R., Traystman, R. J., Herson, P. S. Sexually dimorphic response of TRPM2 inhibition following cardiac arrest-induced global cerebral ischemia in mice. *J Mol Neurosci*, 2013, 51(1): 92–98. https://doi.org/10.1007/s12031-013-0005-9
99. Shimizu, K., Quillinan, N., Orfila, J. E., Herson, P. S. Sirtuin-2 mediates male specific neuronal injury following experimental cardiac arrest through activation of TRPM2 ion channels. *Exp Neurol*, 2016, 275(Pt 1): 78–83. https://doi.org/10.1016/j.expneurol.2015.10.014
100. Jiang, L. H., Li, X., Syed Mortadza, S. A., Lovatt, M., Yang, W. The TRPM2 channel nexus from oxidative damage to Alzheimer's pathologies: An emerging novel intervention target for age-related dementia. *Ageing Res Rev*, 2018, 47: 67–79. https://doi.org/10.1016/j.arr.2018.07.002
101. Doyle, K. P., Simon, R. P., Stenzel-Poore, M. P. Mechanisms of ischemic brain damage. *Neuropharmacology*, 2008, 55(3): 310–318. https://doi.org/10.1016/j.neuropharm.2008.01.005
102. Li, C., Meng, L., Li, X., Li, D., Jiang, L. H. Non-NMDAR neuronal Ca^{2+}-permeable channels in delayed neuronal death and as potential therapeutic targets for ischemic brain damage. *Expert Opin Ther Targets*, 2015, 19(7): 879–892. https://doi.org/10.1517/14728222.2015.1021781
103. Kaneko, S., Kawakami, S., Hara, Y., et al. A critical role of TRPM2 in neuronal cell death by hydrogen peroxide. *J Pharmacol Sci*, 2006, 101(1): 66–76. https://doi.org/10.1254/jphs.fp0060128
104. Sun, Y., Sukumaran, P., Selvaraj, S., et al. TRPM2 promotes neurotoxin MPP^+/MPTP-induced cell death. *Mol Neurobiol*, 2018, 55(1): 409–420. https://doi.org/10.1007/s12035-016-0338-9
105. Stork, C. J., Li, Y. V. Rising zinc: A significant cause of ischemic neuronal death in the CA1 region of rat hippocampus. *J Cereb Blood Flow Metab*, 2009, 29(8): 1399–1408. https://doi.org/10.1038/jcbfm.2009.64
106. Shuttleworth, C. W., Weiss, J. H. Zinc: New clues to diverse roles in brain ischemia. *Trends Pharmacol Sci*, 2011, 32(8): 480–486. https://doi.org/10.1016/j.tips.2011.04.001
107. Aras, M. A., Aizenman, E. Redox regulation of intracellular zinc: Molecular signaling in the life and death of neurons. *Antioxid Redox Signal*, 2011, 15(8): 2249–2263. https://doi.org/10.1089/ars.2010.3607
108. Li, X., Jiang, L. H. Multiple molecular mechanisms form a positive feedback loop driving amyloid β42 peptide-induced neurotoxicity via activation of the TRPM2 channel in hippocampal neurons. *Cell Death Dis*, 2018, 9(2): 195. https://doi.org/10.1038/s41419-018-0270-1
109. Li, X., Jiang, L. H. A critical role of the transient receptor potential melastatin 2 channel in a positive feedback mechanism for reactive oxygen species-induced delayed cell death. *J Cell Physiol*, 2019, 234(4): 3647–3660. https://doi.org/10.1002/jcp.27134
110. Tuo, Q. Z., Zhang, S. T., Lei, P. Mechanisms of neuronal cell death in ischemic stroke and their therapeutic implications. *Med Res Rev*, 2022, 42(1): 259–305. https://doi.org/10.1002/med.21817
111. Mao, R., Zong, N., Hu, Y., Chen, Y., Xu, Y. Neuronal death mechanisms and therapeutic strategy in ischemic stroke. *Neurosci Bull*, 2022, 38(10): 1229–1247. https://doi.org/10.1007/s12264-022-00859-0
112. He, C., Xu, Y., Sun, J., Li, L., Zhang, J. H., Wang, Y. Autophagy and apoptosis in acute brain injuries: From mechanism to treatment. *Antioxid Redox Signal*, 2023, 38(1–3): 234–257. https://doi.org/10.1089/ars.2021.0094

113. Shi, G. S., Qin, Q. L., Huang, C., et al. The pathological mechanism of neuronal autophagy-lysosome dysfunction after ischemic stroke. *Cell Mol Neurobiol*, 2023, 43(7): 3251–3263. https://doi.org/10.1007/s10571-023-01382-0
114. Hu, X., Bao, Y., Li, M., Zhang, W., Chen, C. The role of ferroptosis and its mechanism in ischemic stroke. *Exp Neurol*, 2024, 372: 114630. https://doi.org/10.1016/j.expneurol.2023.114630
115. Wang, Q., Yang, F., Duo, K., et al. The role of necroptosis in cerebral ischemic stroke. *Mol Neurobiol*, 2024, 61(7): 3882–3898. https://doi.org/10.1007/s12035-023-03728-7
116. Hu, X., Wu, L., Liu, X., et al. Deficiency of ROS-activated TRPM2 channel protects neurons from cerebral ischemia-reperfusion injury through upregulating autophagy. *Oxid Med Cell Longev*, 2021, 2021: 7356266. https://doi.org/10.1155/2021/7356266
117. Lai, T. W., Zhang, S., Wang, Y. T. Excitotoxicity and stroke: Identifying novel targets for neuroprotection. *Prog Neurobiol*, 2014, 115: 157–188. https://doi.org/10.1016/j.pneurobio.2013.11.006
118. Jurcau, A., Simion, A. Neuroinflammation in cerebral ischemia and ischemia/reperfusion injuries: From pathophysiology to therapeutic strategies. *Int J Mol Sci*, 2021, 23(1): 14. https://doi.org/10.3390/ijms23010014
119. Ostapchenko, V. G., Chen, M., Guzman, M. S., et al. The transient receptor potential melastatin 2 (TRPM2) channel contributes to β-amyloid oligomer-related neurotoxicity and memory impairment. *J Neurosci*, 2015, 35(45): 15157–15169. https://doi.org/10.1523/JNEUROSCI.4081-14.2015
120. Aminzadeh, M., Roghani, M., Sarfallah, A., Riazi, G. H. TRPM2 dependence of ROS-induced NLRP3 activation in Alzheimer's disease. *Int Immunopharmacol*, 2018, 54: 78–85. https://doi.org/10.1016/j.intimp.2017.10.024
121. Alawieyah Syed Mortadza, S., Sim, J. A., Neubrand, V. E., Jiang, L. H. A critical role of TRPM2 channel in $A\beta_{42}$-induced microglial activation and generation of tumor necrosis factor-α. *Glia*, 2018, 66(3): 562–575. https://doi.org/10.1002/glia.23265
122. Huang, S., Turlova, E., Li, F., et al. Transient receptor potential melastatin 2 channels (TRPM2) mediate neonatal hypoxic-ischemic brain injury in mice. *Exp Neurol*, 2017, 296: 32–40. https://doi.org/10.1016/j.expneurol.2017.06.023
123. Miyanohara, J., Kakae, M., Nagayasu, K., et al. TRPM2 channel aggravates CNS inflammation and cognitive impairment via activation of microglia in chronic cerebral hypoperfusion. *J Neurosci*, 2018, 38(14): 3520–3533. https://doi.org/10.1523/JNEUROSCI.2451-17.2018
124. Cheng, W., Zhao, Q., Li, C., Xu, Y. Neuroinflammation and brain-peripheral interaction in ischemic stroke: A narrative review. *Front Immunol*, 2023, 13: 1080737.
125. Monsour, M., Borlongan, C. V. The central role of peripheral inflammation in ischemic stroke. *J Cereb Blood Flow Metab*, 2023, 43(5): 622–641. https://doi.org/10.1177/0271678X221149509
126. Tiedt, S., Buchan, A. M., Dichgans, M., Lizasoain, I., Moro, M. A., Lo, E. H. The neurovascular unit and systemic biology in stroke - implications for translation and treatment. *Nat Rev Neurol*, 2022, 18(10): 597–612. https://doi.org/10.1038/s41582-022-00703-z
127. Mathias, K., Machado, R. S., Stork, S., et al. Blood-brain barrier permeability in the ischemic stroke: An update. *Microvasc Res*, 2024, 151: 104621. https://doi.org/10.1016/j.mvr.2023.104621
128. Ding, R., Yin, Y. L., Jiang, L. H. Reactive oxygen species-induced TRPM2-mediated Ca^{2+} signalling in endothelial cells. *Antioxidants (Basel)*, 2021, 10(5): 718. https://doi.org/10.3390/antiox10050718
129. Park, L., Wang, G., Moore, J., et al. The key role of transient receptor potential melastatin-2 channels in amyloid-β-induced neurovascular dysfunction. *Nat Commun*, 2014, 5: 5318. https://doi.org/10.1038/ncomms6318
130. Whitehead, B., Karelina, K., Weil, Z. M. Pericyte dysfunction is a key mediator of the risk of cerebral ischemia. *J Neurosci Res*, 2023, 101(12): 1840–1848. https://doi.org/10.1002/jnr.25245
131. Jiang, Q., Gao, Y., Wang, C., et al. Nitration of TRPM2 as a molecular switch induces autophagy during brain pericyte injury. *Antioxid Redox Signal*, 2017, 27(16): 1297–1316. https://doi.org/10.1089/ars.2016.6873
132. Fourgeaud, L., Dvorak, C., Faouzi, M., et al. Pharmacology of JNJ-28583113: A novel TRPM2 antagonist. *Eur J Pharmacol*, 2019, 853: 299–307. https://doi.org/10.1016/j.ejphar.2019.03.043

4 Investigating TRP Channel Function with Tools for High-Precision Manipulation of Gating Mechanism by Light

Denis Krivic, Christina Humer, Christoph Romanin and Klaus Groschner

4.1 INTRODUCTION

4.1.1 Utilizing Light-sensitive Probes to Explore the Function of Signaling Molecules and Pathways

Certain biomolecules undergo defined conformational changes in response to light, thereby transducing optical information to adapt the function, phenotype or behavior of the respective biosystem. Increasing knowledge about biological photosensors and about the chemical features of the involved photoswitches has paved the way towards the related concepts of optogenetics and photopharmacology. The initial goal of these technologies was the high-precision functional interrogation of multicellular, particularly neuronal networks [1, 2]. In this context, the outstanding spatiotemporal precision of optical interventions indeed boosted advances in neuroscience, developmental biology and related fields [3]. In parallel, the therapeutic potential of spatially restricted and temporally defined interventions has been explored [4–6]. To this end, efforts to establish routine clinical applications based on optogenetics or photopharmacology are ongoing while a breakthrough is still awaited. Nonetheless, another promising application for photoswitchable molecules has emerged in the field of basic molecular biosciences. The availability of a wide range of well-characterized, bioactive photoswitches has prompted their use as molecular probes for studies aimed at advancing our insight into the function, dynamics and structure of signaling pathways towards a new, unprecedented level of resolution. Indeed, this vision became feasible with new tools for spatiotemporally defined control over the cellular targeting of signaling elements and the availability of an increasing number of photochromic, light-controllable mimetics of signaling elements. The exceptionally fast conformational transitions triggered in these molecules by light pulses enabled for the first time virtually instantaneous control over the concentrations of active signaling species within defined volume elements of cells. The spatial resolution of this "subcellular concentration-clamp" scenario is merely diffraction-limited, and it opens up the possibility of an exceptionally detailed characterization of the kinetic aspect of signaling processes to a previously unimaginable level. Over the past decade, numerous groundbreaking advances in the understanding of cellular signaling pathways have been achieved by studies involving photopharmacological and optogenetic approaches.

4.1.2 Photochromic Ligands and Optogenetic Tools to Explore TRP Channel Function

The discovery of mammalian homologues of the *Drosophila trp* gene product [7–11], which serve as pivotal elements in a variety of signal transduction cascades, started an ever since expanding

DOI: 10.1201/9781003402732-4

research area. In humans, this gene superfamily comprises 27 genes encoding transmembrane proteins that are understood to function primarily as cation channels involved in cellular Ca^{2+} signaling [12, 13]. Almost 3 decades after their initial discovery, the concepts underlying the function of TRP channels as multimodal cellular sensors for decoding and integration of a range of divergent input information are incompletely understood. The main unknown aspect that remained enigmatic until very recently was the molecular basis by which the channel complexes sense multiple physical- and/or chemical stimuli in their environment. Accordingly, the identification of TRP species as biological heat sensors to provide organisms with information about ambient temperature has been acknowledged with the Nobel Prize in Physiology or Medicine in 2021 [14]. Of note, most TRP channels are subject to cellular regulation by protein-protein and protein-lipid interactions. Moreover, many TRP channels, including also the founding member, the *Drosophila* TRP channel, have been reported to sense information about membrane lipid composition, membrane architecture and mechanical features of the lipid bilayer [15]. Importantly, the interactions between TRP channels and their membrane lipid environment have been recognized as the primary gating mechanism for some subfamilies [16–18]. Such lipid gating processes in ion channels can be envisioned based on i) an interaction between lipid sensing structures and lipids constituting the annular shell of the channels or ii) reversible protein lipidation involving coordination of non-annular lipids within the channel complex. Comprehensive characterization of gating mechanisms in TRP channels has often faced significant challenges, primarily due to the lack of methods that allow precise control over the levels of activating or modulating lipid species in the channels´ membrane surrounding. Conventional protocols for manipulating structural or regulatory membrane lipids have proven inadequate in revealing definitive insights into the underlying mechanism(s). Furthermore, the correlations observed between variations in global cellular lipid levels and the functionality of ion channels are difficult to interpret, given the inherent multifaceted nature of membrane lipids, lipid mediators, and lipid-derived messengers.

4.2 ANALYSIS OF THE MULTIMODAL GATING FEATURES OF TRP CHANNELS – LIPIDS AND OTHER LIGANDS

The experimental hurdles, which complicate the analysis of channel gating by lipids or lipid mediators, are evident from many previous attempts to explore the regulatory role of cholesterol and phosphoinositides, specifically of phosphatidylinositol 4,5-bisphosphate (PIP_2), both of which are recognized as important determinants of TRP channel function [19, 20]. The characterization of functional changes observed in response to a slowly generated, relatively sustained and, in its extent, ill-defined change in the concentration of a membrane lipid does not allow for conclusions about the nature of lipid sensing mechanisms. This is in particular true with respect to a discrimination between direct and indirect mechanisms of channel modulation. Both cholesterol and PIP_2 interact with a plethora of signaling molecules, and they are pivotal for a wide array of cellular pathways in addition to governing membrane architecture and mechanics [21–23]. In most cases, even the identification of a potential lipid recognition/binding element within the channel structure by computational prediction and/or mutagenesis will not help to reveal the nature of the sensing machinery. One promising approach to avoid this dilemma is to use optogenetic or photopharmacological systems, which allow rapid and precisely timed alterations in the presence of a lipid species within the channel's environment. This experimental setup creates a nearly instantaneous concentration clamp scenario for the activating/modulating lipid within the channel's membrane surrounding, thereby enabling a detailed kinetic interrogation. Kinetic features of the "on-" and "off-reactions", representing lipid (ligand) binding/coordination- and conformational changes in the gating structures, provide highly valuable mechanistic information. In particular, when this approach is combined with computational modeling (e.g., MD simulation) and model-driven mutagenesis, unraveling of molecular principles of lipid gating, and similarly channel gating, by lipophilic mediators has proven feasible [16, 24, 25]. The kinetic analysis of channel activation and deactivation processes triggered by "optical lipid clamp" enables more rigorous tests for direct lipid-protein (channel) interactions, and it bears the potential to provide insight into complex stoichiometry and cooperativity in the gating machinery [24]. Importantly, the reliability of conclusions drawn from optogenetics- or photopharmacology-based protocols relies on meeting some general assumptions or at least a thorough understanding and consideration of the potential mechanistic scenarios.

In general, the experimental protocols involve administration of a photoswitchable ligand/modulator (photopharmacology) or, alternatively, introduction of a genetically encoded system for efficient modification of a membrane-resident signaling molecule (optogenetics). It has to be assumed that these actuators are homogeneously distributed and targeted into the membrane system (plasma membrane or organelle), harboring the target channel. Ideally, the photochemical features of the system enable it to adopt a relatively stable "off-state", or essentially low activity state, based on stabilizing the photoswitch in an inactive position. The classical photochromic ligand concept of photopharmacology assumes that this off-state is due to the inability of the inactive conformation of the ligand to bind to its regulatory site on the channel (affinity switch). However, in many cases the photopharmacological concept of an "affinity switch" does not hold and the inactive conformation can indeed bind to its regulatory site though lacking intrinsic activity. The light-induced conformational change then endows the ligand with efficacy to activate or modulate the channel's gating machinery. For both the "efficacy- and affinity switch" scenario (Fig. 4.1) and for optogenetic interventions (outlined below), the light-induced conformational changes are assumed to be sufficiently fast to render the mechanistic steps under investigation (lipid exchange, binding and gating) as rate-limiting.

In addition to careful consideration of the above outlined issues, thorough controls are important to characterize the effects of the light pulses used on the cellular readouts, which are typically membrane currents or ion concentrations [26]. Besides these standard controls, additional tests for enhanced cellular light sensitivity at elevated expression levels of the target channel may need consideration for heterologous overexpression experiments. Moreover, delineation of the

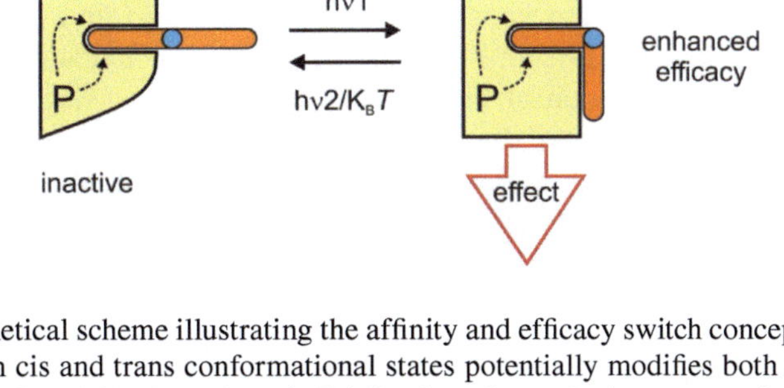

FIGURE 4.1 Hypothetical scheme illustrating the affinity and efficacy switch concepts. Photoisomerization of photolipids between cis and trans conformational states potentially modifies both affinity and efficacy of photoligand action. As found for photochromic DAGs, the scheme depicts a scenario in which a light (h√1)-induced cis isoform exerts higher affinity as well as higher efficacy, thereby initiating a productive change in the target molecule, typically a protein (effect). This is why the "off-kinetics" in terms of returning the target to its inactive state may involve either unbinding of the cis isoform due to its removal from the target's surroundings either by light (h√2)-induced cis-trans isomerization or by thermal relaxation (K_BT) in solution, or, alternatively, due to relaxation of the cis isoform (h√2/K_BT) within its binding pocket (from [25]).

concentration-dependent activity of individual isomers, administered conventionally at photochemical equilibrium, may be required in case photoconversion is potentially insufficient. All these aspects apply to the analysis of regulatory membrane lipids, lipid mediators, as well as synthetic or natural (biogenic) channel modulators. The use of optogenetic systems for efficient local control of the cellular levels of potential channel activators/modulators requires careful control experiments involving nonfunctional actuators. Below we will introduce currently available tools and protocols for both approaches with a focus on the investigation of TRP channel function, highlighting some successful examples along with an outline of experimental limitations and hurdles.

4.3 EXAMPLES HOW TO EXPLORE TRPC AND TRPV FUNCTION BY LIGHT – ADVANTAGES, THREADS AND PROTOCOLS

4.3.1 General Aspects

In this section, we will discuss some showcases of studies that used the concepts of photopharmacology and optogenetics to analyze the structure-function relations in TRP channels. Although these examples were selected specifically from recent work addressing questions of ligand sensing in TRPC and TRPV channels, the outlined experimental approaches are of rather general applicability for ion channel research and beyond. Of note, the gain in knowledge obtained by these studies is for a large part due to combining the light-aided technologies with computational approaches and model-guided mutagenesis [16, 25]. Hence, photopharmacology and optogenetics have recently emerged as experimental concepts, which perfectly complement the computational microscope in structural biology.

The function of both TRPC and TRPV channels is critically controlled by membrane lipids, while the underlying mechanisms are still elusive. Our understanding of the structural basis for lipid sensing in these channels has massively benefited from the recent gain in information on high-resolution 3D structures provided by single-particle cryo-electron microscopy (cryo-EM) studies [27–31]. Many of the reported models of TRPC and TRPV channel structures contained "non-annular" lipids, which may represent either structural elements within the channel complexes or, alternatively, certain regulatory, lipidated states of the channels. In search of experimental protocols that are suitable to perform rigorous tests for direct regulatory protein-lipid interactions and reversible regulatory lipidation processes, lipid photopharmacology was successfully adopted [16, 24]. Unfortunately, the fast, light-controlled generation of bioactive mediators from caged lipids lacked reversibility and, in many cases, also specificity of the intervention. Nonetheless, the development of photochromic lipids [32] provided efficient and reversible tools to address the open questions in a setting termed "optical lipid clamp", in which the relative levels of bioactive lipids are controlled with high spatiotemporal precision. This approach enabled an estimation of the potency of lipid mediators and to explore protein-lipid binding features by mutagenesis. Moreover, as outlined above, kinetic characterization of the on- and off-responses using the "optical concentration clamp" strategy represents a quite direct approach to testing for reversible lipidation. Of note, photoswitchable analogs are not yet available for all lipid species that are considered of regulatory importance. The probably best characterized photochromic lipids are azobenzene-harboring diacylglycerols (photochromic DAGs; [32, 33]). DAGs are of particular relevance for cellular regulation and gating of TRPC channels, and the use of photochromic DAGs has significantly advanced the field. It is important to note that the integration of a photoswitchable chemical moiety into a signaling molecule inevitably alters its biological activity by itself, and these differences need careful consideration in order to correctly interpret the photopharmacological results. Besides DAGs, other membrane lipids such as PIP_2 and cholesterol, which are reportedly relevant for TRP channel function and regulation, have been similarly modified to include azobenzene as a photoswitchable element. However, thorough characterization of these molecule probes for lipid-protein interactions is still awaited. Specifically, the precise manipulation of cellular PIP_2, which is a key regulator of a wide range of ion channels [34, 35], is a long-standing and barely resolved problem of ion channel research. Available tools for experimental control of PIP_2 levels in cell membranes include PIP_2 derivatives for direct, conventional administration as well as synthetic peptides or expressed protein domains, which bind and scavenge the regulatory lipid within the membrane (see in more detail below). In addition, optogenetic approaches for fast and reversible, light-controlled targeting of proteins or peptides into membranes to manipulate PIP_2 availability have been developed, and experimental protocols were established more recently. These methods are primarily based on a cryptochrome-2 dimerization system (CIBN-CYR2), which has proven useful for interrogation of signaling pathways [36–38]. Example protocols for investigation of TRP channel gating are described below.

4.3.2 Exploring TRPC3/6 Function with Photopharmacological Probes

4.3.2.1 Photochromic Probes Available for Interrogation of TRPC Channel Gating

To date, photoswitchable probes for investigation of the TRP channel function have been mainly designed by re-purposing the already established azobenzene photochomic moiety [32]. Currently available probes include synthetic lipids with certain structural features of naturally occurring membrane lipids or lipid mediators such as diacylglycerols (DAGs). These molecules typically harbor an azobenzene group within one or more fatty acid chains, as originally introduced by Dirk Trauner [33]. Moreover, synthetic analogues of TRPC agonists featuring a benzimidazole core structural element [26] have been developed. Photoswitchable DAGs or PhoDAGs were reported to be not only suitable to control TRPC channel activity but also to initiate other DAG-regulated downstream signaling cascades such as protein kinase C or pathways of mechanosensory signaling [25, 33, 39]. Hence, photoswitchable lipids may not be considered as specific TRPC actuators unless such specificity is verified for a certain cellular test system. Shortly after the introduction of PhoDAGs, which contain an azobenzene moiety in a single fatty acid chain of the DAG structure, another actuator (OptoDArG) harboring two photoswitchable moieties (each in a separate arachidonic acid side chain) has been developed and characterized as an actuator of TRPC channels. Of note, photoswitchable DAG probes were found to exert a significant impact on membrane structure and mechanical features, albeit at apparently higher concentrations than those required for gating of TRPC channels. Importantly, kinetic analysis of the changes induced by OptoDArG on TRPC channel activity and membrane mechanics (see Fig. 4.3) demonstrates a lack of correlation between TRPC activity and altered membrane mechanics.

Available data indicate that OptoDArG operates also other DAG-sensing TRPC channels [24, 26]. The reason that OptoDArG outperforms PhoDAG as a TRPC actuator may be due to the presence of two azobenzene rings, which generate higher flexibility of the molecule, thereby enhancing the probability to integrate into the regulatory site of the channel. While PhoDAG-1 and OptoDArG are photoswitchable lipids mimicking the structure of 1,2-SAG, PhoDAG-2 and PhoDAG-3 resemble 1,2-DOG [4]. The azobenzene photoswitch was similarly introduced into the structure of the benzimidazole TRPC agonist GSK1702934A [40] to generate the first small-molecule actuator for photopharmacological manipulation of TRPC3/6 [26]. These compounds, known as OptoBIs and OptoPI, enable lipid-independent precise activation of TRPC3 and TRPC6 channels in heterologous expression systems, such as HEK293 and RBL-2H3, and in native tissues, such as hippocampal neurons and endothelial cells [26, 41]. Although a diverse array of photochromic benzimidazoles is available, only OptoBI-1 was studied thoroughly, while others would require further physiological investigations. Since TRPC channels are multimodal chemical sensors, there are ongoing efforts to identify and synthesize other photochromic molecules, which are potentially relevant for cellular regulation of these channels (such as cholesterol and PIP_2). Of note, this chapter is not aimed at a comprehensive overview of all currently available photochromic TRPC actuators but rather on a more general discussion of typical experimental settings and protocols.

4.3.2.2 Cell Systems, Administration of Probes and Functional Readouts

All the above-mentioned photopharmacological tools are in principle suitable for high precision control over both recombinant and native, endogenous TRPC channels. In this regard, studies in a variety of cell- and tissue models have been published [26, 41, 42]. Meaningful analysis of channel functions requires that the cell model either expresses reasonably well-characterized, endogenous TRPC complexes and/or allows for efficient overexpression of a TRPC channel. When opting for a heterologous expression approach, the expression level of endogenous TRPC proteins and the most likely subunit composition in the genetically modified cells need careful consideration. It is recommended to optimize the gene transfer protocol for the employed cell system and to perform respective control experiments (see below) involving cells that have been exposed to the selected transfection procedures. Of note, transfection reagents not only determine the gene transfer efficiency but may also interfere/modulate the sensitivity of the cells or a certain cellular readout to light exposure. Once a selected cell system is well characterized in terms of endogenous TRPC expression and/or generation of a defined TRPC complex endowment by gene transfer, channel function can be examined with photopharmacological probes. Expression of fluorescent protein-fused TRPC proteins facilitates the experimental approach by providing useful information to judge the expression level in the cells under study. Depending on the type of photochrome and experimental settings, different protocols for administering the photoprobes to cells or tissues are available and need consideration. Photochromic lipids typically require a certain time of preincubation to integrate and equilibrate in the cell membranes. A challenge for all administration protocols is the limited stability of the photochromes in their inactive ("off") conformation, which for most but not all actuators is their *trans* form. Therefore, any light exposure has to be avoided, and for some photochomes, it may be required to include a short pre-illumination phase [16] to ensure that the channels are initially exposed to only negligibly low levels of contaminating *cis* isomer, according to the photochemical equilibrium. Please note that although a specific range of wavelengths is used to operate the photoswitch, isomerization can be promoted by ambient light and is significantly affected by temperature. Particular attention needs to be placed on strict control over the experimental conditions.

Common functional readouts for characterization of TRPC channel function upon photoisomerization and during photocycling of the actuators are electrophysiological recordings of membrane currents (patch clamp) and, more indirectly, the quantification of Ca^{2+} entry through the channel by Ca^{2+}-sensitive dyes (Fluo-4, Fura-2, etc.) or genetically encoded Ca^{2+} reporters (R-GECO1.2, jR-GECO1a, K-GECO, etc.). These fluorescent Ca^{2+} reporters may be used either to monitor Ca^{2+} changes at the global cytosolic level or within a rather local region in the vicinity of the channel by expressing reporter channel fusion constructs [43]. A downside of the strategy using genetically encoded reporters and reporter-channel fusion constructs lies in the fact that the expression of these proteins is often low and the TRPC-reporter fusion channels require careful electrophysiological characterization beforehand. It is also important to keep in mind that monitoring Ca^{2+} entry even when reporters are targeted to a cytosolic domain of the channel is, in most cases, inappropriate as a basis for conclusions on channel gating processes. Moreover, potential hurdles for correct interpretation of data from genetically encoded reporters in photopharmacological or optogenetic experiments arise from the fact that reporters are typically prone to photobleaching and may also give rise to artifacts in certain illumination protocols [44]. Nonetheless, the Ca^{2+} imaging approach is certainly valuable to obtain information at the level of the primary downstream signal of TRPC or generally TRP channels in combination with electrophysiological measurements. Of note, classical Ca^{2+}-sensitive dyes, which may be of advantage in certain aspects, cannot be used in combination with most available photochromic probes because of an incompatible overlap in the excitation spectra of photochromes and actuators.

4.3.2.3 Illumination Protocols

The outstanding benefit of using photochromes over conventional administration of pharmacological tools is their ability to be operated with the precision of light. Typically, UV light triggers *trans-cis* isomerization to generate the activating molecular structure almost instantaneously. The corresponding deactivation process is triggered by blue light (*cis-trans* conversion), as illustrated for the photolipid OptoDArG in Fig. 4.2A. Thereby, the kinetics of the activation and deactivation processes can be precisely studied at a well-controlled, constant level of the activating chemical entity. In addition, the use of a photoswitchable actuator allows, at least to some extent, for control over the actual level of the active signaling species for a defined time. To some extent, photoisomer levels can be modified in a graded manner by changes in illumination intensities and duration. This can be helpful to characterize the actuator-target, e.g., lipid-channel, interaction at a semiquantitative level. Of note, the absolute concentrations of the activating species can hardly be determined unless the approach is combined with an appropriate analytical method. One example of analysis of lipid-channel interactions involving variable illumination intensities is shown in Fig. 4.2B, and an analysis of gating kinetics of a TRPC channel during a constant light pulse that generated a reasonably stable "concentration pulse" (lipid camp approach) is illustrated in Fig. 4.2C.

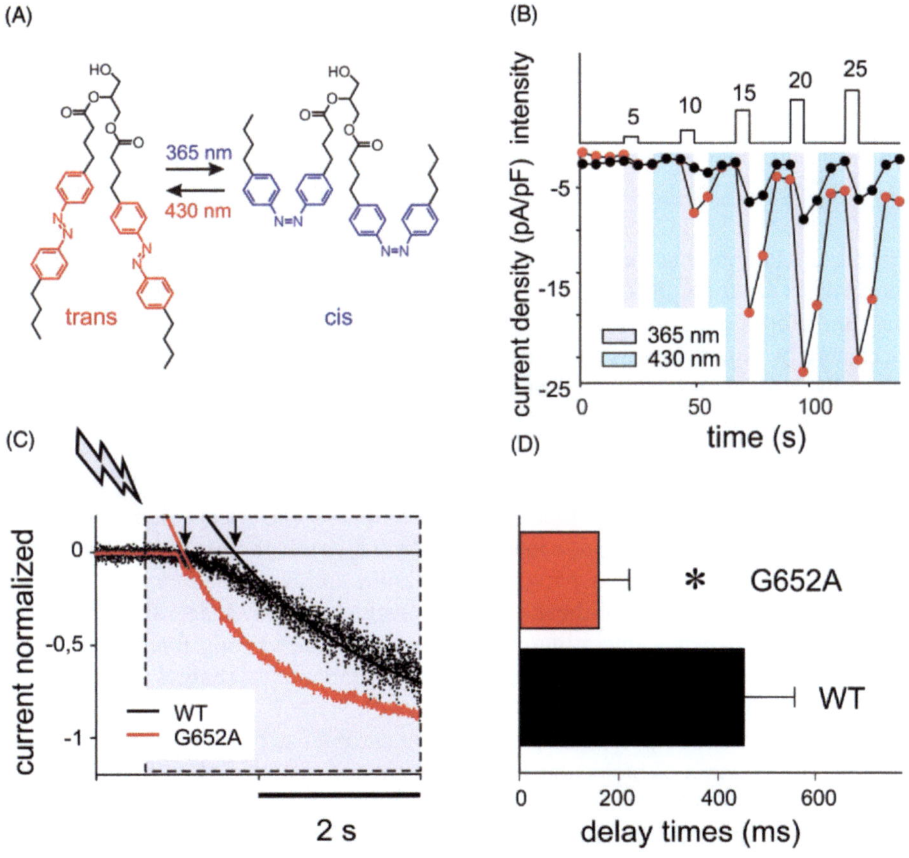

FIGURE 4.2 Photocycling of a DAG photoswitch allows for precise manipulation of TRPC3 gating. OptoDArG as shown in (A) is converted by light to different levels of the activating cis isomer to establish a concentration-response relation as illustrated in (B). (B) Representative responses to increasing light intensity (maximal output power, 670 mW) of TRPC3-WT- (black) and G652A- (a sensitized mutant; red) expressed in HEK293 cells. Currents are induced by photoisomerization of OptoDArG (30 μM) by UV (365 nm, violet, pulse held for 2 s) and blue lights (430 nm, blue, pulse held for 10 s). (C) DAG-induced activation kinetics and its modification by mutagenesis. Solid lines are exponential fits of the initial phase of current activation for TRPC3-WT (black) and G652A (red). Signal was normalized to peak current. The delay of current activation derived from monoexponential fit is indicated by black arrows. (D) Histogram displaying the current activation delay for TRPC3-WT (black) and G652A (red) (modified from [24]).

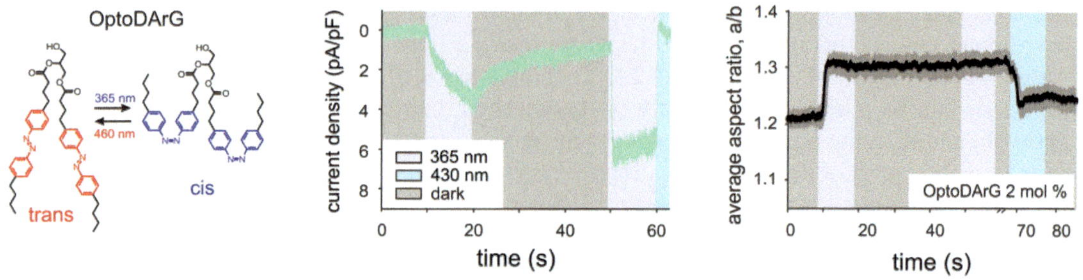

FIGURE 4.3 Comparison of the kinetics of thermal relaxation of OptoDArG within a regulatory channel coordination site and within the lipid bilayer. Photocycling of optoDArG (left) affects TRPC3 membrane currents (middle) and mechanical features of biological membranes (right panel). Biomechanical effects of OptoDArG isomerization are characterized with respect to its kinetics (right) and compared to the observed channel gating processes (middle). Middle panel shows representative traces of the inward currents induced by photoactivation in a whole-cell gap-free recording (holding potential: −40 mV, baseline adjusted to 0 mV, normalized by capacitance) in TRPC3-WT-expressing HEK293 cells. To photoconvert OptoDArG, UV light (365 nm; violet) was applied at 100% intensity for 10 s to switch to active (cis) state, which was followed by dark (UV illumination off) for 30 s (grey) and then blue light (430 nm) for 3 s (blue). Right: shape deformation of giant unilamellar vesicles (GUV), presented in terms of the a/b aspect ratio, due to photoisomerization of OptoDArG; standard error of mean is shown in dark gray (modified from [25]).

It is important to note that the exact kinetics of the isomerization processes and the stability of the photochromes are difficult to determine. It can be assumed that azobenzene switches are converted fast and almost quantitatively from their trans- to cis conformation, and the reverse process is well triggered by blue light while also occurring in the dark by thermal relaxation. Nonetheless, this is typically verified for the photochromes in aqueous solution, while for most experiments these characteristics are primarily relevant for the membrane lipid environment. This aspect has recently been tackled by studies exploring the impact of photolipids on membrane mechanical features, which represent a fairly direct readout for the photoisomerization of lipids in the membrane bilayer [25, 39]. These experiments revealed that at high photolipid levels, effects on membrane mechanics and architecture need consideration as a mechanism by which the photoswitches can affect channels or cellular readouts. By monitoring changes in membrane architecture in parallel, information can be gained about the contribution of these effects to channel regulation and also about the localization of the photoswitch when operating/modifying the channel gating machinery. One example of such parallel measurements of channel gating and membrane mechanics in optical lipid clamp experiments is shown in Fig. 4.3.

Another highly valuable aspect of employing photochromes as biological probes is the option of generating precise patterns of molecular interactions in cells and thereby of cellular activity with high precision in the frequency domain. In view of the key role of oscillatory signaling for the control over cell phenotype and fate, this opens the view on entirely new concepts of pharmacological interventions.

4.3.2.4 Essential Control Experiments, Potential Artifacts and Hurdles

Indispensable control experiments, also mentioned later in Section 4.3.3, include the characterization of cellular test systems for a general sensitivity to the applied light pulses. This characterization will typically focus on the selected readouts but should also embrace other features such as cell viability and structural integrity (morphology). Besides the issue that light pulses may have an impact on the test system itself, controls are needed to test for potential off-target effects of the photoprobes. As outlined already above, most photochromic mediators are not highly selective for the target channels under study. Tests for the effects of the photoprobes in relevant experimental conditions when the cells are lacking the target protein are highly recommended. Moreover, artifacts may be generated by certain illumination protocols when combined with a particular recording technology. Particular attention is needed when genetically encoded reporters are used, as already mentioned above. As an example, the R-GECO fluorescent Ca^{2+} reporters, which, upon illumination with UV light of certain intensity, is prone to generate Ca^{2+}-independent signals in terms of transient recording artifacts [44]. These should be quantified and considered in the data analysis and interpretation.

4.3.3 Exploring TRPV6 Function with Optogenetic Approaches

4.3.3.1 Strategies for Optogenetic Manipulation of PIP_2 Levels

A ubiquitous lipid regulator of cellular ion channel activity is PIP_2, which typically binds target proteins by electrostatic interactions [35]. Conventional strategies and protocols for experimental manipulation of this molecule utilize established pharmacological tools (e.g., PI kinase inhibitors, like Wortmaninn; [45, 46]), genetic manipulation of PIP_2 metabolism (e.g., overexpression of PI5K; [45]), scavenging molecules (e.g., Pal-peptide; [47], anti-PIP_2 antibodies; [48]) or PIP_2 derivatives [23]. Of note, these tools are administered by conventional perfusion systems. The direct exposure of cells or membrane fragments, for example inside-out patches in electrophysiological recordings, to PIP_2 analogs or scavenging molecules enables fairly rapid alterations in PIP_2 levels. However, due to the still limited temporal precision and the lack of reversibility, there was a quest for more refined and precise interventions to analyze biological PIP_2-sensing processes in greater detail.

One important advance towards this goal, albeit limited to voltage-clamp experiments, was the use of voltage-sensitive phosphatases [20]. Recently, a more versatile optogenetic system for reversible protein dimerization, originating from plants and consisting of the photosensitive cryptochrome-2 (CRY2) and the transcription factor cryptochrome interacting basic helix loop helix protein (CIB1), has been established to study membrane signaling processes [36]. This CRY2-CIBN platform enables precise control over cellular targeting of enzymes or scavenging molecules at cellular membranes and thereby manipulation of lipid homeostasis and membrane lipid composition. Reversible targeting of either a PIP_2 hydrolyzing enzyme (5′phosphatase domain of OCRL) or alternatively lipid kinases as well as PIP_2-binding protein domains or peptides can be achieved by blue light-triggered association of actuator-fused CYR2 (CYR2 photolyase homology region) with its membrane-anchored target, the N-terminal region of the transcription factor CIB1 (CIBN). The CYR2-CIBN molecular complex dissociates in the dark, resulting in redistribution of the CYR2 fusion protein towards the cytosol and reversal of changes in PIP_2 membrane levels (see Fig. 4.4). This approach is suitable to characterize the temporal correlation between changes in PIP_2 and channel function as well as reversibility of the regulatory process. As with most biological photoswitches, the light-induced cycling of the CYR2-CIBN photoswitch between off and on-state is reasonably fast but typically incomplete, and the states can be kept stable for only a limited time.

The kinetics of light-induced cycling of this dimerization switch will generate a certain temporal pattern of PIP_2 levels in the membrane. The temporal features of the CYR2 translocation process and also of changes in the lipid (PIP_2) levels can be monitored with fluorescence reporters [36]. This kinetic characterization of the optogenetic protocol appears strongly recommended when a study is focused on kinetic aspects of PIP_2-channel interactions. An example for the characterization of light-induced "on-off" cycling of a CRY2-OCRL-CIBN system in HEK293 cells is illustrated in Fig. 4.5 and discussed below in Section 4.3.3.2. As compared to the photocycling between conformations of a photochromic, azobenzene-based lipid, the light-controlled operation of a CYR2-CIBN system is inherently slower, and lipid manipulation is indirect by interfering with the homeostasis of endogenous PIP_2. Although the photochromic ligand approach outperforms the

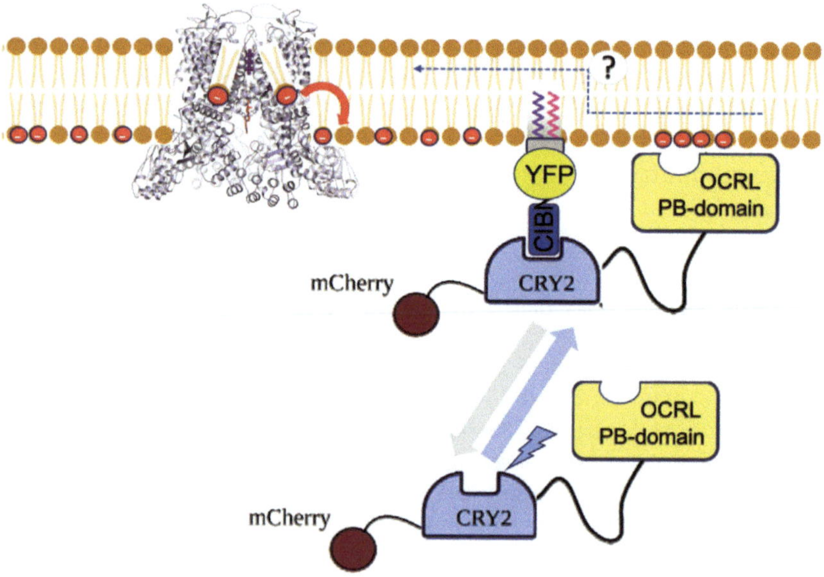

FIGURE 4.4 Cartoon illustrating the reversible targeting of a phosphatase (OCRL) for rapid removal of PIP_2 from the inner leaflet of the plasma membrane to characterize fast regulation of an ion channel by PIP_2.

optogenetic dimerization switch in terms of temporal precision, this strategy offers the potential of achieving substantially enhanced spatial precision by targeting the CIBN anchor into defined regions of the membrane. This strategy might even allow for highly specific manipulation of the vicinity of a target channel, a conceptual advantage, which is considered of high value to advance our understanding of lipid ion channel interactions.

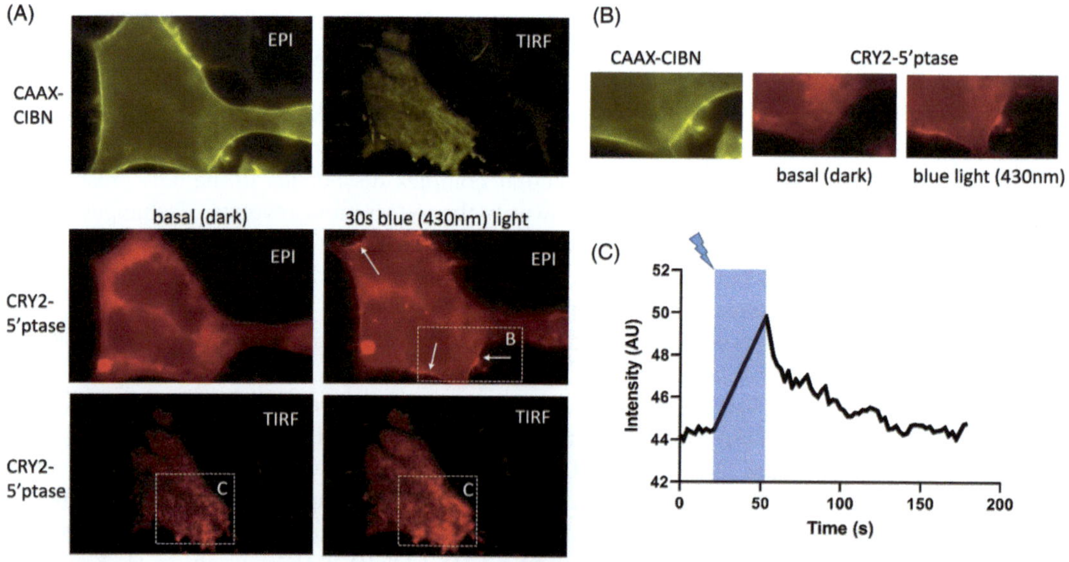

FIGURE 4.5 Reversible targeting of a mCherry-tagged phosphatase (OCRL) to the inner leaflet of the plasma membrane in HEK293 cells. (A) Epifluorescence and TIRF images of the YFP-tagged membrane anchor (CAAX-CIBN). (B) Micrographs illustrating membrane targeting of the constructs. (C) A representative time course of OCRL fluorescence intensity within the membrane (TIRF plane) during the illumination protocol.

4.3.3.2 Cell Systems and Functional Readouts

CRY2-CIBN-mediated targeting of lipid metabolizing enzymes or lipid scavengers to the plasma membrane is in principle feasible for all cell types, which allow for efficient genetic manipulation, ranging from standard expression systems (e.g., HEK293 or CHO cells) to primary cells [36, 38]. The method has so far mainly been utilized to characterize recombinant channels by heterologous overexpression [37, 44, 49], which enables to explore lipid (PIP_2) regulation of defined channel complexes in combination with extensive mutagenesis screens to identify interaction domains and structural elements involved in lipid gating [49, 50]. Of note, the efficiency and kinetic features as well as stability of the off and on states of the CRY2-CIBN photoswitch may significantly vary depending on the cellular background. Therefore, a systematic characterization of these features and extensive controls (see Section 4.3.3.4) are required when establishing the method for a new cell type. Similar to studies on TRPC gating and regulation (above), the functional characterization of TRPV6 is typically performed using classical electrophysiology as the most direct functional readout or by monitoring the Ca^{2+} entry through the channel indirectly in terms of changes in cellular Ca^{2+} levels reported by classical dyes or genetically encoded reporters. Results from an experiment using coexpression of R-GECO1.2 along with TRPV6, a CAAX-CIBN plasma membrane anchor, and the CYR2-OCRL fusion are shown in Fig. 4.6. The use of a genetic reporter for monitoring TRPV6-mediated global cellular Ca^{2+} changes is feasible, though it requires simultaneous expression of multiple constructs, of which usually not all can be visualized as fluorescent fusions. Hence, this approach requires well-established and efficient transfection protocols. It appears important to note that the actual requirements in terms of experimental conditions for different readouts, e.g., cell dialysis or specific temperature limits, may significantly modify the properties of the dimerization photoswitch.

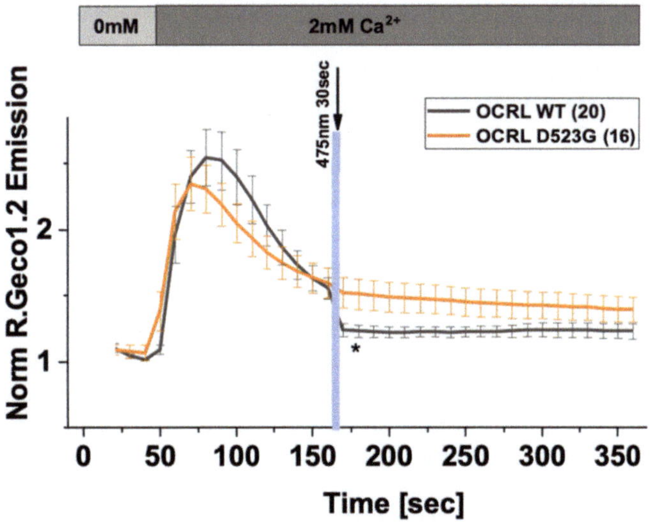

FIGURE 4.6 Light-induced breakdown of membrane PIP_2 reduces TRPV6-mediated Ca^{2+} entry. TRPV6 activity is recorded by monitoring R-GECO1.2 fluorescence during translocation of the OCRL phosphatase (OCRL) to the plasma membrane. Note that TRPV6 constitutive activity gives rise to a transient Ca^{2+} signal upon elevation of extracellular Ca^{2+}. This signal is rapidly suppressed by activating the dimerization photoswitch. A control experiment (orange) in cells expressing a nonfunctional OCRL (D523G) mutant is shown for comparison. Mean values are shown (number of experiments is indicated in brackets, * denotes statistical significance at $P < 0.05$).

4.3.3.3 Illumination Protocols

Fluorescence imaging experiments shown in Fig. 4.5 demonstrate that the light pulses suitable to trigger membrane targeting in a typical experimental setting (illumination by LED; output power ~525 mW) require a duration of approximately 30 sec, with the photoswitch subsequently returning back to its virtual "off state" within approximately 1–2 min. Repetitive cycling is possible. Of note, the transient targeting of the OCRL 5′Ptase will cause a change in homeostasis of endogenous PIP_2, and the time course of changes in this lipid level will depend on the activity of metabolic pathways. Importantly, when characterizing channel activity indirectly via global cation (Ca^{2+} changes), interpretation of the channels' responses to repetitive cycling of PIP_2 levels is barely feasible. Global Ca^{2+} levels may correlate well with channel function during an initial intervention, but cellular Ca^{2+} homeostasis is likely to adapt over time. Moreover, PIP_2 metabolism in the membrane may change during repetitive targeting of the phosphatase. Therefore, characterization of individual illumination protocols for a specific cell type should be performed at the level of the photoswitch (actuator/OCRL translocation; see Fig. 4.5) as well as by monitoring actual PIP_2 changes with fluorescence reporters such as CYPHR [51]. Discrepancies between time courses in phosphatase localization and membrane PIP_2, as well as between TRP channel activity and global cytosolic Ca^{2+}, need careful consideration. Another essential issue is to characterize the basal, background "activity", i.e., the propensity of the optogenetic system to adopt its "on state" (membrane targeting) in the dark (at rest). This feature will be dependent on cell type and experimental conditions. In contrast to photopharmacology (see above), basal activity of the optogenetic switch cannot be effectively modified by altered illumination protocols, except for carefully avoiding background light exposure, but may be improved by genetic modifications to enhance stability of its "off state".

4.3.3.4 Essential Control Experiments, Potential Artifacts and Hurdles

Besides the typical controls for optogenetic and photopharmacological experiments such as tests for a light sensitivity of cellular readouts (membrane currents, cation/Ca^{2+} levels) in the absence of photoswitches, characterization of light effects in cells that express only part of the dimerization switch along with the TRPV channel as well as in cells overexpressing the TRPV channel only are also helpful to understand the cellular model system and optimize protocols. The most important controls are generated by determination of the light responses in cells expressing lack-of-function mutations of the optogenetic switch, such as mutations eliminating enzymatic activity or the systems' capability to dimerize (see experiment in Fig. 4.5; OCRL D523G; [36]).

A series of potential hurdles impairing the value of experiments studying PIP_2 regulation of TRP channels, specifically TRPV6, by CYR2-CIBN switches have already been addressed above. In aggregate, correct interpretation of the function of the optogenetic dimerization switches (specifically CYR2-CIBN) in the cellular test system with respect to its actual impact on lipid metabolism at rest and during a given illumination protocol is a prerequisite for experimental success. Specific hurdles include exceptionally high basal membrane targeting of the CYR2-fusion constructs in the dark (at rest) as well as readout artifacts that are generated with certain illumination protocols (above certain light intensities and pulse duration). Cellular artifacts have been described especially for blue light pulses, and the use of R-GECO constructs as reporters is prone to artifacts [44], which need consideration when characterizing fast responses triggered by light.

4.4 CONCLUSIONS AND OUTLOOK

Regulatory lipidation has emerged as a pivotal process for cellular control for most TRP channels. Our knowledge about structure-function relations underlying these important signaling mechanisms including also their role in human physiopathology is growing. The photopharmacological and optogenetic protocols for manipulation of membrane lipid composition discussed here have significantly contributed to the current understanding and may further pave the way towards therapeutic targeting of lipid-ion channel interactions.

ACKNOWLEDGMENT

The work was supported by FWF funding (P33263 and PAT9543223).

REFERENCES

1. Miesenböck, G. Optogenetic control of cells and circuits. *Annu. Rev. Cell Dev. Biol*, 2011, 27: 731–758. https://doi.org/10.1146/annurev-cellbio-100109-104051.
2. Yizhar, O., Fenno, L. E., Davidson, T. J., Mogri, M., Deisseroth, K. Optogenetics in neural systems. *Neuron*, 2011, 71: 9–34. https://doi.org/10.1016/j.neuron.2011.06.004.
3. Krueger, D., Izquierdo, E., Viswanathan, R., Hartmann, J., Pallares Cartes, C., De Renzis, S. Principles and applications of optogenetics in developmental biology. *Development*, 2019, 146: dev175067. https://doi.org/10.1242/dev.175067.
4. Curcic, S., Tiapko, O., Groschner, K. Photopharmacology and opto-chemogenetics of TRPC channels-some therapeutic visions. *Pharmacol Therap*, 2019, 200: 13–26. https://doi.org/10.1016/j.pharmthera.2019.04.003.
5. Landra-Willm, A., Karapurkar, A., Duveau, A., Chassot, A. A., Esnault, L., Callejo, G., Bied, M., Häfner, S., Lesage, F., Wdziekonski, B., Baron, A., Fossat, P., Marsollier, L., Gasull, X., Boué-Grabot, E., Kienzler, M. A., Sandoz, G. A photoswitchable inhibitor of TREK channels controls pain in wild-type intact freely moving animals. *Nat Commun*, 2023, 14: 1160. https://doi.org/10.1038/s41467-023-36806-4.
6. Berry, M. H., Holt, A., Broichhagen, J., Donthamsetti, P., Flannery, J. G., Isacoff, E. Y. Photopharmacology for vision restoration. *Curr Opin Pharmacol*, 2022, 65: 102259. https://doi.org/10.1016/j.coph.2022.102259.
7. Zhu, X., Chu, P. B., Peyton, M., Birnbaumer, L. Molecular cloning of a widely expressed human homologue for the *drosophila trp* gene. *FEBS Lett*, 1995, 373: 193–198. https://doi.org/10.1016/0014-5793(95)01038-G.
8. Zhu, X., Jiang, M., Peyton, M., Boulay, G., Hurst, R., Stefani, E., Birnbaumer, L. Trp, a novel mammalian Gene family essential for agonist-activated capacitative Ca2+ entry. *Cell*, 1996, 85: 661–671. https://doi.org/10.1016/S0092-8674(00)81233-7.
9. Wes, P. D., Chevesich, J., Jeromin, A., Rosenberg, C., Stetten, G., Montell, C. TRPC1, a human homolog of a drosophila store-operated channel. *Proc. Natl. Acad. Sci. U.S.A*, 1995, 92: 9652–9656. https://doi.org/10.1073/pnas.92.21.9652.
10. Philipp, S., Cavalié, A., Freichel, M., Wissenbach, U., Zimmer, S., Trost, C., Marquart, A., Murakami, M., Flockerzi, V. A mammalian capacitative calcium entry channel homologous to drosophila TRP and TRPL. *EMBO J*, 1996, 15: 6166–6171. https://doi.org/10.1002/j.1460-2075.1996.tb01004.x.
11. Zitt, C., Zobel, A., Obukhov, A. G., Harteneck, C., Kalkbrenner, F., Lückhoff, A., Schultz, G. Cloning and functional expression of a human Ca2+-permeable cation channel activated by calcium store depletion. *Neuron*, 1996, 16: 1189–1196. https://doi.org/10.1016/S0896-6273(00)80145-2.
12. Pan, Z., Yang, H., Reinach, P. S. Transient receptor potential (TRP) gene superfamily encoding cation channels. *Hum Genom*, 2011, 5: 108. https://doi.org/10.1186/1479-7364-5-2-108.
13. Minke, B., Cook, B. TRP channel proteins and signal transduction. *Physiol Rev*, 2002, 82: 429–472. https://doi.org/10.1152/physrev.00001.2002.
14. Gracheva, E. O., Bagriantsev, S. N. Sensational channels. *Cell*, 2021, 184: 6213–6216. https://doi.org/10.1016/j.cell.2021.11.034.
15. Gutorov, R., Katz, B., Rhodes-Mordov, E., Zaguri, R., Brandwine-Shemmer, T., Minke, B. The role of membrane lipids in light-activation of drosophila TRP channels. *Biomolecules*, 2022, 12: 382. https://doi.org/10.3390/biom12030382.
16. Erkan-Candag, H., Clarke, A., Tiapko, O., Gsell, M. A., Stockner, T., Groschner, K. Diacylglycerols interact with the L2 lipidation site in TRPC3 to induce a sensitized channel state. *EMBO Rep*, 2022, 23: e54276. https://doi.org/10.15252/embr.202154276.
17. Storch, U. A greasy business: Identification of a diacylglycerol binding site in human TRPC5 channels by cryo-EM, *Cell Calcium*, 2021, 97: 102414. https://doi.org/10.1016/j.ceca.2021.102414.
18. Storch, U., Forst, A.-L., Pardatscher, F., Erdogmus, S., Philipp, M., Gregoritza, M., Mederos y Schnitzler, M., Gudermann, T. Dynamic NHERF interaction with TRPC4/5 proteins is required for channel gating by diacylglycerol. *Proc Natl Acad Sci USA*, 2017, 114: E37–E46. https://doi.org/10.1073/pnas.1612263114.

19. Graziani, A., Rosker, C., Kohlwein, S. D., Zhu, M. X., Romanin, C., Sattler, W., Groschner, K., Poteser, M. Cellular cholesterol controls TRPC3 function: evidence from a novel dominant-negative knockdown strategy, *Biochem J*, 2006, 396: 147-155. https://doi.org/10.1042/BJ20051246.
20. Mori, M. X., Itsuki, K., Hase, H., Sawamura, S., Kurokawa, T., Mori, Y., Inoue, R. Dynamics of receptor-operated Ca2+ currents through TRPC channels controlled via the PI(4,5)P2-PLC signaling pathway, *Front Pharmacol*, 2015, 6: 22. https://doi.org/10.3389/fphar.2015.00022.
21. Doole, F. T., Kumarage, T., Ashkar, R., Brown, M. F. Cholesterol stiffening of lipid membranes. *J Membrane Biol*, 2022, 255: 385–405. https://doi.org/10.1007/s00232-022-00263-9.
22. Kwik, J., Boyle, S., Fooksman, D., Margolis, L., Sheetz, M. P., Edidin, M. Membrane cholesterol, lateral mobility, and the phosphatidylinositol 4,5-bisphosphate-dependent organization of cell actin. *Proc Natl Acad Sci U S A*, 2003, 100: 13964–13969. https://doi.org/10.1073/pnas.2336102100.
23. Walter, A. M., Müller, R., Tawfik, B., Wierda, K. D., Pinheiro, P. S., Nadler, A., McCarthy, A. W., Ziomkiewicz, I., Kruse, M., Reither, G., Rettig, J., Lehmann, M., Haucke, V., Hille, B., Schultz, C., Sørensen, J. B. Phosphatidylinositol 4,5-bisphosphate optical uncaging potentiates exocytosis. *Elife*, 2017, 6: e30203. https://doi.org/10.7554/eLife.30203.
24. Lichtenegger, M., Tiapko, O., Svobodova, B., Stockner, T., Glasnov, T. N., Schreibmayer, W., Platzer, D., de la Cruz, G. G., Krenn, S., Schober, R., Shrestha, N., Schindl, R., Romanin, C., Groschner, K. An optically controlled probe identifies lipid-gating fenestrations within the TRPC3 channel. *Nat Chem Biol*, 2018, 14: 396–404. https://doi.org/10.1038/s41589-018-0015-6.
25. Erkan-Candag, H., Krivic, D., Gsell, M. A. F., Aleksanyan, M., Stockner, T., Dimova, R., Tiapko, O., Groschner, K. Characterization of DAG binding to TRPC channels by target-dependent cis–trans isomerization of OptoDArG. *Biomolecules*, 2022, 12: 799. https://doi.org/10.3390/biom12060799.
26. Tiapko, O., Shrestha, N., Lindinger, S., Guedes de la Cruz, G., Graziani, A., Klec, C., Butorac, C., Graier, W. F., Kubista, H., Freichel, M., Birnbaumer, L., Romanin, C., Glasnov, T., Groschner, K. Lipid-independent control of endothelial and neuronal TRPC3 channels by light. *Chem Sci*, 2019, 10: 2837–2842. https://doi.org/10.1039/C8SC05536J.
27. Fan, C., Choi, W., Sun, W., Du, J., Lü, W. Structure of the human lipid-gated cation channel TRPC3. *Elife*, 2018, 7: e36852. https://doi.org/10.7554/eLife.36852.
28. Bai, Y., Yu, X., Chen, H., Horne, D., White, R., Wu, X., Lee, P., Gu, Y., Ghimire-Rijal, S., Lin, D. C.-H., Huang, X. Structural basis for pharmacological modulation of the TRPC6 channel. *Elife*, 2020, 9: e53311. https://doi.org/10.7554/eLife.53311.
29. Duan, J., Li, J., Chen, G.-L., Ge, Y., Liu, J., Xie, K., Peng, X., Zhou, W., Zhong, J., Zhang, Y., Xu, J., Xue, C., Liang, B., Zhu, L., Liu, W., Zhang, C., Tian, X.-L., Wang, J., Clapham, D. E., Zeng, B., Li, Z., Zhang, J. Cryo-EM structure of TRPC5 at 2.8-Å resolution reveals unique and conserved structural elements essential for channel function. *Sci Adv*, 2019, 5: eaaw7935. https://doi.org/10.1126/sciadv.aaw7935.
30. Guo, W., Tang, Q., Wei, M., Kang, Y., Wu, J.-X., Chen, L., Structural mechanism of human TRPC3 and TRPC6 channel regulation by their intracellular calcium-binding sites, *Neuron*, 2022, 110: 1023–1035.e5. https://doi.org/10.1016/j.neuron.2021.12.023.
31. Tang, Q., Guo, W., Zheng, L., Wu, J.-X., Liu, M., Zhou, X., Zhang, X., Chen, L. Structure of the receptor-activated human TRPC6 and TRPC3 ion channels. *Cell Res*, 2018, 28: 746–755. https://doi.org/10.1038/s41422-018-0038-2.
32. Morstein, J., Impastato, A. C., Trauner, D. Photoswitchable lipids. *ChemBioChem*, 2021, 22: 73–83. https://doi.org/10.1002/cbic.202000449.
33. Frank, J. A., Yushchenko, D. A., Hodson, D. J., Lipstein, N., Nagpal, J., Rutter, G. A., Rhee, J.-S., Gottschalk, A., Brose, N., Schultz, C., Trauner, D. Photoswitchable diacylglycerols enable optical control of protein kinase c. *Nat Chem Biol*, 2016, 12: 755–762. https://doi.org/10.1038/nchembio.2141.
34. Hammond, G. R. V., Burke, J. E. Novel roles of phosphoinositides in signaling, lipid transport, and disease. *Curr Opin Cell Biol*, 2020, 63: 57–67. https://doi.org/10.1016/j.ceb.2019.12.007.
35. Rosenhouse-Dantsker, A., Logothetis, D. E. Molecular characteristics of phosphoinositide binding. *Pflugers Arch - Eur J Physiol*, 2007, 455: 45–53. https://doi.org/10.1007/s00424-007-0291-6.
36. Idevall-Hagren, O., Dickson, E. J., Hille, B., Toomre, D. K., De Camilli, P. Optogenetic control of phosphoinositide metabolism, *Proc Natl Acad Sci U S A*, 2012, 109: E2316–E2323. https://doi.org/10.1073/pnas.1211305109.
37. Taslimi, A., Zoltowski, B., Miranda, J. G., Pathak, G. P., Hughes, R. M., Tucker, C. L. Optimized second-generation CRY2–CIB dimerizers and photoactivatable Cre recombinase. *Nat Chem Biol*, 2016, 12: 425–430. https://doi.org/10.1038/nchembio.2063.

38. Gada, K. D., Kawano, T., Plant, L. D., Logothetis, D. E. An optogenetic tool to recruit individual PKC isozymes to the cell surface and promote specific phosphorylation of membrane proteins. *J Biologic Chem*, 2022, 298: 101893. https://doi.org/10.1016/j.jbc.2022.101893.
39. Pfeffermann, J., Eicher, B., Boytsov, D., Hannesschlaeger, C., Galimzyanov, T. R., Glasnov, T. N., Pabst, G., Akimov, S. A., Pohl, P. Photoswitching of model ion channels in lipid bilayers. *J Photochem Photobiol B: Biol*, 2021, 224: 112320. https://doi.org/10.1016/j.jphotobiol.2021.112320.
40. Doleschal, B., Primessnig, U., Wolkart, G., Wolf, S., Schernthaner, M., Lichtenegger, M., Glasnov, T. N., Kappe, C. O., Mayer, B., Antoons, G., Heinzel, F., Poteser, M., Groschner, K. TRPC3 contributes to regulation of cardiac contractility and arrhythmogenesis by dynamic interaction with NCX1. *Cardiovasc Res*, 2015, 106: 163–173. https://doi.org/10.1093/cvr/cvv022.
41. Bacsa, B. Pharmaco-optogenetic targeting of TRPC activity allows for precise control over mast cell NFAT signaling. *Frontiers Immunol*, 2020, 11: 8. https://doi.org/10.3389/fimmu.2020.613194.
42. Leinders-Zufall, T., Storch, U., Mederos y Schnitzler, M., Ojha, N. K., Koike, K., Gudermann, T., Zufall, F. A diacylglycerol photoswitching protocol for studying TRPC channel functions in mammalian cells and tissue slices. *STAR Protocols*, 2021, 2: 100527. https://doi.org/10.1016/j.xpro.2021.100527.
43. Curcic, S., Erkan-Candag, H., Pilic, J., Malli, R., Wiedner, P., Tiapko, O., Groschner, K. TRPC3 governs the spatiotemporal organization of cellular Ca2+ signatures by functional coupling to IP3 receptors. *Cell Calcium*, 2022, 108: 102670. https://doi.org/10.1016/j.ceca.2022.102670.
44. Wu, J., Liu, L., Matsuda, T., Zhao, Y., Rebane, A., Drobizhev, M., Chang, Y.-F., Araki, S., Arai, Y., March, K., Hughes, T. E., Sagou, K., Miyata, T., Nagai, T., Li, W., Campbell, R. E. Improved Orange and red ca^{2+} indicators and photophysical considerations for optogenetic applications. *ACS Chem Neurosci*, 2013, 4: 963–972. https://doi.org/10.1021/cn400012b.
45. Velisetty, P., Borbiro, I., Kasimova, M. A., Liu, L., Badheka, D., Carnevale, V., Rohacs, T. A molecular determinant of phosphoinositide affinity in mammalian TRPV channels. *Sci Rep*, 2016, 6: 27652. https://doi.org/10.1038/srep27652.
46. Cai, R., Liu, X., Zhang, R., Hofmann, L., Zheng, W., Amin, M. R., Wang, L., Hu, Q., Peng, J.-B., Michalak, M., Flockerzi, V., Ali, D. W., Chen, X.-Z., Tang, J. Autoinhibition of TRPV6 channel and regulation by PIP2. *iScience*, 2020, 23: 101444. https://doi.org/10.1016/j.isci.2020.101444.
47. Buchmayer, F., Schicker, K., Steinkellner, T., Geier, P., Stubiger, G., Hamilton, P. J., Jurik, A., Stockner, T., Yang, J.-W., Montgomery, T., Holy, M., Hofmaier, T., Kudlacek, O., Matthies, H. J. G., Ecker, G. F., Bochkov, V., Galli, A., Boehm, S., Sitte, H. H. Amphetamine actions at the serotonin transporter rely on the availability of phosphatidylinositol-4,5-bisphosphate. *Proc Natl Acad Sci U S A*, 2013, 110: 11642–11647. https://doi.org/10.1073/pnas.1220552110.
48. Gwanyanya, A., Sipido, K. R., Vereecke, J., Mubagwa, K. ATP and PIP_2 dependence of the magnesium-inhibited, TRPM7-like cation channel in cardiac myocytes. *Am J Physiol-Cell Physiol*, 2006, 291: C627–C635. https://doi.org/10.1152/ajpcell.00074.2006.
49. Abd El-Aziz, T. M., Kaur, A., Shapiro, M. S., Stockand, J. D., Archer, C. R. Optogenetic control of PIP2 interactions shaping ENaC activity. *IJMS*, 2022, 23: 3884. https://doi.org/10.3390/ijms23073884.
50. Humer, C., Radiskovic, T., Horváti, K., Lindinger, S., Groschner, K., Romanin, C., Höglinger, C. Bidirectional allosteric coupling between PIP2 binding and the pore of the oncochannel TRPV6. *Int J Mol Sci* 2024, 25(1): 618. https://doi.org/10.3390/ijms25010618.
51. Violin, J. D., Zhang, J., Tsien, R. Y., Newton, A. C. A genetically encoded fluorescent reporter reveals oscillatory phosphorylation by protein kinase c. *J Cell Biol*, 2003, 161: 899–909. https://doi.org/10.1083/jcb.200302125.

5 Pathophysiological Role of TRPC1 in Cell Function

Ericka Martínez Martínez, Hwei Ling Ong, and Indu S. Ambudkar

5.1 INTRODUCTION

Transient receptor potential canonical 1 (TRPC1) was the first reported human homolog for the *Drosophila melanogaster* TRP channel, which was identified by two separate groups in 1995 (Wes et al., 1995; Zhu et al., 1995). Following its initial discovery, more mammalian homologs of the TRP channel were identified and subdivided into 6 subfamilies: TRPC (canonical), TRPV (vanilloid), TRPM (melastatin), TRPA (ankyrin), TRPP (polycystin), and TRPML (mucolipin) (Lopez et al., 2020a). Further information on other members of the TRPC subfamily as well as other subfamilies can be found in other reviews, which include but are not limited to (Chen et al., 2020; Himmel and Cox, 2020; Ong et al., 2014; Zhang et al., 2023). This chapter highlights what is currently known about the properties and functions of the TRPC1 channel as gleaned from numerous studies using cell lines and animal models (Fig. 5.1).

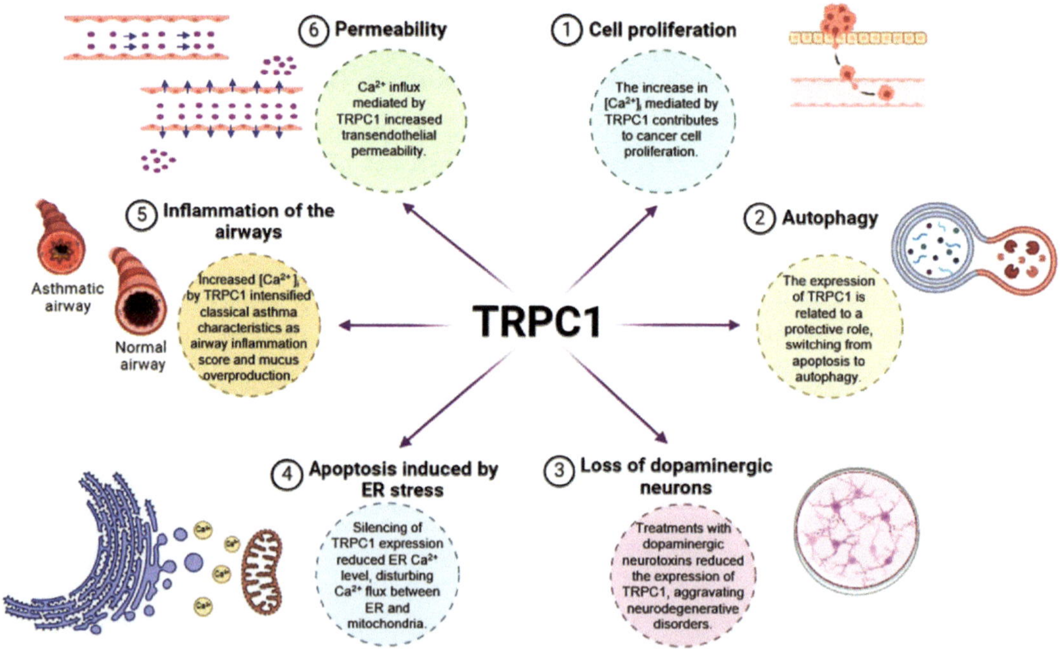

FIGURE 5.1 Diagram shows multiple circles highlighting the contribution of TRPC1 to cellular processes. Increased TRPC1-mediated Ca^{2+} entry contributes to cancer cell proliferation, autophagy, airway inflammation, and increased transendothelial cell permeability. Conversely, silencing TRPC1 expression causes ER stress that consequently induces apoptosis. Reduced TRPC1 expression by dopaminergic neurotoxins aggravates neurodegenerative disorders.

DOI: 10.1201/9781003402732-5

TRPC1 is broadly expressed in many tissues and different cell types. The gene encoding human TRPC1 is found on chromosome 3, and while five mRNA splice variants have been reported, only three variants are functional. TRPC1α has a predicted size of 91 kDa, while TRPC1β is smaller at 88 kDa as it lacks 34 amino acids in the third ankyrin repeat. TRPC1ε has a 34-nucleotide deletion at exon 5, which results in a 7 amino acid deletion downstream from the first transmembrane spanning helix (TM1). TRPC1 has 6 transmembrane-spanning helices (TM1 to TM6), with a pore-forming loop between TM5 and TM6, 4 ankyrin repeats in the N-terminus, and an EWKFAR TRP box plus a calmodulin/IP_3R binding domain in the C-terminal region. The pore-forming loop contains negatively charged glutamate and aspartate and a putative selectivity filter (SYGEE). TRPC1 is predicted to function as a tetramer either as a homomeric channel or in the form of heteromeric channels by interacting with other TRPC subtypes (Barrera et al., 2007; Chen et al., 2020; Ong et al., 2014). Moreover, cryo-EM studies have shown that other TRPC channels assemble as tetramers (Wang et al., 2020a; Zhao et al., 2021).

When TRPC1 was first identified (Wes et al., 1995; Zhu et al., 1995), the search for the store-operated calcium entry (SOCE) channel was being intensely pursued by many investigators. SOCE is a plasma membrane (PM) Ca^{2+} entry process that is controlled by the filling status of the endoplasmic reticulum (ER)-Ca^{2+} stores, namely depletion causing activation and refilling leading to inactivation. The initial reports on TRPC1 cloning proposed that it might contribute to SOCE and could form a component of a store-operated Ca^{2+} channel. Further studies on heterologously expressed and endogenous TRPC1 in CHO, COS, human submandibular gland (HSG) and endothelial cells, plus human platelets, supported this concept (Kim et al., 2009b; Liu et al., 2004; Ma et al., 2011; Zitt et al., 1996; Zhu et al., 1996). However, measurements of currents generated by TRPC1 showed that it was distinct from the store-operated current that had been characterized at the time in T lymphocytes and RBL mast cells, which is now known as I_{CRAC} (Hoth and Penner, 1992; Hoth and Penner, 1993; Prakriya et al., 2006). This led to the suggestion that there may be multiple types of store-operated Ca^{2+} channels. While I_{CRAC} was described as a highly Ca^{2+} selective, inwardly rectifying current, the TRPC1-mediated current (referred to as I_{SOC}, for store-operated Ca^{2+} channel current) was relatively less Ca^{2+} selective and tended to have a linear current-voltage relation. The components of I_{CRAC} were subsequently identified when the SOCE regulatory ER proteins, STIM1 and STIM2, as well as the PM channel Orai1, were discovered (Brandman et al., 2007; Liou et al., 2005; Prakriya et al., 2006).

The interplay between TRPC1, STIM1, and Orai1 is a fascinating study of integrated yet different signaling complexes. Dynamic assembly of the TRPC1-STIM1-Orai1 ternary complex in HSG cells occurs following ER-Ca^{2+} store depletion. An endogenous protein complex was also identified in salivary glands from mice as well as with overexpressed proteins in human embryonic kidney 293 (HEK293) cells (Cheng et al., 2011; Ong et al., 2007). Before stimulation to induce ER-Ca^{2+} store depletion, STIM1 is within the ER but also traffics along the microtubule network, whereas both Orai1 and TRPC1 are diffuse in the PM. Depletion of the ER-Ca^{2+} stores triggers STIM1 aggregation and translocation to ER-PM junctions, where it then recruits Orai1 and TRPC1 into the same junctions (Fig. 5.2). The knockdown of TRPC1 significantly reduced but did not abolish SOCE in these cells, unlike that observed with knockdown of STIM1 or Orai1. These data suggested that TRPC1 depended on STIM1 and Orai1 for function. It was shown that TRPC1 and Orai1 form two distinct channels that are gated by STIM1 (Fig. 5.2) (Cheng et al., 2011; Ong et al., 2012). In a cell line endogenously expressing all three proteins, ER-Ca^{2+} store depletion generated I_{SOC} that was completely abolished by knocking down Orai1. However, knockdown of TRPC1 revealed a residual I_{CRAC} current. Similar data was observed in overexpression studies, where co-expression of Orai1+STIM1 generated I_{CRAC} while co-expression of all three proteins resulted in I_{SOC}. TRPC1 function was drastically reduced when wild-type Orai1 was substituted with the non-functional mutants, Orai1R91W and Orai1E106Q (Cheng et al., 2008; Prakriya et al., 2006). Similarly, substitution of wild-type STIM1 with a mutant that lacks TRPC1 gating ($STIM1^{684}EE^{685}$) or wild-type TRPC1 with a mutant that lacks STIM1 binding ($TRPC1^{639}KK^{640}$) significantly reduced SOCE in HSG cells co-expressing TRPC1+STIM1 and converted I_{SOC} to I_{CRAC} (Pani et al., 2009; Zeng et al.,

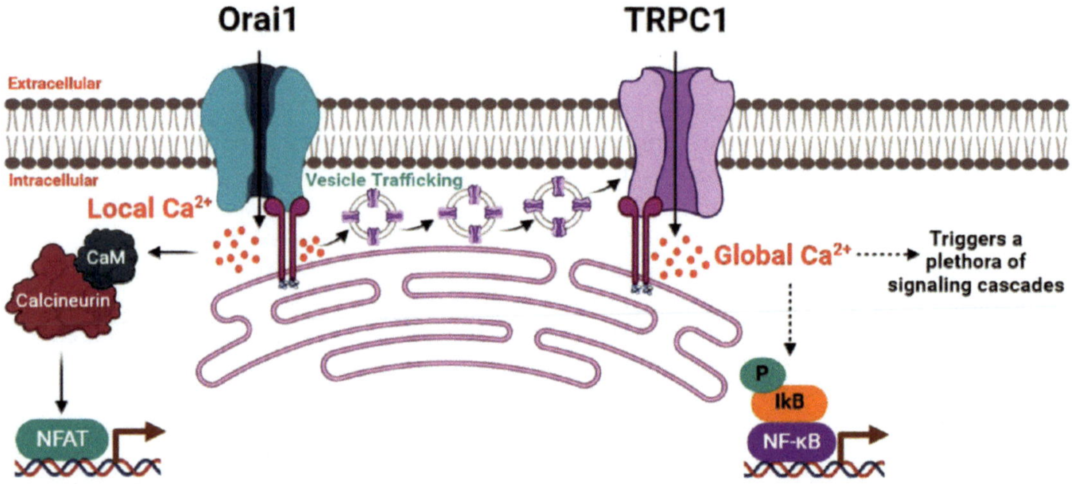

FIGURE 5.2 Model shows how TRPC1 is activated following ER-Ca^{2+} store depletion to mediate SOCE. When STIM1 senses the reduction in $[Ca^{2+}]_{ER}$, it aggregates within the ER-PM junctions, where it activates the Orai1 channel. High local $[Ca^{2+}]$ mediated by Orai1 triggers insertion of TRPC1 into the plasma membrane, where TRPC1 is then activated by STIM1. As the channel mediates a larger magnitude of Ca^{2+} entry, TRPC1 serves to sustain Ca^{2+} influx to regulate a plethora of signaling cascades.

2008; Cheng et al., 2011). However, expression of the STIM-Orai1 activating domain of STIM1 (SOAR) yields a constitutive Ca^{2+} entry and current that was unaffected by knockdown TRPC1 but abolished by loss of Orai1. Moreover, the current observed in HSG cells expressing SOAR alone more closely resembles I_{CRAC} instead of I_{SOC}. These suggest that TRPC1 is not directly gated by the SOAR of STIM1 (Cheng et al., 2011). Additional data showed that expression of TRPC1, Orai1, and STIM1 resulted in I_{SOC}, while that of Orai1+STIM1 generated I_{CRAC} (Cheng et al., 2008; Huang et al., 2006; Yuan et al., 2007), other research have reported TRPC1-associated I_{SOC} (Kim et al., 2009b; Jardin et al., 2008). Together, all these studies show that TRPC1 function is dependent on Orai1+STIM1. The mechanism underlying the functional interaction between TRPC1 and Orai1 was revealed by findings showing that Orai1-mediated Ca^{2+} entry triggers PM insertion of TRPC1, where it is activated by STIM1 (Fig. 5.2) (Cheng et al., 2011). Other studies have also reported similar TRPC1 trafficking to the PM in response to ER-Ca^{2+} store depletion (de Souza et al., 2015).

An interesting aspect of the presence of both Orai1 and TRPC1 in cells is that the two channels generate distinct $[Ca^{2+}]_i$ signals, which are coupled to specific downstream Ca^{2+}-dependent signaling processes. TRPC1-mediated Ca^{2+} entry generates a sustained $[Ca^{2+}]_i$ response when stimulated with high concentrations of carbachol (CCh) but contributes to more oscillatory responses at low concentrations of CCh, which trigger activation of Ca^{2+}-dependent potassium channels and the NFκB signaling pathway (Fig. 5.2). In contrast, Orai1 generates oscillatory responses with a range of CCh concentrations (from high to very low) and contributes to the nuclear translocation of NFAT and activation of NFAT-dependent gene expression (Fig. 5.2). As such, it was proposed that TRPC1-mediated Ca^{2+} entry induces global $[Ca^{2+}]_i$ changes while Orai1-mediated Ca^{2+} entry contributes to more localized $[Ca^{2+}]_i$ signals (Cheng et al., 2011; Ong et al., 2012). Such compartmentalization enables precise spatiotemporal control of SOCE mediated by TRPC1 and Orai1 channels.

5.2 ROLE OF TRPC1 IN DISEASE – STUDIES WITH CELL LINES

Investigation into the development of cancer has gradually shifted its focus from looking at cancer cells alone to studying a tumor microenvironment (TME), where the cooperativity between cancer cells, fibroblasts, vascular cells, and immune cells promotes tumorigenesis. Inflammatory processes during carcinogenesis involve both anti- and pro-tumorigenic processes at the same time. The anti-tumorigenic processes involve immunosurveillance and immunological sculpting of tumor heterogeneity, whereas pro-tumorigenic processes promote a more tumor-permissive state by blocking anti-tumor immunity and exerting direct tumor-promoting signals onto cancer cells. The dysregulated function of TRPC has been shown to contribute to the development and progression of various cancers (Fig. 5.1). As described in a recent review, increased TRPC1 expression is observed in cancer cell lines for pancreas (BxPc3), breast (MDA-MB-468, MCF-7), lung (A549), and colon (HT29), and depending on the type of cancer cells, it may either enhance or inhibit cell motility, proliferation, and migration/invasiveness (Greten and Grivennikov, 2019; Elzamzamy et al., 2020). Such variability can be caused by different interacting partners for TRPC1 in various cell types. For example, TRPC1 is a positive marker for the prognosis of patients with esophageal squamous cell carcinoma (ESCC). A recent clinical study reported a correlation between reduced TRPC1 expression and function and a worse prognosis for ESCC patients. Studies using human ESCC cell lines, EC9706 and KYSE150, showed that downregulating TRPC1 increases cell proliferation, migration, and invasion. Overexpression of TRPC1 had the opposite effect, which is partly mediated via the AKT/p27 pathway (Zeng et al., 2021). However, another study utilizing The Cancer Genome Atlas and The International Cancer Genome Consortium (IGCG) database showed that TRPC1 negatively correlates with the survival rate in patients with liver cancer (Qi et al., 2023).

Breast carcinoma cells, MCF-7, were used to study the contribution of TRPC1 to epirubicin resistance in breast cancer. An epirubicin-resistant cell line, MCF-7/EPI, was made by serially increasing epirubicin concentrations to promote drug resistance in MCF-7 cells. Exposing cells to hypoxia revealed that the microRNA miR-181a-5p modulates the expression of EPDR1 to regulate cell replication, migration, and invasion. MicroRNAs are non-coding RNAs that are approximately 22 nucleotides in length and can function either as oncogenes or tumor suppressors. EPDR1 exerts its effects through TRPC1-mediated modulation of the PI3K/AKT signaling pathway. Overexpression of miR-181a-5p inhibits EPDR1 and TRPC1 expression, leading to the hyperactivation of the PI3K/AKT signaling pathway and diminishing cell sensitivity to epirubicin. Furthermore, overexpression of EPDR1 downregulated the levels of interstitial-related protein breast cancer cells *in vitro* and inhibited tumor growth *in vivo*, suggesting that EPDR1 may inhibit epithelial-mesenchymal transition (EMT) in breast cancer. Thus, drugs modulating the EPDR1/TRPC1 axis may help reduce resistance to epirubicin during chemotherapy (Zhang et al., 2024).

When pancreatic ductal adenocarcinoma cells are exposed to acid adaptation and recovery conditions, TRPC1 expression and function in the PM increase, which correlates with increased cell migration, proliferation, and spheroid growth. The $[Ca^{2+}]_i$ elevation in basal state allows the formation of a complex with the PI3K p85α subunits, calmodulin (CaM), and extracellular signal-regulated kinase 1/2 (ERK1/2), which promotes cell proliferation and faster cell migration (Schnipper et al., 2022). Likewise, the expression of TRPC1 in colorectal cancer (CRC) tissues has been linked with enhanced cell proliferation, migration, and invasion. Loss of TRPC1 decreased the expression of mesenchymal markers such as N-cadherin/CDH2, Snail1, and Slung but increased the expression of epithelial marker E-cadherin/CDH1. TRPC1 may play a scaffolding role in CRC tissues where it promotes the interaction between CaM and the PI3K-p85 subunit and downstream phosphorylation of Akt, which enhances tumor growth, metastasis, and aggressive oncogenic behavior (Sun et al., 2021). In the B-cell leukemia cell line, Jok-1, CD5 promotes Ca^{2+}-dependent constitutive activation of MAPK, phosphorylation of Erk1/2, IL-10 production, and activation of the PI3K/Akt/mTOR pathway. CD5 increased the expression of TRPC1 and TRPV2 channels, which induced a higher basal Ca^{2+} level; this $[Ca^{2+}]_i$ alteration was correlated with a constitutive Erk1/2 activation and an increased IL-10 level. Blocking channel function with La^{3+} decreased the basal phosphorylation of Erk1/2, as similarly observed with the inhibitor of MEK/Erk PD98059. Therefore, it was proposed that TRPC1 activity could be a characteristic of neoplastic B cells resulting from an abnormal Erk1/2 phosphorylation induced by CD5 (Garaud et al., 2018). These studies suggest that TRPC1 may be a useful marker in detecting malignant tumors. In addition, it appears that TRPC1 regulation of cell signaling varies depending on the cell type. The exact basis for this is not yet understood.

There is evidence that TRPC1 is also involved in other cellular processes such as autophagy (Fig. 5.1). In HSG and human neuroblastoma (SHSY-5Y) cells, autophagy was induced by using cell-permeable hypoxia-mimetic agents, DMOG and DFO, or serum starvation. No significant activation of caspase-3, an apoptotic marker, or Ca^{2+} release from the intracellular stores was triggered by the treatment. Hypoxia-mediated autophagy enhanced TRPC1 expression without affecting the expression of other SOCE components, STIM1 and Orai1. As such, SOCE mediated by TRPC1 was also increased, which promoted cell survival by increasing expression of autophagic markers that prevented apoptosis. Silencing TRPC1 (but not TRPC3) function using siRNA or inhibiting autophagy using 3-methyladenine adversely affected cell viability by attenuating hypoxia-induced increase in SOCE, decreasing autophagy, and promoting apoptosis. These findings suggest that TRPC1-mediated Ca^{2+} influx plays a protective role during cell stress by augmenting autophagy to avoid triggering an apoptotic signaling cascade (Sukumaran et al., 2015). Additionally, TRPC1 has been shown to exert other protective effects on SHSY-5Y cells. Treatment with dopaminergic neurotoxins such as 6-hydroxydopamine (6-OHDA), 1-methyl-4-phenylpyridiumions (MPP^+), or 1-methyl-4-phenyl-1, 2, 3, 6-tetrahydropyridine (MPTP) caused neuronal degeneration by reducing SOCE. MPTP or 6-OHDA treatment also decreased TRPC1 expression but did not affect expression levels of other proteins involved in Ca^{2+} influx, such as Orai1, STIM1, TRPC3, or TRPC5. Furthermore, 6-OHDA and MPP^+ increased caspase-3 activity and reduced the levels of an autophagic marker LC3A, which impaired cell survival. These findings present evidence that TRPC1 is the main Ca^{2+} channel in dopaminergic cells, and loss of its function compromises cell viability due to a switch from autophagy to apoptosis (Fig. 5.1) (Sukumaran et al., 2018).

In the myofibroblast cell line CCD-18Co, cyclooxygenase-2 (COX-2), an enzyme that plays an important role in the production of prostaglandin E_2 (PGE_2), is expressed at low levels in unstimulated/resting cells. Following inflammation, tumor necrosis factor (TNF-α) enhances the production of PGE_2 by inducing an abnormal increase in COX-2 expression through tumor promoters. A high concentration of PGE_2 further promotes tumorgenicity. Treatment of CCD-18Co cells with TNF-α increased TRPC1 protein expression and localization in the PM without altering the expression levels of other SOCE proteins such as Orai1 and STIM1. Treatment with the anti-TRPC1 antibody, T1E3, or siRNA specific for *Trpc1* suppressed TRPC1 function and impaired the TNF-α-induced increase in Ca^{2+} influx but potentiated PGE_2 production, suggesting a negative feedback role of TRPC1-dependent Ca^{2+} influx on TNF-α signaling. Suppressing TRPC1-mediated SOCE significantly facilitated TNF-α-induced nuclear translocation of nuclear factor κ B (NF-κB), increasing production of PGE_2 (Hai et al., 2011). It was later found that NF-κB functions together with nuclear factor of activated T cells (NFAT) to shape and decode intracellular Ca^{2+} signals for transmitting the information to downstream cellular effectors (Smedler and Uhlen, 2014). Unlike NF-κB, NFAT3 suppresses COX2 expression. Thus, inhibiting NFAT3 enhanced PGE_2 production without affecting TRPC1 expression and function. As such, NF-κB and NFAT were proposed to be positive and negative transcriptional regulators, respectively, of the TNF-α-induced COX-2 expression that underlies PGE_2 production in pro-inflammatory responses (Hai et al., 2011). Similar dynamic co-regulation by NF-κB and NFAT was recently reported for human T cells, where these transcription factors integrate two key pathways in T cell signaling to regulate distinct modes of gene expression (Huang et al., 2023). Future studies should investigate whether dynamic co-regulation of transcription factors can be found in other types of cancers and the contribution of TRPC1 to the development of these tumors.

Human umbilical vein endothelial cells (HUVECs) were used to study the role of TRPC1 in high glucose-induced oxidative stress. Exposing HUVECs to high glucose conditions significantly reduced TRPC1 expression and impaired cell proliferation and migration. Conversely, increasing TRPC1 expression with adenovirus infection rescued the proliferation and migration of HUVECs exposed to high glucose conditions. Elevated levels of reactive oxygen species (ROS) and malondialdehyde, as well as decreased superoxide dismutase activity, suggest that TRPC1 protects HUVECs against adverse effects caused by high glucose-induced oxidative stress. Thus, drugs targeting TRPC1 in diabetes-induced vascular endothelial cell dysfunction may lead to novel therapeutics for treating diabetic patients (Qu and Du, 2024).

Additionally, TRPC1 has also been shown to contribute to endothelial cell permeability (Fig. 5.1). RhoA is a small GTPase well known to act on the downstream Rho kinase, ROCK, to control vascular tone. Overactivation of the RhoA/ROCK pathway leads to endothelial cell dysfunction and abnormal vascular tone by disrupting the balance in the production of vasodilating and vasoconstricting factors (Yao et al., 2010). It is also reported that RhoA induces TRPC1 association with the inositol 1,4,5-triphosphate receptor (IP_3R) to activate SOCE in human pulmonary arterial (HPAEC) and dermal microvascular (HMEC) cell lines. Overexpression of TRPC1 increased thrombin-induced RhoA-dependent SOCE, suggesting that RhoA regulates TRPC1-mediated SOCE, which is pivotal to thrombin-induced endothelial barrier disruption (Mehta et al., 2003). Vascular endothelial growth factor (VEGF) also increases permeability of the vascular endothelial cells by stimulating SOCE. Angiopoietin-1 (Ang-1) exerts a protective effect by inhibiting VEGF-induced Ca^{2+} influx mediated by TRPC1, TRPC1-IP_3R interaction, and then hyperpermeability (Jho et al., 2005). TRPC1 forms a heteromeric channel with TRPC4 in endothelial cells, where one subunit of TRPC1 interacts with at least two subunits of TRPC4 (Cioffi et al., 2012). While Orai1 interacts with TRPC1 and contributes to its function in HSG cells (Cheng et al., 2011), thus far, only Orai1 interaction with TRPC4 has been reported in endothelial cells, where loss of Orai1 impaired heteromeric TRPC1/TRPC4 channel activation and improved endothelial cell barrier function (Cioffi et al., 2012). As recently reviewed (Moccia et al., 2023), SOCE in endothelial cells may be mediated by different channels: I_{CRAC} by Orai1 and STIM1; I_{SOC} by TRPC1, TRPC4, and STIM1; and a moderately Ca^{2+}-selective, I_{CRAC}-like current by TRPC1, TRPC4, Orai1, and STIM1. Further studies are required to clarify how each channel type contributes to and regulates endothelial cell barrier function.

Parkinson's disease (PD) is a neurodegenerative disorder characterized by a gradual loss of dopaminergic neurons (DN), causing involuntary movements (motor deficits) and confusion or memory loss (cognitive decline). Current PD treatments do not adequately counter these symptoms of PD and have serious side effects, which greatly diminish patients' quality of life. TRPC1 is widely expressed in the brain and has been found in the substantia nigra pars compacta and reticulata, plus the striatum and subthalamic nuclei. Expression of TRPC1 in these areas is generally reduced in PD (Vaidya and Sharma, 2020). In a cellular model where mesenchymal stem cells (MSCs) were differentiated into DN to mimic native neurons, treatment with MPP^+ reduced not only extracellular Ca^{2+} influx but in the long term also Ca^{2+} release triggered by thapsigargin (a blocker of the sarco/endoplasmic reticulum Ca^{2+}-ATPase pump). Expression of TRPC1, but not Orai1, was substantially reduced after the MPP^+ treatment (Sun et al., 2018). Dysregulation of Ca^{2+} influx alters Ca^{2+} homeostasis in multiple organelles and induces cellular stress (Fig. 5.1). For instance, decreased $[Ca^{2+}]_{ER}$ triggers an unfolded proteins response (UPR) that results in the accumulation of unfolded protein in the ER lumen (Chami and Checler, 2020). Silencing of TRPC1 using siRNAs reduced both the $[Ca^{2+}]_{ER}$ and extracellular Ca^{2+} influx while increasing the expression of chaperones such as GRP78, GRP94, and CHOP, and enhancing phosphorylation of PERK and its effector eIF2α (Selvaraj et al., 2012). In agreement with this, another study reported that deleting TRPC1 in striatal neuronal cells modifies the expression of NDUV2, DJ-1 oxidative stress-related proteins, such as p-PERK, p-eIF2α, and CHOP ER stress response-related proteins (Wang et al., 2018). On the other hand, alterations in mitochondrial Ca^{2+} homeostasis and modifications of Ca^{2+} fluxes from the ER to the mitochondria have been associated with various pathological conditions, namely excessive production of ROS that can lead to programmed cell death or apoptosis (Guerrero-Hernandez et al., 2020; Sanchez-Vazquez et al., 2023; Singh et al., 2024; Sun et al., 2018). In PD, changes in mitochondria and ER Ca^{2+} homeostasis are linked to neuronal survival, while A549 (human lung epithelial) cells under oxidative stress conditions have been reported to exhibit an increased expression of apoptotic markers such as caspases-3 and -9 (Selvaraj et al., 2012; Shin et al., 2023). Collectively, these findings highlight the relevance of TRPC1 channel activity to maintain the Ca^{2+} levels not only in the cytoplasm but also in intracellular organelles.

A few studies showed that TRPC1 may also function as a non-Ca^{2+} selective channel activated by mechanical stress. Asthma is a disease characterized by chronic inflammation of the airways (Liu et al., 2021), which has been correlated with an increased Ca^{2+} influx mediated by TRPC1 (Fig. 5.1). Human bronchial epithelial (16HBE) cells subjected to mechanical stress were used to evaluate the role of the TRPC1 channel. Knockdown of TRPC1 reduced the eosinophils concentration, IL-13, TGF-β1, and MMP-9 expression levels, and airway inflammation score (Li et al., 2019). Another characteristic of asthma is mucus overproduction, which has been linked to IL-4 (Kim et al., 2009a). In the bronchial epithelial cell (BEC) line, IL-4 increases mucus production due to overexpression of the mucin glycoprotein MUC5AC (Xia et al., 2017). The molecular basis for this was correlated with clusters of caveolin-1 in lipid rafts and TRPC1, where increased Ca^{2+} influx mediated by TRPC1 may activate NF-κB to induce MUC5AC overexpression (Xia et al., 2017). Further studies are required to evaluate the suitability of TRPC1 as a candidate target of future therapies for airway inflammatory diseases.

5.3 REVEALING TRPC1 CHANNEL FUNCTION USING ANIMAL MODELS

Disease- and protein-specific animal models are broadly utilized to study the contribution of TRPC1 to disease development and elucidate physiological functions of the channel. Studies with TRPC1$^{-/-}$ mice showed that loss of TRPC1 substantially decreases acinar cell fluid section from salivary glands and pancreas (Ambudkar, 2018). TRPC1 plays a key role in regulating SOCE and fluid secretion in acinar cells of salivary glands. Briefly, stimulation of cells by neurotransmitters leads to the generation of IP_3, which then binds to its receptor on the ER membrane to induce Ca^{2+} release. Depletion of the ER Ca^{2+} stores is sensed by the STIM proteins, STIM1 and STIM2, resulting in their aggregation and translocation to ER-PM junctions. Both STIMs then recruit the Orai1 and TRPC1 channels to mediate extracellular Ca^{2+} influx into the acinar and ductal cells. Spatiotemporal regulation of these Ca^{2+} signals enables precise control of Ca^{2+}-dependent downstream effectors, thereby ensuring their specificity and efficacy. Sustained increases in $[Ca^{2+}]_i$ in acinar cells are required to activate ion transporters and channels, such as the $Na^+/K^+/2Cl^-$ cotransporter 1 (NKCC1), Anoctamin 1 (ANO1), and Ca^{2+}-dependent K^+ (K_{Ca}) channel. The coordinated actions of NKCC1, ANO1, and K_{Ca} channel promote vectorial transport of Cl^- from the basolateral to luminal membranes of polarized salivary gland epithelial cells, generating an osmotic gradient across the luminal membrane that drives water secretion via the water channel, Aquaporin 5 (AQP5). Loss of TRPC1 causes significant loss of SOCE, Ca^{2+}-activated K^+ current, and saliva secretion in the salivary gland (Ambudkar, 2018; Liu et al., 1999; Liu et al., 2018). TRPC1 requires both Orai1 and STIM1 for its function. Yet these proteins are differentially localized in pancreatic acinar cells. Orai1 is found in the apical pole of the lateral membrane, with its localization unchanged by cell stimulation. Conversely, STIM1 was recruited to both apical and lateral regions, where TRPC1 is also localized. Pancreatic acinar cells from TRPC1$^{-/-}$ mice showed reduced Ca^{2+} influx, Ca^{2+} oscillation frequency, and Ca^{2+}-activated Cl^- current (Hong et al., 2011).

Primary Sjögren's Syndrome (pSS) is a systemic autoimmune disease characterized by loss of fluid secretion from salivary glands as well as immune cell infiltration of salivary glands. Immunomodulatory therapies that aimed to ameliorate abnormal immune regulation have yielded little or no benefits in pSS patients (Sun et al., 2023). However, oral pilocarpine treatment improves the symptoms of dry mouth and dry eyes in these cohorts (Tsifetaki et al., 2003; Vivino et al., 1999). Pilocarpine induces Ca^{2+} signaling by activating muscarinic acetylcholine receptors (mAChRs) in salivary gland cells. Activation of the M1 and M4 mAChR subtypes induced ER-Ca^{2+} release and extracellular Ca^{2+} influx, resulting in a net increase in $[Ca^{2+}]_i$ that is mediated by TRPC1 (as shown by studies in TRPC1$^{-/-}$ and Cav1$^{-/-}$ mice) (Busch and Borda,

2007; Liu et al., 2007; Nezu et al., 2015; Pani et al., 2013; Sun et al., 2023). Using the IL14a transgenic (Tg) mouse model for pSS, loss of TRPC1-mediated SOCE resulted in loss of salivary gland acinar cell function and cell death. This preceded immune cell infiltration of the salivary glands. TRPC1 expression was decreased in salivary glands from pSS patients and the IL14a Tg mouse model. Those obtained from TRPC1$^{-/-}$ mice showed abnormal immune cell detection in salivary glands due to infiltration. Loss of TRPC1-mediated SOCE in siRNA-treated cell lines induced an ER stress response, characterized by increased expression of CHOP, an ER stress marker, and increased apoptosis. Similarly, salivary tissues obtained from pSS patients showed elevated levels of CHOP in both acinar and ductal cells. Aberrant Ca^{2+} signaling in diseased salivary glands caused the release of damage-associated molecular patterns (DAMPs), such as histone-3, HMGBI, HSP70, and alarmins, all of which promoted immune cell infiltration (Sun et al., 2023). Elevated IL-17 levels have been observed in pSS patients, and IL-17-deficient mice were resistant to experimental Sjögren's Syndrome (ESS) induction, exhibiting normal salivary gland function. Treatment of epithelial cells from the human labial gland and murine salivary gland with IL-17 significantly reduced TRPC1 mRNA transcript levels. Similar treatment of human salivary gland (SG) cell line A253 and primary murine SG epithelial cells downregulated TRPC1 protein expression. Interestingly, during ESS development, IL-17-deficient mice had higher channel expression when compared to WT mice. IL-17 promotes *Nfkbiz* gene expression that encodes the IkB-z protein, both of which were significantly increased during ESS development. IkB-z inhibited NF-κB in the nucleus, which in turn downregulated *Trpc1* levels and protein expression (Xiao et al., 2021).

Neurodegenerative diseases characterized by abnormal aggregates of tau in neuronal and glial cells are called tauopathies. Tau is one of the most abundant microtubule-associated proteins in neurons, where it contributes to the structure of the cytoskeleton network by catalyzing microtubule polymerization and assembly. Additionally, tau is also important for intracellular axonal transport (Devi, 2023). In hippocampal neurons, elevated levels of tau upregulate the C/EBPβ/TRPC1 pathway to increase TRPC1 expression and enhance SOCE. These in turn induce ER stress and create an imbalance in the activities of protein kinases and phosphatases, which aggravates tau hyperphosphorylation, creating a vicious feed-forward cycle. TRPC1 is the main channel contributing to neuronal SOCE, as pharmacologically inhibiting or genetically knocking out TRPC1 attenuates tau phosphorylation, rescuing synaptic dysfunction and memory deficits in human tau-expressing mice (Ye et al., 2020). Intriguingly, TRPC1$^{-/-}$ mice did not exhibit learning and memory impairment until they were injected with amyloid-β (Aβ), which stimulates the formation of neurofibrillary tangles in the brain. Aβ-induced neuronal apoptosis was also enhanced by the loss of TRPC1, as shown by a significant reduction in immunostaining for astrocyte and microglia markers in the hippocampal CA3 of TRPC1$^{-/-}$ mice. TRPC1 competes with γ-secretase for interaction with the Aβ precursor protein (AβPP), which prevents γ-secretase-mediated cleavage of AβPP to generate Aβ. Thus, increasing TRPC1 interaction with AβPP may also have a neuroprotective effect by preventing the formation of Aβ plaques (Li et al., 2018). TRPC1 has also been implicated in PD, where channel activation confers a protective effect on neuronal cells (Gao et al., 2021). PD patients exhibit significant neuronal loss in the substantia nigra, coupled with decreased TRPC1 function and enhanced unfolded protein response (UPR), suggesting that post-translational processing and targeting of TRPC1 were impaired. Neurotoxin treatment of TRPC1$^{-/-}$ mice produced similar results, which were ameliorated by overexpression of TRPC1 as Ca^{2+} entry mediated by the channel activates the neuro-protective Akt/mTOR pathway (Selvaraj et al., 2012). Dopaminergic neurons (DNs) derived from mesenchymal stem cells (MSCs) are an attractive therapeutic option to ameliorate neuronal loss in PD patients. One study using MSC-derived DNs showed that following treatment with 1-Methyl-4-phenylpyridinium (MPP^+), TRPC1 expression and function were significantly reduced. Addition of SKF96365, a TRPC channel blocker, to MSC-derived DNs significantly reduced dopamine release but increased cell death (Sun et al., 2018). Collectively, these data highlight the functional importance of TRPC1 to physiological neuronal activities.

Traumatic brain injuries (TBIs) and diverse stress stimulus over astrocytes can provoke astrogliosis, which is a reactive state where astrocytes increase their proliferation and grow toward the injury site. Astrocyte activity can be protective by generating a glial scar to cover the damaged area. On the other hand, the glial scar hinders axonal regeneration. Astrocytes from TRPC1$^{-/-}$ mice showed an increased Ca^{2+} influx when stimulated with 1-oleoyl-2-acetyl-sn-glycerol (OAG), an analogue of DAG. However, astrocytes from TRPC3$^{-/-}$ mice elicited a substantial reduction of Ca^{2+} influx. Interestingly, *in vitro* migration assays revealed that TRPC1 negatively regulates astrocyte movement, whereas TRPC3 promotes cell migration. Further studies are required to elucidate how TRPC1 co-regulates astrocyte Ca^{2+} signals with TRPC3 (Belkacemi et al., 2017). Neutrophil infiltration of the inflamed brain area following ischemic stroke has been linked to worse prognosis and clinical outcomes for patients. While it is known that there is crosstalk between the microglia and infiltrating immune cells, little is known about the underlying molecular mechanisms for such crosstalk. In a recent study, microglial *Trpc1* conditional knockdown mice were generated and used in experiments with middle cerebral artery occlusion reperfusion (MCAO/R) and *in vitro* oxygen-glucose deprivation/reoxygenation (OGD/R). In both conditions, TRPC1 expression within the microglia was significantly upregulated. Knockdown of microglial TRPC1 markedly reduced neutrophil infiltration primed pro-inflammatory phenotype with less chemokine secretion, SOCE, and NFATc1 levels. Conversely, overexpression of *Nfatc1* ameliorated the chemokine reducing effect of *Trpc1* knockdown. The authors proposed that upregulated expression of TRPC1 stimulates chemokine production via the Ca^{2+}/NFATc1 pathway, leading to increased neutrophil infiltration in the brain to aggravate reperfusion injury (Qian et al., 2024).

Airway smooth muscle cells (ASMCs) from asthmatic mouse models exhibited upregulated expression of TRPC1, TRPC3, and TRPC6, together with elevated $[Ca^{2+}]_i$ and enhanced cell viability, all of which were blocked by SKF96365 treatment and siRNA-mediated knockdown of TRPC1 or TRPC3 (Zhang et al., 2018). Additionally, sustained levels of cytosolic $[Ca^{2+}]_i$ in pulmonary epithelial (PE) cells have been proposed to promote bronchial wall thickening and airway constriction in respiratory disease states, such as chronic obstructive pulmonary disease (COPD). Continuous production of ROS increased the levels of various markers associated with fibrosis, inflammation, and apoptosis. PE cells isolated from COPD patients show elevated protein levels of the TRPC1 channel, in addition to TRPC3, STIM1, and Orai1. TRPC1 worked in tandem with TRPC3 to mediate extracellular Ca^{2+} influx that increased cytosolic $[Ca^{2+}]_i$, triggered ER stress responses, and induced apoptosis, all of which were inhibited by knockdown of both channels (Shin et al., 2023). Chronic airway inflammation induced by house dust mite (HDM) challenge triggers a severe inflammatory response and airway remodeling. Increased levels of $[Ca^{2+}]_i$ and pro-inflammatory cytokines such as IL-6, TNF-α, and IL-4, plus enhanced activation of STAT3 and NF-κB, were observed in control mice, which were ameliorated in TRPC1$^{-/-}$ mice. This research showed that eosinophils in the lung, inflammation index, and lipid peroxidation were reduced in TRPC1$^{-/-}$ mice compared with control mice. The authors proposed that TRPC1 facilitates EMT and enhances STAT3/NF-κB signaling to induce a severe asthmatic response triggered by house dust mites (Pu et al., 2019). The Schisandrin B (SB) plant is used in traditional Chinese medicine for its anti-inflammatory and antioxidant properties. Treatment of Sprague-Dawley rats with ovalbumin (OVA) triggered an asthmatic response and induced inflammation in the lungs, as shown by increased expression of the NLRP3 inflammasome, caspase-1, and GSDMD-N, as well as higher levels of IL-1β and IL-18. Treatment with SB attenuates the expression of these inflammatory proteins and reduces the levels of IL-1β and IL-18. TRPC1 facilitates activation of the STAT3/NF-κB signaling complex, and SB treatment was proposed to inhibit TRPC1 function to ameliorate OVA-induced lung inflammation (Chen et al., 2021a). In a BALB/c mouse model of asthma induced by OVA, it was also found that SB reduced both the level of inflammatory factors such as eotaxin, IL-13, IL-5, and IL-4 and the concentration of inflammatory cells such as neutrophils, macrophages, lymphocytes, and eosinophils from bronchoalveolar lavage fluids. In this research, it was proposed that SB improves asthmatic complications caused by inflammation through inhibiting the NF-κB pathway and activating the Nrf2 signaling pathway (Chen et al., 2021b). While TRPC1-mediated Ca^{2+} influx has been shown to be crucial for activating NF-κB signaling (Ong et al., 2012), it remains to be conclusively shown how SB directly affects TRPC1 channel function.

Familial hypocalciuric hypercalcemia (FHH) is a genetic condition associated with inactivating mutations in genes for the Ca^{2+}-sensing receptor (*CaSR*) and G protein α subunit 11 (Gα11; encoded by *GNA11*). Clinical symptoms shown by FHH patients include hypocalciuria, hypercalcemia, and abnormally high levels of parathyroid hormone (PTH). High serum [Ca^{2+}] has a negative feedback effect on PTH by suppressing its production and/or secretion from the parathyroid gland (Onopiuk et al., 2020). The CaSR on parathyroid cells senses small changes in serum [Ca^{2+}] and modifies the release of PTH accordingly. Activation of CaSR by extracellular Ca^{2+} leads to the generation of IP_3, which binds to its receptor on the ER membrane and triggers Ca^{2+} release from the ER lumen. STIMs in the ER membrane sense depletion of the ER-Ca^{2+} stores and activate PM Ca^{2+} channels, such as TRPC1 and Orai1. CaSR has been shown to activate TRPC1 in colonic epithelial, endothelial, and breast cancer cells, as well as rabbit mesenteric arteries (Chen and Goodman, 2004; Onopiuk et al., 2020). Interestingly, CaSR is found abundantly within the PM caveolae, where TRPC1 is also enriched. The TRPC1$^{-/-}$ mouse model mimicked human FHH by developing hypercalcemia and hypocalciuria and exhibited elevated PTH levels. Although TRPC1 has been shown to act downstream of CaSR, channel activation by store depletion may not account for its role in PTH release. An interaction of Gα11 with both N- and C-termini of TRPC1 has been proposed to activate the channel in parathyroid gland cells. It remains to be seen whether co-expression of TRPC1 and STIM1 in PTH-C1 (a cell line derived from rat parathyroid gland) will confer ER-Ca^{2+} store dependence of CaSR-mediated PTH release (Chen and Goodman, 2004; Onopiuk et al., 2020).

While TRPC1 is not a voltage-gated channel, many studies have reported its contribution to calcium signaling in excitable cells. Skeletal muscles of TRPC1$^{-/-}$ mice exhibit smaller muscle mass, increased muscle fatigue, and resting muscle stiffness, as well as slower regeneration after muscle injury (Zanou et al., 2010). While an increased protein expression of TRPC1 has been reported in vascular smooth muscles from *mdx* mice (a murine model for Duchenne muscular dystrophy) (Lopez et al., 2020b), it is unclear how TRPC1 contributes to the $[Ca^{2+}]_i$ and $[Na^+]_i$ dyshomeostasis observed in these cells. Interestingly, enhanced TRPC1 channel function has been observed in Homer 1-deficient and defective dystrophin-expression mice, which exhibit higher proteolysis, muscle necrosis, and myopathy. TRPC1 was hypothesized to form a putative "costameric" macromolecular complex together with Homer 1, dystrophin, and dystroglycans in skeletal muscle sarcolemma, regulating the development of muscle atrophy (Gorza et al., 2021).

While expression of TRPC1 is sparse in mouse and rat retinas, the channel has been found in retinal glial cells, where Ca^{2+} signaling contributes significantly to various synaptic activities. The α and β isoforms of TRPC1 are expressed in dorsal root ganglions (DRGs), and decreased channel activity impaired neuronal sensitivity to mechanical stimuli (Staaf et al., 2009; Thebault, 2021). TRPC1 has also been found in human retinal vascular endothelial cells, where it may contribute to retinal blood circulation (Thebault, 2021). Degranulation of mast cells is an early event following antigen stimulation, resulting in anaphylaxis. Mast cells derived from the bone marrow of TRPC1$^{-/-}$ mice point to a protective effect as loss of TRPC1 delayed recovery from the induction of antigen-mediated anaphylaxis. Interestingly, calcium signaling was enhanced in these cells with increased Ca^{2+}-dependent activation of NFAT and AP1 transcriptional pathways but little effect on degranulation. Furthermore, the levels of inflammatory cytokines such as TNF-α, IL-6, and GM-CSF were also elevated. The authors proposed that TRPC1 functions as a negative regulator of SOCE in mast cells by suppressing mast cell production of TNF-α following antigen stimulation (Medic et al., 2013). Nonetheless, the contribution of Orai1 to SOCE and the NFAT signaling pathway in mast cells (Clemens and Lowell, 2019; Park et al., 2020) cannot be discounted. The close correlation between TRPC1 and Orai1 makes it challenging to elucidate the contribution of each channel when present together in any cell (Ambudkar et al., 2017).

TRPC1 may contribute to other diseases via its interaction with other TRP channels. For example, the binding site for TRPC1 in the TRPP1 protein is mutated in patients with autosomal dominant polycystic kidney disease (AKPKD). TRPC1 also interacts with TRPC3 and TRPC6 in human and rat mesangial cells, where impaired TRPC1 channel activity adversely affects glomerular mesangial function. Proteinuria (high protein concentration in urine) is a characteristic symptom of diabetic nephropathy that results in end-stage renal disease, and genetic polymorphisms in the *TRPC1* gene have been linked to patients of Han Chinese descent (Chen et al., 2013). Vascular smooth muscle cell (VSMC) plasticity allows for reversible phenotypic switching between highly proliferative (synthetic) and fully differentiated (contractile) phenotypes. Tonic regulation of vascular homeostasis is controlled by contractive VSMCs. Vascular disorders, such as atherosclerosis and hypertension, are caused by abnormal phenotypic switching of VSMCs. TRPC1 is ubiquitously expressed in the cardiovascular system, with studies reporting channel contribution to vasoconstriction, neointimal hyperplasia, and cardiac hypertrophy. While the expression of TRPC1 and TRPC6 is upregulated in pathological VSMCs, more studies are required to elucidate how channel function has been altered by pathophysiological conditions (Nishida et al., 2020). TRPC1 may also form heteromeric channels with TRPC3 and TRPC6 to allow chronic Ca^{2+} and Na^{+} influx in VSMCs from the *mdx* mice (Lopez et al., 2020b). Spontaneously hypertensive rats (SHRs) have been used to study brain damage caused by arterial hypertension. While the rats were normotensive at birth, they developed sustained hypertension by 6 months. SHRs exhibit neurological and behavioral changes similar to those reported in patients showing vascular brain damage and attention deficit with hyperactivity disorder (ADHD) (Tayebati et al., 2012). Muscarinic receptor expression and function is altered in adrenal medullary chromaffin (AMC) cells from SHRs (*c.f.* normotensive rats). TRPC1, TRPC4, and STIM1 are expressed in SHR AMC cells, where heteromeric TRPC1/TRPC4 channels have been proposed to contribute to non-selective cation channel activity that mediates the effects of enhanced muscarinic receptor activation (Inoue and Harada, 2023). Both TRPC1 and TRPC6 have been shown to contribute to hypoxia-induced pulmonary hypertension in neonatal rats (Huang et al., 2022) and 8- to 12-week-old Wistar albino rats (Kaymak et al., 2021). Expression of both channels was also upregulated in the forebrain of rats with diet-induced obesity (DIO), but it remains to be elucidated how this affects long-term neuronal health (Roy et al., 2021).

EMT is a key step in cancer cell proliferation and metastasis. Pancreatic cancer cells expressed high levels of TRPC1 protein, in addition to TRPC4 and TRPC6. Knockdown of TRPC1, but not TRPC4 and TRPC6, significantly reduced TGF-β-induced cell motility and migration. Similarly, reduced TRPC1 expression and function decreased the proliferative and invasive capacities of breast, lung, and colon cancer cells, as well as that of malignant glioblastoma (Elzamzamy et al., 2020). TRPC1 has also been proposed to function as a mechanosensitive channel in airway epithelial cells, mediating non-selective cation influx in response to mechanical stress. As for cancer cells, EMT has been proposed to be a key factor for airway remodeling in COPD. Aberrant repair of the airway epithelial layer and fibroblast accumulation leads to irreversible airway obstruction and progressive upregulation of airway pressure. TRPC1 expression was upregulated in cells obtained from human lung tissues of COPD patients and mechanically stretched 16HBE cells. The occurrence of EMT was shown by decreased expression of cytokeratin 9 and E-cadherin together with a simultaneous increase in α-SMA expression in mechanically stretched 16HBE cells. siRNA-medicated knockdown of TRPC1 abolished changes in the expression of EMT markers (Wang et al., 2020b; Wu et al., 2023). Further studies are required to confirm the mechano-sensitivity of TRPC1 and how it impacts channel function.

Herpes viruses cause a variety of diseases in the human population and can persist in the human body. Lifelong latent infections mean that the herpesviruses can be reactivated at a later stage, causing recurrent disease manifestations. The herpes simplex virus-1 (HSV-1) belongs to the subfamily of Alphaherpesvirinae within the Herpesviridae family (Zhao et al., 2023). HSV-1 infections have been implicated in clinical conditions such as oral herpes (Petti and Lodi, 2019) and vestibular neuritis (acute and persistent peripheral vertigo) (Zhao et al., 2023). In the HEp-2 cell line (HeLa derivative, cancerous), HSV-1 infection induced SOCE that is blocked by inhibitors such as HSK-7975A and 2-aminoethoxydiphenyl borate (2-APB). HEp-2 cells pre-infected with HSV-1 produced greater thapsigargin-induced SOCE, showing that HSV-1 infection enhances SOCE in infected cells. Extracellular Ca^{2+} entry was crucial for cell infection by HSV-1, as blocking SOCE significantly reduced cytoplasmic accumulation of the HSV-1 antigen. Knockdown of Orai1, STIM1, and TRPC1 using siRNAs abrogated HSV-1-induced SOCE, whereas knockdown of TRPCs 3–6 had no effect. Even though treatment with siTRPC1 only partially reduced SOCE (*c.f.* siOrai1, siSTIM1), all knockdowns equally inhibited virus entry into the cell. HSV-1 infection of mouse embryonic fibroblasts (MEFs) from TRPC1$^{-/-}$ mice revealed significant reductions in both HSV-1 viral entry and cytoplasmic Ca^{2+} levels. An interesting observation from this study was with CHO cells overexpressing TRPC1. While native CHO cells are resistant to HSV-1 infection due to the absence of known entry receptors on the cell surface, TRPC1-expressing CHO cells showed significant HSV-1 entry following inoculation. This suggested that TRPC1 can function as an entry receptor to facilitate HSV-1 viral entry into the cell. TIRF imaging of HSV-1-infected HEp-2 cells showed that PM localization of TRPC1 was increased following cell infection. The interaction between TRPC1 and HSV-1 is mediated by the third ectodomain of TRPC1 and the viral glycoprotein of HSV-1 at the cell surface. Further evidence for the role of TRPC1 in HSV-1 infections came from studies with mouse models and buccal biopsies from human patients. Ocular infection of TRPC1$^{-/-}$ mice using two HSV-1 strains revealed that in the absence of TRPC1, there was mild ocular abnormality, almost no retinal detachment, significantly slower herpes stromal keratitis development, and higher survival rates than similarly infected wild-type mice. Buccal biopsies from healthy human patients showed dispersed localization of TRPC1 in the cytoplasm with low PM expression, while those from patients diagnosed with herpes oral lesions had significantly higher TRPC1 expression in the PM (*c.f.* cytoplasm). The data show that TRPC1 can be a vital target for the development of new therapies to prevent and mitigate the severity of HSV-1 infection (He et al., 2020). Whether TRPC1 plays a similar role in HSV-2 infection (genital herpes) remains to be investigated.

5.4 RESEARCH CHALLENGES

Several isoforms of TRPC1 caused by variable splicing of the *Trpc1* gene have been reported, but only the α, β, and ε isoforms are functional. Not many studies have looked at the specific involvement of each splice variant in SOCE in cells where more than one variant can be found. Additionally, TRPC1 can form heteromeric channels with TRPCs (except for TRPC2), which complicates efforts to elucidate TRPC1 channel function *per se* in these complexes. Moreover, TRPC1 activity can be modulated and/or regulated by protein-protein interactions with other channels (e.g., TRPP1, TRPV4, and TRPV6) and signaling proteins (e.g., CaM, Homer 1, and IP_3R) (Ong et al., 2014; Yuan et al., 2003). The ability to specifically detect protein expression is crucial when determining cell and/or tissue expression of TRPC1, as well as functional effects from overexpression, knockdown (using siRNA or shRNA), and knockout (gene deletion) studies. While anti-TRPC1 antibodies are commercially available, questions have been raised regarding the specificity of these antibodies in detecting TRPC1. A band within the predicted size of TRPC1 (60-100 kDa) was detected using these antibodies in Western blots of protein lysates from TRPC1$^{-/-}$ mice in several studies, despite RT-PCR confirming *Trpc1* gene deletion (Kim et al., 2014; Liu et al., 2007; Ong et al., 2013; Ong et al., 2002; Sakura and Ashcroft, 1997). Ideally, each anti-TRPC1 antibody should be confirmed for each cell type and tissue used in a study, with data showing that not only does the antibody specifically detect overexpressed protein but also partial/complete loss of the antibody-recognized protein band(s) in Western blots following *Trpc1* gene knockdown/knockout. Some studies also utilize anti-TRPC1 antibodies to inhibit channel function, which makes it important to be certain of the antibody's specificity in detecting TRPC1 alone (Elzamzamy et al., 2020).

As previously highlighted (Ong et al., 2014), TRPC1 has multiple interacting partners that regulate protein trafficking to and insertion into the PM and also modulate channel function.

Orai1-mediated Ca^{2+} influx is required to drive TRPC1 channel insertion into the PM, and both Orai1 isoforms, Orai1α and Orai1β, were interchangeable for TRPC1 channel function (Cheng et al., 2011; Desai et al., 2015; Zeng et al., 2008). Activation of TRPC1 requires channel interaction with and gating by STIM1 in the ER-PM junctions (Cheng et al., 2011; Huang et al., 2006; Kim et al., 2009b; Lee et al., 2010; Ong et al., 2007; Zeng et al., 2008). While the polybasic domain of STIM1 interacts with TRPC1 electrostatically, how TRPC1 is activated by STIM1 has not been fully elucidated, although the ezrin/radixin/moesin domain of STIM1 may be involved (Huang et al., 2006; Yuan et al., 2007; Zeng et al., 2008). In contrast, many studies have elucidated the stoichiometry of CRAC channels formed by the Orai1/STIM1 signaling complex and how the respective domains contribute (Qiu and Lewis, 2019; Yen and Lewis, 2019). So, a vexing conundrum for heterologous expression studies using cell lines is how much TRPC1 should be co-expressed with STIM1 and Orai1 to recapitulate TRPC1 channel function and current. Moreover, TRPC1 also interacts with other TRPC channels and forms heteromeric channels, which further muddles efforts to study channel function in this manner. *In vitro* experiments using various cell lines and different *in vivo* models where more than one TRPC channel is present can make it difficult to conceptualize an overall role for the channel. For example, upregulation of TRPC6 expression was positively correlated with increased coronary artery contractility in metabolic syndrome pigs. However, TRPC6 was shown to facilitate phenotypic switching in VSMCs from contractile to highly proliferative/synthetic phenotype. Inhibition of TRPC6 facilitated the contractile phenotype, while inhibition of TRPC1 and TRPC3 decreased cell proliferation. It is unclear why *in vitro* experiments revealed a greater contribution of TRPC1 and TRPC3 to cell proliferation while the animal model points to TRPC6 (Chen et al., 2020).

Many pharmacological compounds activate or block TRPC1 together with TRPC4 and TRPC5 (Bon et al., 2022; Rubaiy, 2019; Wang et al., 2020a). While maitotoxin has been reported to potentially be a selective activator of TRPC1, there have not been many studies to clarify how the compound acts on the channel (Brereton et al., 2001; Chen and Barritt, 2003; Flores et al., 2017; Woods et al., 1999). Elucidating the structure of TRPC1 using cryo-EM and other techniques will allow us to extract novel channel features that can be used to develop highly specific drugs. These can then be used to pinpoint the exact contribution of TRPC1 when the channel is found in heteromeric channel complexes or works in tandem with other channels to mediate SOCE. High resolution (near 3 Å) cryo-EM structures of homomeric TRPCs 3–6 have been reported in the past few years. The protein structures of these TRPCs are similar, with six TM α helices in each subunit, where the first four helices form a voltage sensor-like domain (VSLD) and the last two forming a structurally conserved pore domain (PD). Both the N- and C-termini for each subunit are in the cytoplasm. In a homomeric tetramer, the VSLD from one subunit interacts with the PD from an adjacent subunit, creating a "domain-swapped" conformation. The ion permeation pore of the tetrameric complex is formed by four PDs of each subunit at the center. The relative motions of the first C-terminal helix with the TRP domain and linker between the fourth and fifth TM α helices have been proposed to determine the overall protein shape. TRPC4 and C5 are wider (100–105 Å) when compared to TRPC3 and C6 (75–85 Å). The third TM α helix of TRPC3 and C6 channels is three turns longer than that of TRPC4 and C5 channels. Furthermore, two conserved cysteine residues found only in TRPC4 and C5 form a disulfide bond at the loop connecting the fifth TM α helix and the pore helix, stabilizing the channel pore and functioning as a redox sensor. Unique structural features that distinguish TRPC4 and C5 from TRPC3 and C6 underlie distinct properties attributed to each TRPC, such as channel activation, regulation, and function. It would be exciting to see how the structures of TRPC1 and C7 compare with these TRPCs. Additionally, heteromeric channels with different ratios of TRPC subunits may exhibit different structural conformations, providing further insight into how each subunit contributes to the make-up of a heteromultimeric channel complex. Understanding the correlation between protein conformational changes and channel function will provide further insights into the physiological properties of each TRPC. This will enable the development of highly specific compounds to either activate or inhibit distinct TRPCs, as well as novel therapies to ameliorate pathophysiological conditions caused by aberrant TRPC function (Wang et al., 2020a; Zhao et al., 2021).

The TRPC1$^{-/-}$ mice have proved useful to correlate TRPC1 expression and function with a number of diseased states. Nonetheless, additional evidence supporting this can be provided by animal models used to study a specific disease. For example, multiple genetic and inducible mouse models for Sjögren's Syndrome have been developed. Even though TRPC1 is a crucial channel mediating SOCE required for saliva secretion, channel expression and function have not been fully elucidated in these models that show variable clinical manifestations and immunological profiles of Sjögren's Syndrome (Gao et al., 2020). Many studies used the TRPC1$^{-/-}$ mouse model developed by Lutz Birnbaumer at NIEHS, NIH, which had a mixed 129Sv:C57BL/6J background (Dietrich et al., 2007), to study TRPC1 alone or intercross the TRPC1$^{-/-}$ mice to generate multiple knockouts. For example, the TRPC1/4/5$^{-/-}$ mice were generated after backcrossing three different mouse lines with the C57Bl6/N strain (Broker-Lai et al., 2017; Camacho Londono et al., 2015). Additionally, the TRPC1$^{-/-}$ mouse model is a general knockout of TRPC1 in all tissues. It is unknown whether long-term knockout of TRPC1 protein induces any compensatory responses by other proteins in the model. Tissue-specific and conditional knockouts of the *Trpc1* gene may help allay this concern. The development of specific antagonists of TRPC1 would be useful in determining whether inhibiting channel function in wild-type animal models is sufficient to induce a diseased state. Such tools will be most useful in studies on disease progression in animal models.

Several studies have proposed that TRPC1 may be activated by mechanisms other than ER-Ca^{2+} store depletion, refuting the contribution of TRPC1 to SOCE. However, the underlying mechanisms in these cases, including mechano- or stretch-induced activation of TRPC1, have not been fully elucidated. As reported in several literature reviews (Patel et al., 2010, Sharif-Naeini et al., 2008, Yin and Kuebler, 2010), TRPC1 has been proposed to be a mechano-sensitive channel that transduces mechanical force into a biological response, given its abundance in endothelial cells and myocytes of the heart, arteries, and skeletal muscle. More recently, TRPC1 was proposed to function as a mechanosensitive channel contributing to pressure-mediated activation of airway remodeling (Li et al., 2019), morphological polarity during directory migration of human bone osteosarcoma U2OS cells (Huang et al., 2015), Müller cells (Jo et al., 2022), and human bronchial epithelial (16HBE) cells (Wang et al., 2020b). Yet the molecular mechanisms underlying the mechano-sensitivity of TRPC1 have not been fully elucidated, which makes it difficult to evaluate its contribution to the physiological function of various mechano-sensitive cells.

5.5 CONCLUSIONS

Dysregulation of TRPC1 expression is highly related to several diseases and their stages. Overexpression of TRPC1 enhances SOCE and increases the basal $[Ca^{2+}]_i$, which triggers a plethora of downstream effectors to induce pathophysiological states. TRPC1 may be modifying the cellular behavior in several ways, either as a non-selective cation channel or as a protein able to regulate specific cellular targets. The participation of TRPC1 depends strongly on the cellular type or diseases; e.g., overexpression of TRPC1 has been related to a worse prognosis in many diseases, while in others, low expression has been linked to a better prognosis. Increasing our understanding of TRPC1 pathophysiological functions will allow us to employ it as a target of treatments for several human diseases.

5.6 ACKNOWLEDGMENTS

Fig. 5.1 and Fig. 5.2 were partially created with BioRender. Funding for this work was provided by NIDCR-DIR, NIH (Z01-DE00438-35) for I.S.A.

REFERENCES

Ambudkar, I. Calcium signaling defects underlying salivary gland dysfunction. *Biochim Biophys Acta Mol Cell Res*, 2018, 1865: 1771–1777.

Ambudkar, I. S., De Souza, L. B., Ong, H. L. TRPC1, Orai1, and STIM1 in SOCE: Friends in tight spaces. *Cell Calcium*, 2017, 63: 33–39.

Barrera, N. P., Shaifta, Y., Mcfadzean, I., Ward, J. P., Henderson, R. M., Edwardson, J. M. AFM imaging reveals the tetrameric structure of the TRPC1 channel. *Biochem Biophys Res Commun*, 2007, 358: 1086–1090.

Belkacemi, T., Niermann, A., Hofmann, L., Wissenbach, U., Birnbaumer, L., Leidinger, P., Backes, C., Meese, E., Keller, A., Bai, X., Scheller, A., Kirchhoff, F., Philipp, S. E., Weissgerber, P., Flockerzi, V., Beck, A. TRPC1- and TRPC3-dependent Ca^{2+} signaling in mouse cortical astrocytes affects injury-evoked astrogliosis in vivo. *Glia*, 2017, 65: 1535–1549.

Bon, R. S., Wright, D. J., Beech, D. J., Sukumar, P. Pharmacology of TRPC channels and its potential in cardiovascular and metabolic medicine. *Annu Rev Pharmacol Toxicol*, 2022, 62: 427–446.

Brandman, O., Liou, J., Park, W. S., Meyer, T. STIM2 is a feedback regulator that stabilizes basal cytosolic and endoplasmic reticulum Ca^{2+} levels. *Cell*, 2007, 131: 1327–1339.

Brereton, H. M., Chen, J., Rychkov, G., Harland, M. L., Barritt, G. J. Maitotoxin activates an endogenous non-selective cation channel and is an effective initiator of the activation of the heterologously expressed hTRPC-1 (transient receptor potential) non-selective cation channel in H4-IIE liver cells. *Biochim Biophys Acta*, 2001, 1540: 107–126.

Broker-Lai, J., Kollewe, A., Schindeldecker, B., Pohle, J., Nguyen Chi, V., Mathar, I., Guzman, R., Schwarz, Y., Lai, A., Weissgerber, P., Schwegler, H., Dietrich, A., Both, M., Sprengel, R., Draguhn, A., Kohr, G., Fakler, B., Flockerzi, V., Bruns, D., Freichel, M. Heteromeric channels formed by TRPC1, TRPC4 and TRPC5 define hippocampal synaptic transmission and working memory. *EMBO J*, 2017, 36: 2770–2789.

Busch, L., Borda, E. Signaling pathways involved in pilocarpine-induced mucin secretion in rat submandibular glands. *Life Sci*, 2007, 80: 842–851.

Camacho Londono, J. E., Tian, Q., Hammer, K., Schroder, L., Camacho Londono, J., Reil, J. C., He, T., Oberhofer, M., Mannebach, S., Mathar, I., Philipp, S. E., Tabellion, W., Schweda, F., Dietrich, A., Kaestner, L., Laufs, U., Birnbaumer, L., Flockerzi, V., Freichel, M., Lipp, P. A background Ca^{2+} entry pathway mediated by TRPC1/TRPC4 is critical for development of pathological cardiac remodelling. *Eur Heart J*, 2015, 36: 2257–2266.

Chami, M., Checler, F. Alterations of the endoplasmic reticulum (ER) calcium signaling molecular components in Alzheimer's disease. *Cells*, 2020, 9: 2577.

Chen, J., Barritt, G. J. Evidence that TRPC1 (transient receptor potential canonical 1) forms a Ca^{2+}-permeable channel linked to the regulation of cell volume in liver cells obtained using small interfering RNA targeted against TRPC1. *Biochem J*, 2003, 373: 327–336.

Chen, K., Jin, X., Li, Q., Wang, W., Wang, Y., Zhang, J. Association of TRPC1 gene polymorphisms with type 2 diabetes and diabetic nephropathy in Han Chinese population. *Endocr Res*, 2013, 38: 59–68.

Chen, R. A., Goodman, W. G. Role of the calcium-sensing receptor in parathyroid gland physiology. *Am J Physiol Renal Physiol*, 2004, 286: F1005–F1011.

Chen, X., Sooch, G., Demaree, I. S., White, F. A., Obukhov, A. G. Transient receptor potential canonical (TRPC) channels: Then and now. *Cells*, 2020, 9: 1983.

Chen, X., Xiao, Z., Jiang, Z., Jiang, Y., Li, W., Wang, M. Schisandrin b attenuates airway inflammation and airway remodeling in asthma by inhibiting NLRP3 inflammasome activation and reducing pyroptosis. *Inflammation*, 2021a, 44: 2217–2231.

Chen, Y., Kong, Y., Wang, Q., Chen, J., Chen, H., Xie, H., Li, l. Schisandrin b attenuates airway inflammation by regulating the NF-κB/Nrf2 signaling pathway in mouse models of asthma. *J Immunol Res*, 2021b, 2021: 8029963.

Cheng, K. T., Liu, X., Ong, H. L., Ambudkar, I. S. Functional requirement for Orai1 in store-operated TRPC1-STIM1 channels. *J Biol Chem*, 2008, 283: 12935–12940.

Cheng, K. T., Liu, X., Ong, H. L., Swaim, W., Ambudkar, I. S. Local Ca^{2+} entry via Orai1 regulates plasma membrane recruitment of TRPC1 and controls cytosolic Ca^{2+} signals required for specific cell functions. *PLoS Biol*, 2011, 9: e1001025.

Cioffi, D. L., Wu, S., Chen, H., Alexeyev, M., St Croix, C. M., Pitt, B. R., Uhlig, S., Stevens, T. Orai1 determines calcium selectivity of an endogenous TRPC heterotetramer channel. *Circ Res*, 2012, 110: 1435–1444.

Clemens, R. A., Lowell, C. A. CRAC channel regulation of innate immune cells in health and disease. *Cell Calcium*, 2019, 78: 56–65.

De Souza, L. B., Ong, H. L., Liu, X., Ambudkar, I. S. Fast endocytic recycling determines TRPC1-STIM1 clustering in ER-PM junctions and plasma membrane function of the channel. *Biochim Biophys Acta*, 2015, 1853: 2709–2721.

Desai, P. N., Zhang, X., Wu, S., Janoshazi, A., Bolimuntha, S., Putney, J. W., Trebak, M. Multiple types of calcium channels arising from alternative translation initiation of the Orai1 message. *Sci Signal*, 2015, 8: ra74.

Devi, G. The tauopathies. *Handb Clin Neurol*, 2023, 196: 251–265.

Dietrich, A., Kalwa, H., Storch, U., Mederos Y Schnitzler, M., Salanova, B., Pinkenburg, O., Dubrovska, G., Essin, K., Gollasch, M., Birnbaumer, L., Gudermann, T. Pressure-induced and store-operated cation influx in vascular smooth muscle cells is independent of TRPC1. *Pflugers Arch*, 2007, 455: 465–477.

Elzamzamy, O. M., Penner, R., Hazlehurst, L. A. The role of TRPC1 in modulating cancer progression. *Cells*, 2020, 9: 388.

Flores, P. L., Rodriguez, E., Zapata, E., Carbo, R., Farias, J. M., Martinez, M. Maitotoxin is a potential selective activator of the endogenous transient receptor potential canonical type 1 channel in *xenopus laevis* oocytes. *Mar Drugs*, 2017, 15: 198.

Gao, Y., Chen, Y., Zhang, Z., Yu, X., Zheng, J. Recent advances in mouse models of Sjögren's syndrome. *Front Immunol*, 2020, 11: 1158.

Gao, Y. Y., Tian, W., Zhang, H. N., Sun, Y., Meng, J. R., Cao, W., Li, X. Q. Canonical transient receptor potential channels and their modulators: Biology, pharmacology and therapeutic potentials. *Arch Pharm Res*, 2021, 44: 354–377.

Garaud, S., Taher, T. E., Debant, M., Burgos, M., Melayah, S., Berthou, C., Parikh, K., Pers, J. O., Luque-Paz, D., Chiocchia, G., Peppelenbosch, M., Isenberg, D. A., Youinou, P., Mignen, O., Renaudineau, Y., Mageed, R. A. CD5 expression promotes IL-10 production through activation of the MAPK/Erk pathway and upregulation of TRPC1 channels in B lymphocytes. *Cell Mol Immunol*, 2018, 15: 158–170.

Gorza, L., Sorge, M., Secli, L., Brancaccio, M. Master regulators of muscle atrophy: Role of costamere components. *Cells*, 2021, 10: 61.

Greten, F. R., Grivennikov, S. Inflammation and cancer: Triggers, mechanisms, and consequences. *Immunity*, 2019, 51: 27–41.

Guerrero-Hernandez, A., Sanchez-Vazquez, V. H., Martinez-Martinez, E., Sandoval-Vazquez, L., Perez-Rosas, N. C., Lopez-Farias, R., Dagnino-Acosta, A. Sarco-endoplasmic reticulum calcium release model based on changes in the luminal calcium content. *Adv Exp Med Biol*, 2020, 1131: 337–370.

Hai, L., Kawarabayashi, Y., Imai, Y., Honda, A., Inoue, R. Counteracting effect of TRPC1-associated Ca^{2+} influx on TNF-α-induced COX-2-dependent prostaglandin E2 production in human colonic myofibroblasts. *Am J Physiol Gastrointest Liver Physiol*, 2011, 301: G356–G367.

He, D., Mao, A., Li, Y., Tam, S., Zheng, Y., Yao, X., Birnbaumer, L., Ambudkar, I. S., Ma, X. TRPC1 participates in the HSV-1 infection process by facilitating viral entry. *Sci Adv,*, 2020, 6: eaaz3367.

Himmel, N. J., Cox, D. N. Transient receptor potential channels: Current perspectives on evolution, structure, function and nomenclature. *Proc Biol Sci*, 2020, 287: 20201309.

Hong, J. H., Li, Q., Kim, M. S., Shin, D. M., Feske, S., Birnbaumer, L., Cheng, K. T., Ambudkar, I. S., Muallem, S. Polarized but differential localization and recruitment of STIM1, Orai1 and TRPC channels in secretory cells. *Traffic*, 2011, 12: 232–245.

Hoth, M., Penner, R. Depletion of intracellular calcium stores activates a calcium current in mast cells. *Nature*, 1992, 355: 353–356.

Hoth, M., Penner, R. Calcium release-activated calcium current in rat mast cells. *J Physiol*, 1993, 465: 359–386.

Huang, G. N., Zeng, W., Kim, J. Y., Yuan, J. P., Han, L., Muallem, S., Worley, P. F. STIM1 carboxyl-terminus activates native SOC, I_{crac} and TRPC1 channels. *Nat Cell Biol*, 2006, 8: 1003–1010.

Huang, W., Lin, W., Chen, B., Zhang, J., Gao, P., Fan, Y., Lin, Y., Wei, P. NFAT and NF-κB dynamically co-regulate TCR and CAR signaling responses in human T cells. *Cell Rep*, 2023, 42: 112663.

Huang, W., Liu, N., Tong, X., Du, Y. Sildenafil protects against pulmonary hypertension induced by hypoxia in neonatal rats via activation of PPARγ-mediated downregulation of TRPC. *Int J Mol Med*, 2022, 49: 19.

Huang, Y. W., Chang, S. J., Harn, H. I., Huang, H. T., Lin, H. H., Shen, M. R., Tang, M. J., Chiu, W. T. Mechanosensitive store-operated calcium entry regulates the formation of cell polarity. *J Cell Physiol*, 2015, 230: 2086–2097.

Inoue, M., Harada, K. Enhancement of muscarinic receptor-mediated excitation in spontaneously hypertensive rat adrenal medullary chromaffin cells. *Auton Neurosci*, 2023, 248: 103108.

Jardin, I., Lopez, J. J., Salido, G. M., Rosado, J. A. Orai1 mediates the interaction between STIM1 and hTRPC1 and regulates the mode of activation of hTRPC1-forming Ca^{2+} channels. *J Biol Chem*, 2008, 283: 25296–25304.

Jho, D., Mehta, D., Ahmmed, G., Gao, X. P., Tiruppathi, C., Broman, M., Malik, A. B. Angiopoietin-1 opposes VEGF-induced increase in endothelial permeability by inhibiting TRPC1-dependent Ca^{2+} influx. *Circ Res*, 2005, 96: 1282–1290.

Jo, A. O., Lakk, M., Rudzitis, C. N., Krizaj, D. TRPV4 and TRPC1 channels mediate the response to tensile strain in mouse Muller cells. *Cell Calcium*, 2022, 104: 102588.

Kaymak, E., Akin, A. T., Tufan, E., Basaran, K. E., Taheri, S., Ozdamar, S., Yakan, B. The effect of chloroquine on the TRPC1, TRPC6, and CaSR in the pulmonary artery smooth muscle cells in hypoxia-induced experimental pulmonary artery hypertension. *J Biochem Mol Toxicol*, 2021, 35: e22636.

Kim, C. H., Kim, K. E., Yoon, J. H., Song, K. S. Upregulation of MUC5AC gene expression by IL-4 through CREB in human airway epithelial cells. *J Cell Biochem*, 2009a, 108: 974–981.

Kim, J., Kwak, M., Jeon, J. P., Myeong, J., Wie, J., Hong, C., Kim, S. Y., Jeon, J. H., Kim, H. J., So, I. Isoform- and receptor-specific channel property of canonical transient receptor potential (TRPC)1/4 channels. *Pflugers Arch*, 2014, 466: 491–504.

Kim, M. S., Zeng, W., Yuan, J. P., Shin, D. M., Worley, P. F., Muallem, S. Native store-operated Ca^{2+} influx requires the channel function of Orai1 and TRPC1. *J Biol Chem*, 2009b, 284: 9733–9741.

Lee, K. P., Yuan, J. P., So, I., Worley, P. F., Muallem, S. STIM1-dependent and STIM1-independent function of transient receptor potential canonical (TRPC) channels tunes their store-operated mode. *J Biol Chem*, 2010, 285: 38666–38673.

Li, M., Liu, E., Zhou, Q., Li, S., Wang, X., Liu, Y., Wang, L., Sun, D., Ye, J., Gao, Y., Yang, X., Liu, J., Yang, Y., Wang, J. Z. TRPC1 null exacerbates memory deficit and apoptosis induced by amyloid-β. *J Alzheimers Dis*, 2018, 63: 761–772.

Li, N., He, Y., Yang, G., Yu, Q., Li, M. Role of TRPC1 channels in pressure-mediated activation of airway remodeling. *Respir Res,*, 2019, 20: 91.

Liou, J., Kim, M. L., Heo, W. D., Jones, J. T., Myers, J. W., Ferrell, J. E. Jr., Meyer, T. STIM is a Ca^{2+} sensor essential for Ca^{2+}-store-depletion-triggered Ca^{2+} influx. *Curr Biol*, 2005, 15: 1235–1241.

Liu, G., Philp, A. M., Corte, T., Travis, M. A., Schilter, H., Hansbro, N. G., Burns, C. J., Eapen, M. S., Sohal, S. S., Burgess, J. K., Hansbro, P. M. Therapeutic targets in lung tissue remodelling and fibrosis. *Pharmacol Ther*, 2021, 225: 107839.

Liu, X., Cheng, K. T., Bandyopadhyay, B. C., Pani, B., Dietrich, A., Paria, B. C., Swaim, W. D., Beech, D., Yildrim, E., Singh, B. B., Birnbaumer, L., Ambudkar, I. S. Attenuation of store-operated Ca^{2+} current impairs salivary gland fluid secretion in TRPC1(-/-) mice. *Proc Natl Acad Sci U S A*, 2007, 104: 17542–17547.

Liu, X., Groschner, K., Ambudkar, I. S. Distinct Ca^{2+}-permeable cation currents are activated by internal Ca_{2+}-store depletion in RBL-2H3 cells and human salivary gland cells, HSG and HSY. *J Membr Biol,*, 2004, 200: 93–104.

Liu, X., Ong, H. L., Ambudkar, I. TRP channel involvement in salivary glands-some good, some bad. *Cells,* 2018, 7: 74.

Liu, X., Singh, B. B., Ambudkar, I. S. ATP-dependent activation of K_{Ca} and ROMK-type K_{ATP} channels in human submandibular gland ductal cells. *J Biol Chem*, 1999, 274: 25121–25129.

Lopez, J. J., Jardin, I., Sanchez-Collado, J., Salido, G. M., Smani, T., Rosado, J. A. TRPC channels in the SOCE scenario. *Cells*, 2020a, 9: 126.

Lopez, J. R., Uryash, A., Faury, G., Esteve, E., Adams, J. A. Contribution of TRPC channels to intracellular Ca^{2+} dyshomeostasis in smooth muscle from mdx mice. *Front Physiol*, 2020b, 11: 126.

Ma, X., Cheng, K. T., Wong, C. O., O'neil, R. G., Birnbaumer, L., Ambudkar, I. S., Yao, X. Heteromeric TRPV4-C1 channels contribute to store-operated Ca^{2+} entry in vascular endothelial cells. *Cell Calcium*, 2011, 50: 502–509.

Medic, N., Desai, A., Olivera, A., Abramowitz, J., Birnbaumer, L., Beaven, M. A., Gilfillan, A. M., Metcalfe, D. D. Knockout of the *Trpc1* gene reveals that TRPC1 can promote recovery from anaphylaxis by negatively regulating mast cell TNF-alpha production. *Cell Calcium*, 2013, 53: 315–326.

Mehta, D., Ahmmed, G. U., Paria, B. C., Holinstat, M., Voyno-Yasenetskaya, T., Tiruppathi, C., Minshall, R. D., Malik, A. B. RhoA interaction with inositol 1,4,5-trisphosphate receptor and transient receptor potential channel-1 regulates Ca^{2+} entry. Role in Signaling Increased Endothelial Permeability. *J Biol Chem*, 2003, 278: 33492–33500.

Moccia, F., Brunetti, V., Perna, A., Guerra, G., Soda, T., Berra-Romani, R. The molecular heterogeneity of store-operated Ca^{2+} entry in vascular endothelial cells: The different roles of Orai1 and TRPC1/TRPC4 channels in the transition from Ca^{2+}-selective to non-selective cation currents. *Int J Mol Sci,* 2020, 24: 3259.

Nezu, A., Morita, T., Tojyo, Y., Nagai, T., Tanimura, A. Partial agonistic effects of pilocarpine on Ca^{2+} responses and salivary secretion in the submandibular glands of live animals. *Exp Physiol*, 2015, 100: 640–651.

Nishida, M., Tanaka, T., Mangmool, S., Nishiyama, K., Nishimura, A. Canonical transient receptor potential channels and vascular smooth muscle cell plasticity. *J Lipid Atheroscler*, 2020, 9: 124–139.

Ong, E. C., Nesin, V., Long, C. L., Bai, C. X., Guz, J. L., Ivanov, I. P., Abramowitz, J., Birnbaumer, L., Humphrey, M. B., Tsiokas, L. A TRPC1 protein-dependent pathway regulates osteoclast formation and function. *J Biol Chem*, 2013, 288: 22219–22232.

Ong, H. L., Chen, J., Chataway, T., Brereton, H., Zhang, L., Downs, T., Tsiokas, L., Barritt, G. Specific detection of the endogenous transient receptor potential (TRP)-1 protein in liver and airway smooth muscle cells using immunoprecipitation and Western-blot analysis. *Biochem J*, 2002, 364: 641–648.

Ong, H. L., Cheng, K. T., Liu, X., Bandyopadhyay, b., Paria, C., Soboloff, C., Pani, J., Gwack, B., Srikanth, Y., Singh, S., Gill, B. B., Ambudkar, D. L. Dynamic assembly of TRPC1-STIM1-Orai1 ternary complex is involved in store-operated calcium influx. Evidence for similarities in store-operated and calcium release-activated calcium channel components. *J Biol Chem*, 2007, 282: 9105–9116.

Ong, H. L., De Souza, L. B., Cheng, K. T., Ambudkar, I. S. Physiological functions and regulation of TRPC channels. *Handb Exp Pharmacol*, 2014, 223: 1005–1034.

Ong, H. L., Jang, S. I., Ambudkar, I. S. Distinct contributions of Orai1 and TRPC1 to agonist-induced $[Ca^{2+}]_i$ signals determine specificity of Ca^{2+}-dependent gene expression. *PLoS One*, 2012, 7: e47146.

Onopiuk, M., Eby, B., Nesin, V., Ngo, P., Lerner, M., Gorvin, C. M., Stokes, V. J., Thakker, R. V., Brandi, M. L., Chang, W., Humphrey, M. B., Tsiokas, L., Lau, K. Control of PTH secretion by the TRPC1 ion channel. *JCI Insight*, 2020, 5: e132496.

Pani, B., Liu, X., Bollimuntha, S., Cheng, K. T., Niesman, I. R., Zheng, C., Achen, V. R., Patel, H. H., Ambudkar, I. S., Singh, B. B. Impairment of TRPC1-STIM1 channel assembly and AQP5 translocation compromise agonist-stimulated fluid secretion in mice lacking caveolin1. *J Cell Sci*, 2013, 126: 667–675.

Pani, B., Ong, H. L., Brazer, S. C., Liu, X., Rauser, K., Singh, B. B., Ambudkar, I. S. Activation of TRPC1 by STIM1 in ER-PM microdomains involves release of the channel from its scaffold caveolin-1. *Proc Natl Acad Sci U S A*, 2009, 106: 20087–20092.

Park, Y. J., Yoo, S. A., Kim, M., Kim, W. U. The role of calcium-calcineurin-NFAT signaling pathway in health and autoimmune diseases. *Front Immunol*, 2020, 11: 195.

Patel, A., Sharif-Naeini, R., Folgering, J. R., Bichet, D., Duprat, F., Honore, E. Canonical TRP channels and mechanotransduction: From physiology to disease States. *Pflugers Arch*, 2010, 460: 571–581.

Petti, S., Lodi, G. The controversial natural history of oral herpes simplex virus type 1 infection. *Oral Dis*, 2019, 25: 1850–1865.

Prakriya, M., Feske, S., Gwack, Y., Srikanth, S., Rao, A., Hogan, P. G. Orai1 is an essential pore subunit of the CRAC channel. *Nature*, 2006, 443: 230–233.

Pu, Q., Zhao, Y., Sun, Y., Huang, T., Lin, P., Zhou, C., Qin, S., Singh, B. B., Wu, M. TRPC1 intensifies house dust mite-induced airway remodeling by facilitating epithelial-to-mesenchymal transition and STAT3/NF-κB signaling. *FASEB J*, 2019, 33: 1074–1085.

Qi, H., Wu, F., Wang, H. Function of TRPC1 in modulating hepatocellular carcinoma progression. *Med Oncol*, 2023, 40: 97.

Qian, H., Zhang, H. N., Gao, T., Wang, X. S., Wang, X., Yu, M. Y., Li, M. K., Huang, J. Upregulation of TRPC1 in microglia promotes neutrophil infiltration after ischemic stroke. *Brain Res Bull*, 2024, 208: 110894.

Qiu, R., Lewis, R. S. Structural features of STIM and Orai underlying store-operated calcium entry. *Curr Opin Cell Biol*, 2019, 57: 90–98.

Qu, M., Du, L. Upregulation of TRPC1 protects against high glucose-induced HUVECs dysfunction by inhibiting oxidative stress. *Biochem Biophys Res Commun*, 2024, 699: 149560.

Roy, P., Martinelli, I., Moruzzi, M., Maggi, F., Amantini, C., Micioni Di Bonaventura, M. V., Cifani, C., Amenta, F., Tayebati, S. K., Tomassoni, D. Ion channels alterations in the forebrain of high-fat diet fed rats. *Eur J Histochem*, 2021, 65: 3305.

Rubaiy, H. N. Treasure troves of pharmacological tools to study transient receptor potential canonical 1/4/5 channels. *Br J Pharmacol*, 2019, 176: 832–846.

Sakura, H., Ashcroft, F. M. Identification of four trp1 gene variants murine pancreatic beta-cells. *Diabetologia*, 1997, 40: 528–532.

Sanchez-Vazquez, V. H., Martinez-Martinez, E., Gallegos-Gomez, M. L., Arias, J. M., Pallafacchina, G., Rizzuto, R., Guerrero-Hernandez, A. Heterogeneity of the endoplasmic reticulum Ca^{2+} store determines colocalization with mitochondria. *Cell Calcium*, 2023, 109: 102688.

Schnipper, J., Kouba, S., Hague, F., Girault, A., Telliez, M. S., Guenin, S., Ahidouch, A., Pedersen, S. F., Ouadid-Ahidouch, H. Acid adaptation promotes TRPC1 plasma membrane localization leading to pancreatic ductal adenocarcinoma cell proliferation and migration through Ca^{2+} entry and interaction with PI3K/CaM. *Cancers (Basel)*, 2022, 14: 4946.

Selvaraj, S., Sun, Y., Watt, J. A., Wang, S., Lei, S., Birnbaumer, L., Singh, B. B. Neurotoxin-induced ER stress in mouse dopaminergic neurons involves downregulation of TRPC1 and inhibition of AKT/mTOR signaling. *J Clin Invest*, 2012, 122: 1354–1367.

Sharif-Naeini, R., Dedman, A., Folgering, J. H., Duprat, F., Patel, A., Nilius, B., Honore, E. TRP channels and mechanosensory transduction: Insights into the arterial myogenic response. *Pflugers Arch*, 2008, 456: 529–540.

Shin, S., Gombedza, F. C., Awuah Boadi, E., Yiu, A. J., Roy, S. K., Bandyopadhyay, B. C. Reduction of TRPC1/TRPC3 mediated Ca^{2+}-signaling protects oxidative stress-induced COPD. *Cell Signal*, 2023, 107: 110681.

Singh, R., Kaur, N., Choubey, V., Dhingra, N., Kaur, T. Endoplasmic reticulum stress and its role in various neurodegenerative diseases. *Brain Res*, 2024, 1826: 148742.

Smedler, E., Uhlen, P. Frequency decoding of calcium oscillations. *Biochim Biophys Acta*, 2014, 1840: 964–969.

Staaf, S., Maxvall, I., Lind, U., Husmark, J., Mattsson, J. P., Ernfors, P., Pierrou, S. Down regulation of TRPC1 by shRNA reduces mechanosensitivity in mouse dorsal root ganglion neurons *in vitro*. *Neurosci Lett*, 2009, 457: 3–7.

Sukumaran, P., Sun, Y., Antonson, N., Singh, B. B. Dopaminergic neurotoxins induce cell death by attenuating NF-κB-mediated regulation of TRPC1 expression and autophagy. *FASEB J*, 2018, 32: 1640–1652.

Sukumaran, P., Sun, Y., Vyas, M., Singh, B. B. TRPC1-mediated Ca^{2+} entry is essential for the regulation of hypoxia and nutrient depletion-dependent autophagy. *Cell Death Dis,*, 2015, 6: e1674.

Sun, Y., Nascimento Da Conceicao, V., Chauhan, A., Sukumaran, P., Chauhan, P., Ambrus, J. L., Vissink, A., Kroese, F. G. M., Muniswamy, M., Mishra, B. B., Singh, B. B. Targeting alarmin release reverses Sjogren's syndrome phenotype by revitalizing Ca^{2+} signalling. *Clin Transl Med*, 2023, 13: e1228.

Sun, Y., Selvaraj, S., Pandey, S., Humphrey, K. M., Foster, J. D., Wu, M., Watt, J. A., Singh, B. B., Ohm, J. E. MPP^+ decreases store-operated calcium entry and TRPC1 expression in mesenchymal stem cell derived dopaminergic neurons. *Sci Rep,*, 2018, 8: 11715.

Sun, Y., Ye, C., Tian, W., Ye, W., Gao, Y. Y., Feng, Y. D., Zhang, H. N., Ma, G. Y., Wang, S. J., Cao, W., Li, X. Q. TRPC1 promotes the genesis and progression of colorectal cancer via activating CaM-mediated PI3K/AKT signaling axis. *Oncogenesis,*, 2021, 10: 67.

Tayebati, S. K., Tomassoni, D., Amenta, F. Spontaneously hypertensive rat as a model of vascular brain disorder: Microanatomy, neurochemistry and behavior. *J Neurol Sci*, 2012, 322: 241–249.

Thebault, S. Minireview: Insights into the role of TRP channels in the retinal circulation and function. *Neurosci Lett*, 2021, 765: 136285.

Tsifetaki, N., Kitsos, G., Paschides, C. A., Alamanos, Y., Eftaxias, V., Voulgari, P. V., Psilas, K., Drosos, A. A. Oral for the treatment of ocular symptoms in patients with Sjögren's syndrome: A randomised 12 week controlled study. *Ann Rheum Dis*, 2003, 62: 1204–1207.

Vaidya, B., Sharma, S. S. Transient receptor potential channels as an emerging target for the treatment of Parkinson's disease: An insight into role of pharmacological interventions. *Front Cell Dev Biol*, 2020, 8: 584513.

Vivino, F. B., Al-Hashimi, I., Khan, Z., Leveque, F. G., Salisbury, P. L. 3rd, Tran-Johnson, T. K., Muscoplat, C. C., Trivedi, M., Goldlust, B., Gallagher, S. C. Pilocarpine tablets for the treatment of dry mouth and dry eye symptoms in patients with Sjögren syndrome: A randomized, placebo-controlled, fixed-dose, multicenter trial. P92-01 study group. *Arch Intern Med*, 1999, 159: 174–181.

Wang, D., Yu, H., Xu, B., Xu, H., Zhang, Z., Ren, X., Yuan, J., Liu, J., Guo, Y., Spencer, P. S., Yang, X. TRPC1 deletion causes striatal neuronal cell apoptosis and proteomic alterations in mice. *Front Aging Neurosci*, 2018, 10: 72.

Wang, H., Cheng, X., Tian, J., Xiao, Y., Tian, T., Xu, F., Hong, X., Zhu, M. X. TRPC channels: Structure, function, regulation and recent advances in small molecular probes. *Pharmacol Ther*, 2020a, 209: 107497.

Wang, J., He, Y., Yang, G., Li, N., Li, M., Zhang, M. Transient receptor potential canonical 1 channel mediates the mechanical stress-induced epithelial-mesenchymal transition of human bronchial epithelial (16HBE) cells. *Int J Mol Med*, 2020b, 46: 320–330.

Wes, P. D., Chevesich, J., Jeromin, A., Rosenberg, C., Stetten, G., Montell, C. TRPC1, a human homolog of a drosophila store-operated channel. *Proc Natl Acad Sci U S A*, 1995, 92: 9652–9656.

Woods, N. M., Dixon, C. J., Yasumoto, T., Cuthbertson, K. S., Cobbold, P. H. Maitotoxin-induced free Ca changes in single rat hepatocytes. *Cell Signal*, 1999, 11: 805–811.

Wu, X., Jia, B., Luo, X., Wang, J., Li, M. Glucocorticoid alleviates mechanical stress-induced airway inflammation and remodeling in COPD via transient receptor potential canonical 1 channel. *Int J Chron Obstruct Pulmon Dis*, 2023, 18: 1837–1851.

Xia, Y., Cai, P. C., Yu, F., Xiong, L., He, X. L., Rao, S. S., Chen, F., Yang, X. P., Ma, W. L., Ye, H. IL-4-induced caveolin-1-containing lipid rafts aggregation contributes to MUC5AC synthesis in bronchial epithelial cells. *Respir Res*, 2017, 18: 174.

Xiao, F., Du, W., Zhu, X., Tang, Y., Liu, L., Huang, E., Deng, C., Luo, C., Han, M., Chen, P., Ding, L., Hong, X., Wu, L., Jiang, Q., Zou, H., Liu, D., Lu, L. IL-17 drives salivary gland dysfunction via inhibiting TRPC1-mediated calcium movement in Sjögren's syndrome. *Clin Transl Immunology*, 2021, 10: e1277.

Yao, L., Romero, M. J., Toque, H. A., Yang, G., Caldwell, R. B., Caldwell, R. W. The role of RhoA/Rho kinase pathway in endothelial dysfunction. *J Cardiovasc Dis Res*, 2010, 1: 165–170.

Ye, J., Yin, Y., Yin, Y., Zhang, H., Wan, H., Wang, L., Zuo, Y., Gao, D., Li, M., Li, J., Liu, Y., Ke, D., Wang, J. Z. Tau-induced upregulation of C/EBPbeta-TRPC1-SOCE signaling aggravates tauopathies: A vicious cycle in Alzheimer neurodegeneration. *Aging Cell*, 2020, 19: e13209.

Yen, M., Lewis, R. S. Numbers count: How STIM and Orai stoichiometry affect store-operated calcium entry. *Cell Calcium*, 2019, 79: 35–43.

Yin, J., Kuebler, W. M. Mechanotransduction by TRP channels: General concepts and specific role in the vasculature. *Cell Biochem Biophys*, 2010, 56: 1–18.

Yuan, J. P., Kiselyov, K., Shin, D. M., Chen, J., Shcheynikov, N., Kang, S. H., Dehoff, M. H., Schwarz, M. K., Seeburg, P. H., Muallem, S., Worley, P. F. Homer binds TRPC family channels and is required for gating of TRPC1 by IP3 receptors. *Cell*, 2003, 114: 777–789.

Yuan, J. P., Zeng, W., Huang, G. N., Worley, P. F., Muallem, S. STIM1 heteromultimerizes TRPC channels to determine their function as store-operated channels. *Nat Cell Biol*, 2007, 9: 636–645.

Zanou, N., Shapovalov, G., Louis, M., Tajeddine, N., Gallo, C., Van Schoor, M., Anguish, I., Cao, M. L., Schakman, O., Dietrich, A., Lebacq, J., Ruegg, U., Roulet, E., Birnbaumer, L., Gailly, P. Role of TRPC1 channel in skeletal muscle function. *Am J Physiol Cell Physiol*, 2010, 298: C149–C162.

Zeng, W., Yuan, J. P., Kim, M. S., Choi, Y. J., Huang, G. N., Worley, P. F., Muallem, S. STIM1 gates TRPC channels, but not Orai1, by electrostatic interaction. *Mol Cell*, 2008, 32: 439–448.

Zeng, Y. Z., Zhang, Y. Q., Chen, J. Y., Zhang, L. Y., Gao, W. L., Lin, X. Q., Huang, S. M., Zhang, F., Wei, X. L. TRPC1 inhibits cell Proliferation/Invasion and is predictive of a better prognosis of esophageal squamous cell carcinoma. *Front Oncol*, 2021, 11: 627713.

Zhang, M., Ma, Y., Ye, X., Zhang, N., Pan, L., Wang, B. TRP (transient receptor potential) ion channel family: Structures, biological functions and therapeutic interventions for diseases. *Signal Transduct Target Ther*, 2023, 8: 261.

Zhang, X., Zhao, Z., Ma, L., Guo, Y., Li, X., Zhao, L., Tian, C., Tang, X., Cheng, D., Chen, Z., Zhang, L. The effects of transient receptor potential channel (TRPC) on airway smooth muscle cell isolated from asthma model mice. *J Cell Biochem*, 2018, 119: 6033–6044.

Zhang, Y., Guan, Y., Zheng, X., Li, C. Hypoxia-induced miR-181a-5p up-regulation reduces epirubicin sensitivity in breast cancer cells through inhibiting EPDR1/TRPC1 to activate PI3K/AKT signaling pathway. *BMC Cancer,*, 2024, 24: 167.

Zhao, Y., Mcveigh, B. M., Moiseenkova-Bell, V. Y. Structural pharmacology of TRP channels. *J Mol Biol*, 2021, 433: 166914.

Zhao, Z., Liu, X., Zong, Y., Shi, X., Sun, Y. Cellular processes induced by HSV-1 infections in vestibular neuritis. *Viruses,* 2023, 16: 12.

ZHU, X., CHU, P. B., PEYTON, M., BIRNBAUMER, L. Molecular cloning of a widely expressed human homologue for the trp gene. *FEBS Lett*, 1995, 373: 193–198.

Zhu, X., Jiang, M., Peyton, M., Boulay, G., Hurst, R., Stefani, E., Birnbaumer, L. Trp, a novel mammalian gene family essential for agonist-activated capacitative Ca^{2+} entry. *Cell*, 1996, 85: 661–671.

Zitt, C., Zobel, A., Obukhov, A. G., Harteneck, C., Kalkbrenner, F., Luckhoff, A., Schultz, G. Cloning and functional expression of a human Ca^{2+}-permeable cation channel activated by calcium store depletion. *Neuron*, 1996, 16: 1189–1196.

6 TRPM4 and TRPM5 Channels as Chemical-to-Electrical Signal Converter

Simon Vu and Jie Zheng

6.1 INTRODUCTION

The possible earliest electrophysiological recordings of TRPM4 or TRPM5 preceded their molecular cloning by several decades. The first prospective evidence of TRPM4 came from observations that calcium ions (Ca^{2+}) played a noticeable role in the generation of transient depolarizations (TDs) in canine Purkinje tissue (Ferrier and Moe, 1973). The underlying change in current was investigated and found to be generated by a transient inward (TI) current (Lederer and Tsien, 1976). Initially, the hypothesis was that the observed TI was a Ca^{2+} current (Ferrier and Moe, 1973; Lederer and Tsien, 1975; Lederer and Tsien, 1976). However, follow-up studies revealed that this current could not be carried by Ca^{2+} ions, as reversal potential (E_{rev}) measurements showed no change to differing concentrations of Ca^{2+}; rather, Na^+ ions were concluded to play a major role in TI current (Kass et al., 1978a; Kass et al., 1978b). Using this simple but elegant test of ionic dependence of E_{rev}, Kass, Tsien, and Weingart argued further that Na^+ could not be the sole permeant ion because the observed reversal potential for TI was near 0 mV, far more negative than the equilibrium potential for Na^+ (+70 mV); involvement of K^+ was suggested (Kass et al., 1978b). Thus, the existence of a Ca^{2+}-activated nonselective cation (CAN) channel became known.

Single-channel recordings of CAN channels were reported soon after the revolutionary study of improved patch-clamp techniques (Hamill et al., 1981). This channel, having a single-channel conductance of ~30 pS, was activated by micromolar concentrations of intracellular Ca^{2+} and exhibited little selectivity between monovalent cations in neonatal myocytes (Colquhoun et al., 1981) (Fig. 6.1A). In addition, the channel displayed weak voltage dependence and current desensitization upon activation. Yellen reported that CAN channels in neuroblastomas have almost equal permeability to various monovalent cations (Na^+, K^+, Li^+, Cs^+), suggesting that they may be responsible for the TI currents previously mentioned in cardiac cells (Yellen, 1982). With this, it was suggested that CAN channels have widespread expression throughout the body. These Ca^{2+}-sensitive channels were suggested to play a role in the regulation of electrical activity; in cardiac pacemaker cells, they could provide a sustained depolarizing current whereas other known channels would inactivate too quickly (Colquhoun et al., 1981; Partridge and Swandulla, 1988; Reuter et al., 1984). Indeed, CAN channels possess a unique capability of linking intracellular Ca^{2+} fluctuations to changes in membrane potential, whereas many ion channels are known to directly or indirectly convert membrane potential changes into changes in intracellular Ca^{2+} (Hille et al., 2001; Zheng and Trudeau, 2023).

6.2 CLONING OF TRPM4 AND TRPM5 SPLICE VARIANTS

More than a decade after the extensive electrophysiological characterizations of CAN channels, a new ion channel named TRPM4 was cloned from human cDNA libraries (Xu et al., 2001). The channel belongs to the transient receptor potential (TRP) channel family, which is a large group of cation channels that exhibit the common feature as cellular sensors for diverse environmental stimuli as well as internal signaling (Clapham, 2003). TRPM4 fits in

DOI: 10.1201/9781003402732-6

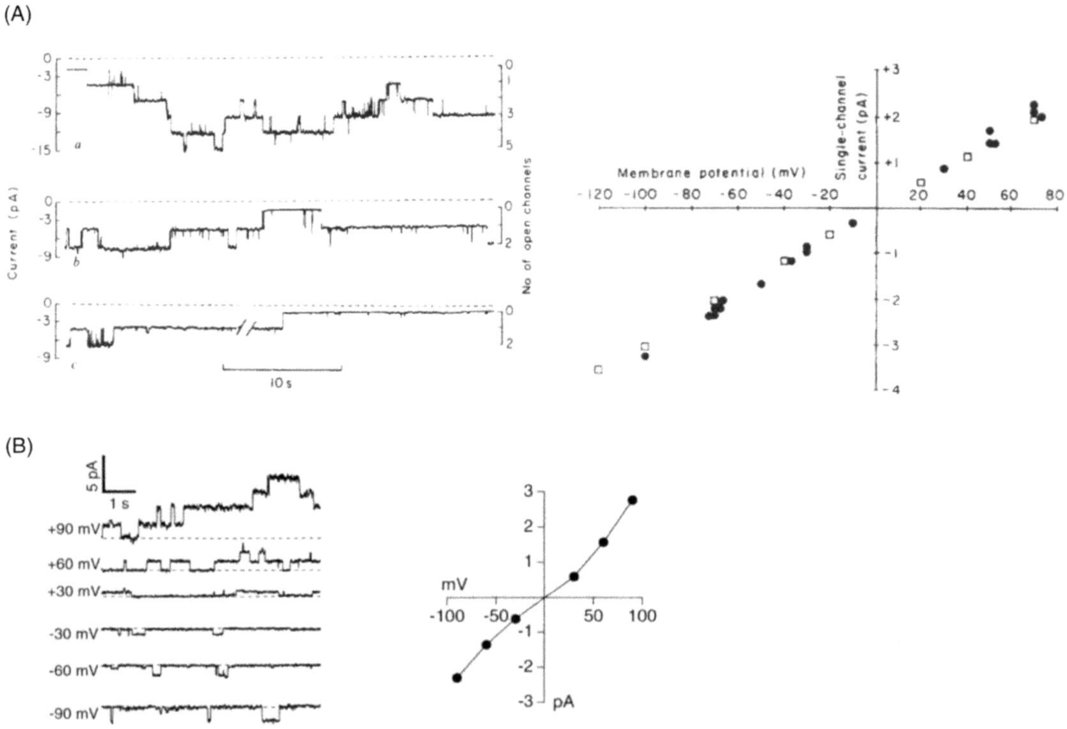

FIGURE 6.1 (A) Representative CAN channel current traces recorded from rat ventricular muscle cells (left panel) at 6 μM (top), 1.5 μM (middle), and zero (bottom) intracellular calcium, and the I-V plots (right panel) at 20 μM (filled curcles) or 1 μM (open squares) calcium. Reproduced from Colquhoun et al. (1981). (B) Representative TRPM4 current traces (left panel) recorded from a TRPM4b-transfected HEK293 cell at 300 nM calcium and the associated I-V plot (right panel). (Reproduced from Launay et al., 2002.)

the TRPM (M for melastatin) subfamily according to sequence homology (Zheng et al., 2013). The first cloned TRPM4 (TRPM4a) is a short splice variant. A year later, a second splice variant (TRPM4b) was cloned, which contained an additional 174 amino acids at the N-terminal (Launay et al., 2002). This longer variant of TRPM4 was confirmed to be Ca^{2+}-activated and permeable to Na^+ and K^+ but not Ca^{2+} or other divalent cations (Fig. 6.1B). Later, a third splice variant, TRPM4c, which lacked 537 amino acids at the N-terminal, was discovered (Nilius et al., 2003). In addition, TRPM4a and TRPM4b isoforms were identified and cloned from rats with differing properties compared to the human homologs (Yoo et al., 2010). It might be difficult and speculative to assume that TRPM4 is the CAN channel many others previously characterized, but it is amazing to see the strikingly similar properties. Indeed, with the knowledge of the cloned TRPM4, many of the original observations of the CAN channel are now confirmed. Though the physiological significance of each isoform is still not fully understood, TRPM4b is the most widely studied isoform of this channel and will be simply referred to as TRPM4 for the remainder of this review.

Around the same time, a close homolog of TRPM4, now named TRPM5, was cloned (Enklaar et al., 2000; Prawitt et al., 2000). TRPM5 is also Ca^{2+}-activated and monovalent cation-selective (Prawitt et al., 2003). It has a relatively restricted expression pattern compared to TRPM4, as reviewed elsewhere (Liman et al., 2014; McKemy et al., 2015). High-resolution cryo-EM structure of TRPM5 is now available (Ruan et al., 2021). Given the similarity in functional properties, this review mainly focuses on TRPM4 and refers to TRPM5, where key additional information is known.

6.3 TRPM4 AND TRPM5 CA^{2+}-DEPENDENT ACTIVATION AND DESENSITIZATION

Activities of native CAN and cloned TRPM4 are dependent on intracellular Ca^{2+}. Early studies of the native CAN channels noted that channel activation was dependent on elevated intracellular Ca^{2+} concentration and readily reversible when Ca^{2+} was removed (Colquhoun et al., 1981; Yellen, 1982). Upon the discovery of TRPM4, it was estimated that Ca^{2+}-dependent half-maximal activation concentration (EC_{50}) was 300-to-400 nM in whole-cell recordings of heterolgously expressed TRPM4 channels in HEK293 cells (Launay et al., 2002). Proceeding studies reported that no noticeable activation was observed until concentrations of Ca^{2+} reached around 1 μM or higher (Nilius et al., 2003, Nilius et al., 2004, Nilius et al., 2004, Nilius et al., 2005; Nilius et al., 2005; Ullrich et al., 2005; Nilius et al., 2006; Yarishkin et al., 2008; Amarouch et al., 2013). In contrast, the EC_{50} of Ca^{2+}-dependent activation of TRPM4 in inside-out patches is substantially higher; several groups reported values in the range of 150-to-1000 μM (Earley et al., 2007; Nilius et al., 2004; Ullrich et al., 2005; Yarishkin et al., 2008; Zhang et al., 2005). TRPM5, on the other hand, exhibits more than 10 times higher sensitivity to Ca^{2+} (Ullrich et al. 2005).

There are several potential reasons for the observed differences in TRPM4 Ca^{2+} sensitivities. It has been shown that TRPM4 activity is regulated by various cytoplasmic factors such as phosphatidylinositol 4,5-bisphosphate (PIP2), protein kinase C (PKC), calmodulin, and ATP. It is possible for important regulatory factors to be lost during patch excision, hence the differences in calcium sensitivity. The difficulty in studying these channels was evident in earlier CAN channel studies when a continued application of Ca^{2+} resulted in a reduction of channel activity (Yellen, 1982). Single-channel recordings suggested a complex gating scheme that could not be characterized by a simple first-order transition. Observations of unusually long open dwells and extremely long closing dwells (sometimes in the order of minutes) in combination with clusters of much shorter openings and closings revealed the complex nature of channel gating kinetics (Colquhoun et al., 1981; Yellen, 1982). Even today, the complexity of TRPM4 gating kinetics presents itself as a challenge.

Several groups observed that, upon prolonged Ca^{2+}-activation, TRPM4 undergoes rapid desensitization (Earley et al., 2007; Gonzales and Earley, 2012; Maruyama and Petersen, 1984; Nilius et al., 2003; Nilius et al., 2005; Nilius et al., 2006; Simon et al., 2010; Yamaguchi et al., 2019; Zhang et al., 2005). The first single-channel recordings of CAN channels also exhibited this phenomenon, which is one of the many complexities in TRPM4 gating (Colquhoun et al., 1981; Yellen, 1982). Nilius et al. showed that this desensitization process occurred at various Ca^{2+} concentrations but more rapidly at higher concentrations (Nilius et al., 2005). This may in part explain why such discrepancies are observed with the Ca^{2+}-dependent half-maximal activation concentration. Indeed, it was observed that the EC_{50} shifted 5-fold in a desensitized state (Zhang et al., 2005). It has been shown that application of hydrogen peroxide to the intracellular face of TRPM4 eliminated channel desensitization where Cys1093 on the C-terminal domain helix was suggested to be involved (Simon et al., 2010).

6.4 PIP2 REGULATION

Given the pronounced, rapid desensitization and the differences in Ca^{2+}-sensitivity observed in whole-cell and cell-free patch configurations, understanding these processes would offer insight into the physiological function of TRPM4. PIP2 is one key factor known to dynamically regulate TRPM4 activity. PIP2 is a phospholipid that plays a role in modulating a vast number of receptors and ion channels, including many TRP channels. For example, among the TRPM channels, breakdown of PIP2 has been shown to induce TRPM7 channel closure and underlie desensitization of TRPM5 and TRPM8 (Liu and Liman, 2003; Liu and Qin, 2005; Rohacs et al., 2005; Runnels et al., 2002). PIP2 positively regulates TRPM4 channel activity (Nilius et al., 2006; Zhang et al., 2005). More specifically, TRPM4 channel desensitization could be fully reversed with intracellular application of PIP2. In the presence of PIP2, Ca^{2+} sensitivity is increased, which also shifts the voltage dependence towards negative potentials (Nilius et al., 2006; Zhang et al., 2005). Mg_2ATP, which activates lipid kinases to replenish PIP2, promoted similar current recovery as intracellular application of PIP2; however, free ATP^{4-} has been shown to inhibit TRPM4 activity (Nilius et al., 2005; Zhang et al., 2005). In accordance, inhibition of phospholipase C (PLC) or PI4K activity and scavenging of PIP2 down-regulate channel activity, promote TRPM4 currents, consistent with PIP2 being a positive modulator.

Due to the clear importance of PIP2 in the regulation of TRPM4 and TRPM5 channels, its binding sites were searched for. In TRPM4, a binding site was initially hypothesized to be at either the N-terminus or C-terminus (Bousova et al., 2015; Bousova et al., 2020; Nilius et al., 2006). Mutations to positively charged residues on the C-terminal TRP box resulted in reduced sensitivity to PIP2 along with rapid desensitization (Nilius et al., 2006). In contrast, using surface plasmon resonance to study direct binding of PIP2 to TRPM4 channels in liposomes, Teisinger and colleagues proposed that the binding site was located at the N-terminus; PIP3 was also proposed to bind to a similar site (Bousova et al., 2015). Later, using peptides corresponding to the N- or C-terminus, molecular dynamics, and mutational analysis, it was observed that PIP2 has the potential to bind to both termini (Bousova et al., 2018). Given the observed importance of PIP2 for TRPM4 and TRPM5 activity, structural evidence of these channels in complex with PIP2 is desired, as it would not only offer a more direct clue of PIP2 binding location but also provide a better understanding of the mechanism of positive regulatory action.

6.5 VOLTAGE DEPENDENCE

Membrane potential is another regulating factor for TRPM4 and TRPM5 activity. Many TRP channels have been shown to be voltage-dependent, though their voltage sensitivity is much less than classical voltage-dependent channels such as the Shaker K^+ channel (Sigworth et al., 1994; Zheng et al., 2013). The molecular mechanisms underlying TRP channel voltage dependence are still not fully understood but are likely diverse. Whereas some TRP channels appear to utilize the classic S4 voltage sensor mechanism (Voets et al., 2007), TRPV1 is recently found to rely on an unorthodox mechanism of voltage-sensing, using the outer pore domain as the voltage sensor (Yang et al., 2020). Early studies of CAN channels noted that these channels showed weak voltage dependence with nearly linear macroscopic I-V relationships (Lindau and Fernandez, 1986; Marty et al., 1984; Petersen and Maruyama, 1983). However, studies in neutrophils and Schwann cells reported the macroscopic current to be slightly outward rectifying, with the open probability increasing with membrane depolarization (Bevan et al., 1984; von Tscharner et al., 1986). Following the cloning and identification of TRPM4, it was confirmed that the channel is voltage-dependent with a sigmoidal I-V relationship and increased open probability and open times at depolarizing potentials (Launay et al., 2002). More in-depth studies noted that, in the presence of intracellular Ca^{2+}, TRPM4 currents deactivated at negative potentials and activated at positive potentials with outwardly rectifying macroscopic currents (Nilius et al., 2003). The half-maximal activation voltage was variable within the range of –20 mV and +60 mV depending on factors such as Ca^{2+} concentration, PIP2, and temperature, suggesting the possibility of dynamic voltage regulation under varying physiological and pathological conditions (Nilius et al., 2006). In accordance with the weak voltage sensitivity, the channel displayed a small gating charge of 0.7 e_0, which is similar to other TRP channels such as TRPM8 and TRPV1 (Nilius et al., 2005; Yang et al. 2020). Like TRPM4, TRPM5 is also weakly voltage-sensitive (Hofmann et al., 2003; Ullrich et al., 2005).

6.6 TEMPERATURE DEPENDENCE

Several TRP channels are highly temperature sensitive and are known to serve as temperature sensors. A common method for measuring and comparing temperature sensitivity is the Q_{10} value (fold increase in current amplitude due to a change of 10°C in temperature). For example, heat activation of TRPV1 exhibits a Q_{10} value above 20 and that of the cold-activated TRPM8 channel is about 24 (Benham et al., 2003; Brauchi et al., 2004; Yang and Zheng, 2014). In contrast, simple chemical reactions and protein functions often display Q_{10} values in the range of 1–4. Studies of the single-channel current amplitude of CAN channels determined the Q_{10} to be ~2.3 (Colquhoun et al., 1981). However, studies on the macroscopic currents from heterologously expressed TRPM4 and TRPM5 yielded much higher Q_{10} values, suggesting that the gating process may have substantial sensitivity to heat. TRPM5 displayed a Q_{10} of 10.3 and that of TRPM4 was reported to be 8.5 between 15°C and 25°C (Talavera et al., 2005). In addition, it was shown that temperature had a minor effect on Ca^{2+} dependence as the EC_{50} only slightly decreased from 15°C to 37°C; elevating temperature shifted the voltage-dependent curve to more negative potentials. It remains unclear how TRPM4 and TRPM5 channels sense temperature changes and how their apparent high temperature sensitivity contributes to physiology.

6.7 TRPM4 AND TRRPM5 PHARMACOLOGY

Whereas Ca^{2+} is required to activate TRPM4, there are a handful of known modulators that facilitate channel activation (Table 6.1). Over the years, only a couple of positive modulators of TRPM4 have been discovered. As discussed earlier, PIP2 and Mg_2ATP positively modulate TRPM4, thus

TABLE 6.1
TRPM4 Modulators

Molecule	EC_{50}/IC_{50} (μM)	Affects CAN Channels?	References
Activator/Potentiator			
Intracellular Ca^{2+}	0.3–1000	+	Launay, 2002; Nilius et al., 2003; Nilius et al., 2004; Nilius et al., 2005; Ullrich et al., 2005; Yarishkin et al., 2008; Amarouch et al., 2013; Winkler et al., 2017; Autzen et al., 2018; Colquhoun et al., 1981; Yellen, 1982
BTP2	0.008		Takezawa et al., 2006
PIP2	4–5		Zhang et al., 2005; Nilius et al., 2006; Guo et al., 2017
Decavanadate	1.9		Nilius et al., 2004; Winkler et al., 2017
Mg_2ATP	ND		Zhang et al., 2005; Nilius et al., 2006; Guo et al., 2017
Inhibitor			
AMP	19	+	Nilius et al., 2004; Sturgess et al., 1986
ADP	2.2	+	Nilius et al., 2004; Sturgess et al., 1986
ATP	1.7	+	Nilius et al., 2004; Guo et al., 2017; Sturgess et al., 1986
Adenosine	630	+	Nilius et al., 2004; Sturgess et al., 1986
AMP-PNP	19	+	Nilius et al., 2004; Thorn and Petersen, 1992
Spermine	61		Nilius et al., 2004
Flufenamic acid	3–5	+	Guinamard et al., 2006; Demion etal., 2007; Morita et al., 2007
Clotrimazole	ND		Ullrich et al., 2005
Glibenclamide	≥ 10000		Guinamard et al., 2006; Demion et al., 2007; Grand et al., 2008; Sala-Rabanal et al., 2012
Quinine	100–400	+	Talavera et al., 2008; Gray and Argent, 1990
9-phenanthrol	10–30		Grand et al., 2008; Gonzales et al., 2010
CBA (Compound 5)	1.5		Ozhathil et al., 2018

ND, not determined.

antagonizing desensitization. Decavanadate was found to potentiate Ca^{2+} activation, though it lacks specificity (Nilius et al., 2004). This vanadate oligomer has been proposed to be an inhibitor of several ATPases and to interact with ATP binding sites (Pezza et al., 2002; Toyoshima et al., 2000). Decavanadate was observed to affect channel gating, where it rendered TRPM4 to be nearly voltage-independent over a voltage range between -180 mV and +140 mV (Nilius et al., 2004). In addition, decavanadate was shown to increase the sensitivity of TRPM4 to block by ATP^{4-}. The binding site for decavanadate was proposed to be at the C-terminal coiled-coil region, which was confirmed in structural studies to be one of the sites (Nilius et al., 2004; Winkler et al., 2017); this will later be discussed in further detail. BTP2 (3,5-Bis(trifluoromethyl)pyrazole derivative) was also observed to enhance TRPM4 activation. BTP2 is an immunosuppressive compound that was found to inhibit interleukin-2 (IL-2) production in lymphocytes (Takezawa et al., 2006; Zitt et al., 2004). It was demonstrated that BTP2 enhances Ca^{2+}-activated current in a concentration-dependent manner at physiological membrane potentials. It is hypothesized that BTP2 shifts the voltage-dependence toward more negative potentials, thus increasing the open probability of TRPM4 in a physiological setting.

Similarly, there are only a few known inhibitors of TRPM4 (Table 6.1), with the majority being nonspecific. Early studies of CAN channels observed that adenosine and its derivatives AMP, ADP, and ATP inhibit currents (Sturgess et al., 1986). Later, it was verified that TRPM4 current was indeed reversibly inhibited by ATP, ADP, AMP, AMP-PNP, adenosine, spermine, and quinine, with ADP being the most potent (Nilius et al., 2004; Talavera et al., 2008). Two of the first described nonspecific inhibitors of TRPM4, flufenamic acid and chlotrimazole, were noted to be gating modifiers (Demion et al., 2007; Guinamard et al., 2006; Morita et al., 2007; Ullrich et al., 2005). Glibenclamide, which inhibits K_{ATP} channels through interactions with the SUR1 subunit, was also found to inhibit TRPM4, though these results have been disputed (Demion et al., 2007; Grand et al., 2008; Guinamard et al., 2006; Sala-Rabanal et al., 2012). Further studies identified a more specific and potent inhibitor, 9-phenanthrol, which had no effect on its closest homolog, TRPM5, or other TRP channels such as TRPC3, TRPC6, and TRPM7 (Gonzales et al., 2010; Grand et al., 2008). However, 9-phenanthrol was reported to activate the intermediate conductance calcium-activated potassium channel at low micromolar concentrations (Garland et al., 2015). Recently, a more potent and specific inhibitor of TRPM4 was discovered through virtual screening methods and verified functionally (Ozhathil et al., 2018). This inhibitor, an anthranilic amide analog of flufenamic acid that the authors called Compound 5, displayed selectivity over other TRP channels, including TRPM5, TRPM7, TRPM8, TRPV1, TRPV3, and TRPV6. The IC_{50} was measured to be ~1 μM, which was several folds lower compared to previous inhibitors, thus being the most selective and potent TRPM4 inhibitor to date. As discussed later, all the known human disease-causing hereditary TRPM4 mutations produce a gain-of-function phenotype, highlighting the need to identify specific and effective channel inhibitors.

Besides sharing the Ca^{2+}-sensitivity feature with TRPM4, TRPM5 is sensitive to PIP2 (Liu and Liman, 2003), arachidonic acid (Oike et al., 2006), extracellular protons (Liu et al., 2005), and zinc (Uchida and Tominaga, 2013), but is not sensitive to 9-phenanthrol which inhibits TRPM4 (Grand et al., 2008). Recently, the structure of TRPM5 in complex with a potent antagonist, *N′*-(3,4-dimethoxybenzylidene)-2-(naphthalen-1-yl)acetohydrazide (NDNA), was resolved (Ruan et al., 2021). This ligand was bound between the S1–S4 domain and the pore domain, also known as the "vanilloid binding pocket". Interestingly, several other TRP channels contain this binding pocket, suggesting that many TRP channels may exhibit a shared gating mechanism. This functional conservation has been confirmed in TRPV2 (Yang et al., 2016; Zhang et al., 2016) and TRPV3 (Zhang et al., 2019).

6.8 TRPM4 AND TRPM5 STRUCTURES

Like most other TRP channels, TRPM4 and TRPM5 are homotetramers, with each subunit having six transmembrane segments (S1–S6) and cytosolic N- and C-termini. Similar to voltage-gated potassium channels, the S1–S4 domain constitutes a peripheral voltage sensor-like domain

(VSLD), and the S5–S6 the pore-forming unit. Earlier reports of the low-resolution structures of TRPM4 revealed a tetrameric subunit organization of 100-to-120 Å diameters (Clemencon et al., 2014; Constantine et al., 2016). Within a short time during 2017 and 2018, four independent groups reported high-resolution structures of TRPM4 (Autzen et al., 2018; Duan et al., 2018; Guo et al., 2017; Winkler et al., 2017). Indeed, TRPM4 was shown to be tetrameric with its transmembrane core adopting a domain-swapped architecture similar to many voltage-gated cation channels and all known TRP channels (Fig. 6.2). The cytoplasmic domain consists of the N-terminal melastatin homology regions (MHRs) and a C-terminal coiled-coil domain. Several different states were resolved, including the apo state, Ca^{2+}-bound state, Ca^{2+}-and-decavanadate-bound state, and ATP-bound state, though all of the resolved structures are presumed to be closed (Autzen et al., 2018; Duan et al., 2018; Guo et al., 2017; Winkler et al., 2017).

Within an intracellular cavity formed by the S1-S4 domain, a density located in this hydrophilic pocket was proposed to represent bound Ca^{2+} (Fig. 6.2) (Autzen et al., 2018). A similar Ca^{2+}-binding site has so far been identified also in several other TRPM channels including TRPM2, TRPM5, and TRPM8 (Fig. 6.2B) (Duan et al., 2018; Huang et al., 2019; Ruan et al., 2021; Yin et al., 2019). In TRPM4, the calcium ion was observed to be coordinated by four residues: E828, Q831, N865, and D868 (residue numbers are according to the human homolog), at a position close to the intracellular side of the membrane. The binding site is large and hydrophilic enough to accommodate multiple water molecules, offering an environment supportive of Ca^{2+} occupation. Indeed, in the new TRPM5 cryo-EM structures, solved at resolutions up to 2.3 Å, two water molecules are identified as Ca^{2+}-coordinating ligands together with four channel residues to satisfy the octahedral Ca^{2+} coordination

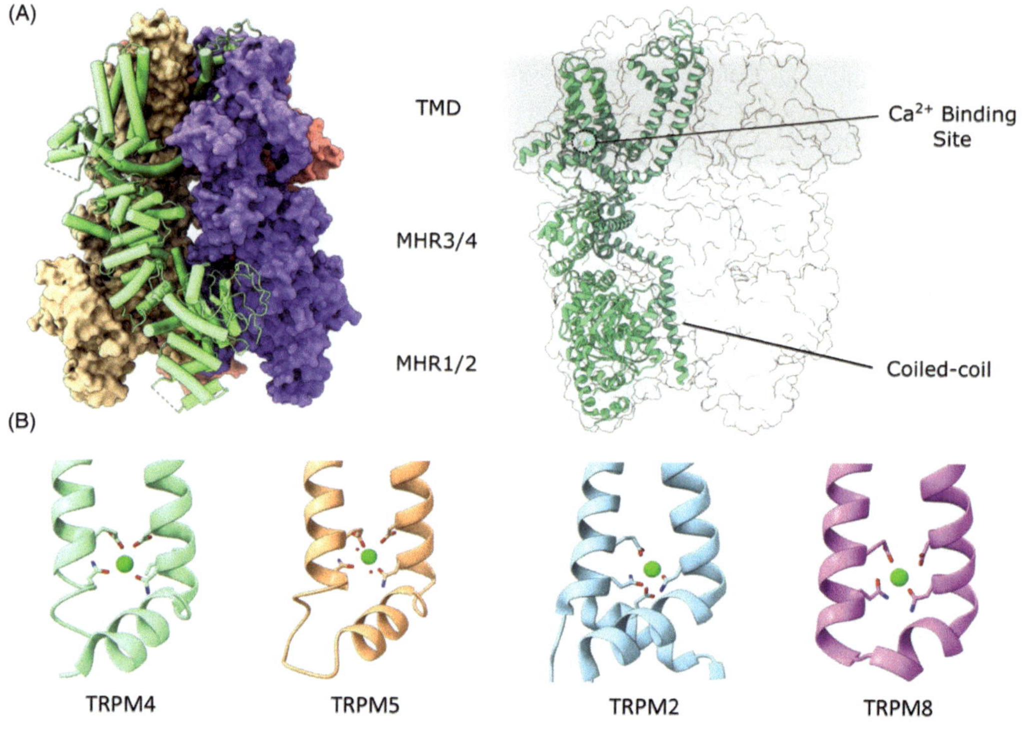

FIGURE 6.2 (A) Overall architecture of human TRPM4, with one subunit, shown in green, highlighted for key structural domains. The location of the cell membrane in the right panel is indicated by the shade. (B) Zoom-in view of calcium binding sites in TRPM4, TRPM5, TRPM2, and TRPM8 (PDB: 6BQV, 7MBS, 6PUS, and 6NR3, respectively). Calcium coordination residues are highlighted; calcium ion is shown as a green sphere. For TRPM5, red spheres represent water molecules.

geometry (Fig. 6.2B) (Ruan et al., 2021). Mutations to any of the four proposed residues that interact with Ca^{2+} or nearby residues either shifted the concentration-response curve or eliminated sensitivity to Ca^{2+}, offering supportive functional evidence of this binding site (Yamaguchi et al., 2019). Previous functional studies reported that in TRPM4 the E1068Q mutation, located on the TRP helix near the bound Ca^{2+}, reduced Ca^{2+} sensitivity (Yamaguchi et al., 2014). In the new TRPM5 structure, the equivalent residue E994 is seen to coordinate one of the two water molecules that directly interact with the bound Ca^{2+} (Ruan et al., 2021).

The intracellular N-terminal segment of TRPM4 contains four melastatin homology regions (MHR1-4), with MHR1 and MHR2 forming an intact domain structure and MHR3 connecting MHR1/2 and MHR4 (Fig. 6.2A, right) (Winkler et al., 2017). MHR1/2 was seen to have extensive interactions with MHR3, which consists of ankyrin repeats, within the same subunit but also interactions with MHR3 from the adjacent subunit. Within this domain, densities corresponding to ATP and decavanadate (one of two sites) were independently observed (Guo et al., 2017; Winkler et al., 2017). There are two opposing effects of ATP—Mg_2ATP induces recovery from desensitization and ATP^{4-} inhibits TRPM4. Studies of mutations within this binding pocket determined that this site exhibited inhibitory properties and had no effect on the recovery from desensitization; thus, the observed binding pocket was proposed to be of the inhibitory state (Guo et al., 2017).

The C-terminal domain consists of a coiled-coil helix with a segment running horizontal, which then bends ~120° into vertical helices that are parallel to the vertical helices of the other subunits. The C-terminus has been previously proposed to contain modulation sites for PIP2, calmodulin binding, and phosphorylation and is important for tetramer formation (Launay et al., 2004; Nilius et al., 2005; Nilius et al., 2006). In addition, functional studies suggested that decavanadate binds to the C-terminal domain. Indeed, structural studies observed a second decavanadate-binding site at the positively charged hinge of the C-terminal coiled-coil domain (Winkler et al., 2017). Previous functional studies observed that the presence of decavanadate increased the sensitivity of TRPM4 to ATP^{4-} inhibition. Thus, the binding of decavanadate to the C-terminal domain is consistent with this site being involved in the positive modulation effects, as it should not competitively antagonize N-terminal ATP^{4-} binding site. How decavanadate increases sensitivity to ATP^{4-} inhibition is still a question.

The new TRPM5 structures (Ruan et al., 2021) offer exciting new insights into the Ca^{2+} activation mechanism. Unlike all the available TRPM4 structures, the Ca^{2+}-bound TRPM5 structure appears to represent an open state, with the hydrophobic gate opening up substantially compared to the apo (closed) state. An allosterically coupled regulatory site for Ca^{2+} is also identified. It is located intracellularly at the interface between the MHR1/2 and MHR3/4 domains; binding of Ca^{2+} to this site causes local conformational rearrangements of the MHR domains. Interestingly, the effect of Ca^{2+} binding to the intracellular site can be felt at the transmembrane Ca^{2+}-binding site some 70 Å away, yielding a substantially increased apparent binding affinity and positive allosteric regulation of channel activation. TRPM4 is considered to be devoid of the intracellular Ca^{2+}-binding site. The absence of positive allosteric coupling may help explain the much lower Ca^{2+} sensitivity observed in TRPM4 channels.

6.9 TRPM4 AND TRPM5 EXPRESSION AND PHYSIOLOGICAL FUNCTION

The first CAN currents described were in cardiac Purkinje fibers (Kass et al., 1978a). In subsequent years, CAN channels have been shown to exist in various tissues including heart (Colquhoun et al., 1981), skeletal muscle (Salkoff and Wyman, 1983; Sheu and Blaustein, 1983), neuroblastoma (Yellen, 1982), Schwann cells (Bevan et al., 1984), pituitary cells (Adler et al., 1983), neurons (Swandulla and Lux, 1985), pancreas (Gallacher et al., 1984; Maruyama and Petersen, 1984; Maruyama and Peterson, 1982; Petersen and Maruyama, 1983), thyroid (Maruyama et al., 1985), lens epithelial cells (Rae and Levis, 1984), insulinoma (Sturgess et al., 1986), macrophages (Lipton et al., 1986), neutrophils (von Tscharner et al., 1986), mast cells (Lindau and Fernandez, 1986), and dorsal root

TABLE 6.2
TRPM4 Distribution

Tissue/Organ	TRPM4	CAN Channel	References
heart	+	+	Colquhoun et al., 1981; Fonfria et al., 2006
pituitary	+		Fonfria et al., 2006
brain	+	+	Yellen, 1982; Partridge and Swandulla, 1987; Bevan et al., 1984; Fonfria et al., 2006
skin	+	+	Galietta et al., 1991; Fonfria et al., 2006; Wang etal., 2019
lung	+	+	Feng et al., 1993; Fonfria et al., 2006
liver	+	+	Fitz et al., 1994; Fonfria et al., 2006
kidney	+	+	Matsunaga et al., 1991; Teulon et al., 1987; Nonaka et al., 1995; Ono et al., 1994; Fonfria et al., 2006
stomach	+		Fonfria et al., 2006
intestine	+	+	Nonaka, 1995; Fonfria et al., 2006
spleen	+		Fonfria et al., 2006
blood	+	+	von Tscharner et al., 1986; Lindau and Fernandez, 1986; Fonfria et al., 2006
macrophages	+		Fonfria et al., 2006
adipose	+	+	Siemen and Reuhl, 1987; Koivisto, 1992; Fonfria et al., 2006
pancreas	+	+	Maruyama and Peterson, 1982; Maruyama and Petersen, 1984; Thorn and Petersen, 1993; Fonfria et al., 2006
prostate	+		Fonfria et al., 2006
placenta	+		Fonfria et al., 2006
bone	+		Fonfria et al., 2006
salivary gland	+	+	Holzer, 2011

ganglia (Simonneau et al., 1987) (Table 6.2). Given the wide expression profile, the physiological functions of CAN channels are likely to be diverse. CAN channels have been proposed to play roles in stimulation-secretion coupling and regulation of electrical activity in excitable cells. In the early days, the most characterized function of these channels was explored in neurons. Here, after Ca^{2+} activation, they are thought to be responsible for the sustained depolarization during action potentials and identified as pacemaker currents (Partridge and Swandulla, 1987; Swandulla and Lux, 1985). Though first discovered in cardiac cells, TRPM4's cellular function was debated. As previously noted, in Purkinje fibers, a TI current was observed (Kass et al., 1978a; Kass et al., 1978b); however, others proposed that the inward current was mainly produced by the Na^+/Ca^{2+} exchanger (Sipido et al., 1995).

The cloned TRPM4 has been shown to be widely expressed in various organ types, which include but are not limited to the brain, pituitary, heart, skin, lung, liver, kidney, skeletal muscle, stomach, intestine, spleen, peripheral blood mononuclear cells, macrophages, adipose, pancreas, prostate, placenta, salivary gland, and bone, with the intestine and prostate showing the highest expression levels (Fonfria et al., 2006; Holzer et al., 2011). As expected, several of the cell types that were found to exhibit CAN currents in earlier studies were also found to express TRPM4, as summarized in Table 6.2. With the emergence of new tools, our understanding of the physiological role of CAN channels, and more specifically TRPM4 and TRPM5, has significantly advanced along with new insights into the channels' biophysical properties. It is now clear that the primary function of these two TRPM channels is to depolarize the membrane upon an increase of intracellular calcium, serving as a converter of intracellular chemical signaling to electrical signaling. This will be further discussed below in a few systems where a clearer understanding has emerged.

6.9.1 Cardiovascular Physiology

TRPM4 expression has been found in the sino-atrial node, atrium, ventricles, and smooth muscle cells (Demion et al., 2007; Earley et al., 2013; Guinamard et al., 2006; Liu et al., 2010; Mathar et al., 2014; Piao et al., 2015). In sino-atrial pacemaker cells, spontaneous diastolic depolarization is mediated by a number of ion channels and transporters, such as the hyperpolarization-activated cyclic nucleotide-gated (HCN) channel and an inward current activated by Ca^{2+}. This inward current was originally thought to be carried mostly by the Na^+/Ca^{2+} exchanger (NCX); however, it was observed that the current was still present when Na^+ was replaced with the NCX-impermanent ion K^+ or Li^+, suggesting that TRPM4 may be involved (Guo et al., 1996; Sasse et al., 2007). In addition, 9-phenathrol reduced the beating rate with a half-maximal inhibition concentration similar to what was observed to inhibit TRPM4. In TRPM4 knockout mice, this effect was not observed, providing further evidence of the involvement of TRPM4 (Hof et al., 2013). Of note, these Ca^{2+}-activated inward currents were observed in the early studies of CAN channels that suggested their involvement in producing the pacemaker current.

In addition to the contribution to pacemaker current, TRPM4 has been suggested to play a role in ventricular and atrial action potentials. It can be inferred that the increase in intracellular calcium during an action potential (estimated to reach 30 μM locally due to ryanodine receptor 2 activation (Dedkova and Blatter, 2008; Szalai et al., 2000; Dorn and Maack, 2013) may activate TRPM4 in the late diastolic phase to prolong the plateau phase in cardiomyocytes (Guinamard et al., 2015). Indeed, inhibition of TRPM4 by 9-phenanthrol led to a reduction in atrial action potential duration, an effect not observed in TRPM4 knockout mice; moreover, TRPM4 knock-out mice displayed a shortened action potential duration (faster repolarization) compared to wild-type atria (Simard et al., 2013). In the ventricles, similar results were observed; TRPM4 knock-out mice displayed shortened ventricular action potential durations compared to wild-type mice (Mathar et al., 2014). However, others noted that 9-phenanthrol had no effect on ventricular myocytes even though it affected atria as previously reported (Simard et al., 2013). Thus, the role of TRPM4 in ventricles is still not fully understood. Future studies will be needed to delineate the importance of TRPM4 in the cardiac system.

6.9.2 Immune Physiology

Changes in intracellular calcium, often in the form of calcium oscillations, are important for signaling in the propagation and maintenance of immune responses. TRPM4 has been suggested to play a role in calcium signaling in T cells, mast cells, and dendritic cells (Barbet et al., 2008; Launay et al., 2004; Vennekens and Nilius, 2007). Calcium influx through calcium release-activated calcium (CRAC) channels, driven by both concentration gradient and negative membrane potential, leads to an increase in intracellular Ca^{2+} concentration in these cells. It was suggested that TRPM4 could control calcium oscillations through membrane depolarization to limit calcium entry (Launay et al., 2004). In this sense, TRPM4 activity serves as the negative feedback mechanism in calcium oscillation, whereas activities of Kv1.3 and KCa3.1 serve as the positive feedback mechanism, as they hyperpolarize membrane potential and increase the driving force for calcium entry. Indeed, TRPM4 activation by BTP2 led to a decrease in calcium influx, resulting in the suppression of cytokine release (Takezawa et al., 2006). In dendritic cells, TRPM4 has been proposed to be important for the migration, but not maturation, to secondary lymphoid organs by controlling calcium influx (Barbet et al., 2008). In mast cells, activation of the high-affinity IgE receptor leads to rapid hyperpolarization and calcium influx, whereas sustained depolarization leads to much shorter peak calcium levels. However, in mast cells with TRPM4 knocked-out, calcium influx is dramatically increased following the activation of the IgE receptor due to prolonged negative membrane potentials (Vennekens and Nilius, 2007). Thus, TRPM4 plays a key role in regulating IgE-mediated calcium influx through regulation of membrane potentials, which in turn results in the control of inflammatory responses.

6.9.3 Central Nervous System

According to the human protein atlas (Uhlen et al., 2015), TRPM4 is expressed in the olfactory region, cerebral cortex, hippocampal formation, amygdala, basal ganglia, hypothalamus, thalamus, midbrain, pons and medulla, cerebellum, and pituitary gland. The general hypothesis, similar to cardiac and immune cells, is that TRPM4 participates in burst firing, persistent activity, and rhythmic firing (Riquelme et al., 2018). TRPM4 has been shown to be involved in the plateau potential and neuronal bursting in response to postsynaptic currents by glutamatergic activation (Mironov et al., 2008; Mironov et al., 2013). More recently, TRPM4 has been suggested to physically interact with N-methyl-D-aspartate receptors, which leads to excitotoxicity (Yan et al., 2020). In the hypothalamus, TRPM4 and TRPM5 were suggested to be involved in fast depolarizing after potentials in the generation of phasic bursts (Teruyama et al., 2011).

Of particular note was the involvement of TRPM4 in depolarization-induced slow currents in cerebellar Purkinje cells (Shin et al., 2008). It was observed that brief depolarization of the spontaneously firing Purkinjie cells induced a slow inward current (time-to-peak was measured to be 1.8 sec), which was found to be dependent on intracellular calcium (Kim et al., 2013). The current was found to be Na^+ and K^+ permeable but Ca^{2+} impermeable. Further characterizations using TRPM4 antagonists observed that these slow inward currents could be partially attenuated (Kim et al., 2013). Early studies of CAN channels also noted similar observations in cardiac Purkinje fibers (Kass et al., 1978b). These slow transient inward currents were observed to be Na^+ and K^+ permeable, Ca^{2+} impermeable, intracellular Ca^{2+}-dependent, and a time-to-peak of about 1.5 seconds at −80 mV. Given the similar observations, it is probable that the CAN channels observed decades ago were in fact TRPM4 channels.

Compared to TRPM4, TRPM5 appears to have a narrower tissue distribution. It is known to function in taste cells (Perez et al., 2002; Zhang et al., 2003), pancreatic islets (Brixel et al., 2010; Colsoul et al., 2010), gastrointestinal tract (Bezencon et al., 2007), hypothalamus (Teruyama et al., 2011), and other tissues or cell types. In chemosensing neurons, TRPM5 is found to be important for both generating taste responses (Damak et al., 2006; Kaske et al., 2007) and sensing pheromones (Lin et al., 2007; Lopez et al., 2014).

6.10 TRPM4 AND TRPM5 IN HUMAN DISEASES

Given the physiological roles of TRPM4 and TRPM5 in regulating membrane potential and their broad distribution, it can be anticipated that dysfunction of these two channels would have severe consequences. Indeed, a large number of TRPM4 hereditary mutations have been linked to disorders in the heart and, more recently, in the skin (Fig. 6.3). Interestingly, most TRPM4 hereditary mutations identified so far are gain-of-function in nature, though the causes of such gain-of-function are different.

The first study to link TRPM4 with a hereditary disease discovered an E7K mutation in a family with cardiac bundle branch conduction block (Kruse et al., 2009). It was reported that this mutation did not affect the biophysical properties of TRPM4. Rather, the mutation led to increased expression levels at the plasma membrane due to altered channel SUMOylation and consequently impaired endocytosis, resulting in increased TRPM4 current amplitudes, hence amplified responses to Ca^{2+}. Later studies discovered additional mutations that have similar phenotypes (Bianchi et al., 2018; Liu et al., 2010; Stallmeyer et al., 2012). Though the underlying physiological consequence is not fully understood, it can be inferred that an increased expression of TRPM4 would lead to prolonged action potential durations and thus disrupt normal cardiac rhythm. In addition, an increased number of channels could lead to a more depolarized resting membrane potential (which is observed in keratinocytes and will be discussed in detail later) and in turn inactivate voltage-gated sodium channels, thus affecting conduction. Within patients with Brugada syndrome, several mutations were found throughout the TRPM4 channel that displayed different phenotypes (Duthoit et al., 2012; Liu et al., 2013; Stallmeyer et al., 2012). For example, K914X resulted in a truncated nonfunctional channel, and P779R was observed to have a reduction in current density, which could be explained by a

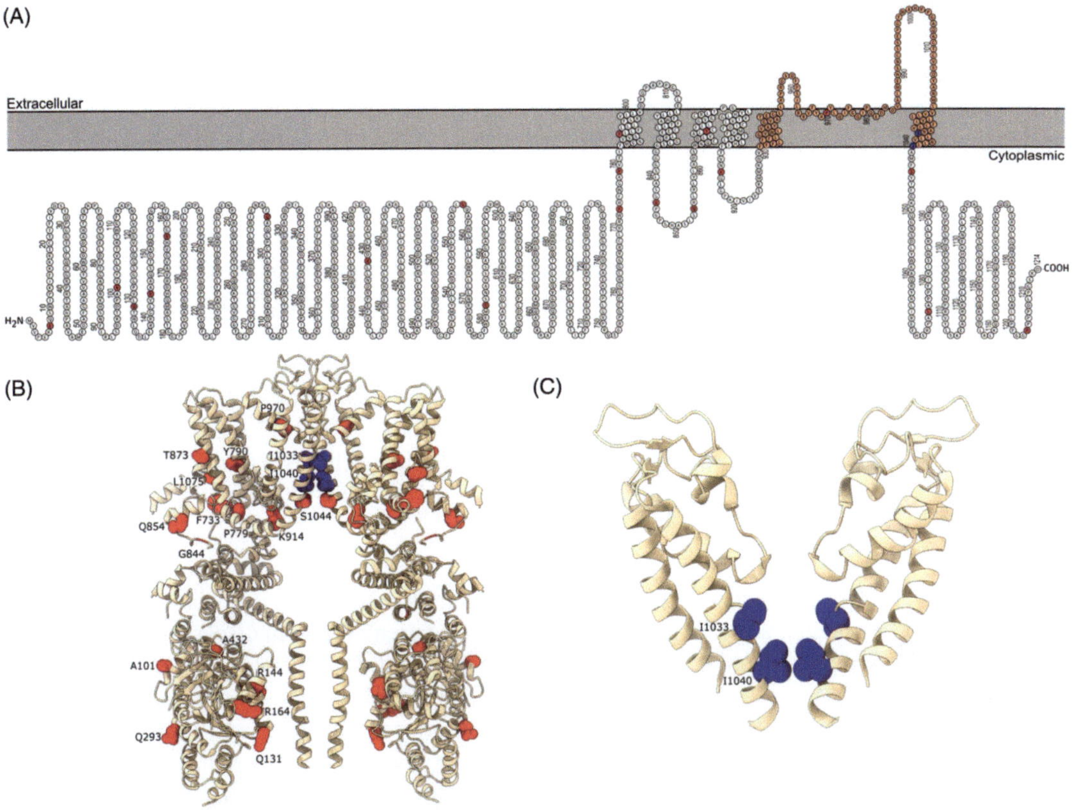

FIGURE 6.3 (A) Topology of TRPM4 with known mutants colored in red. Pore domain is in orange, with pore mutants colored blue. Designed with Protter (Omasits et al. 2014). (B) Known mutants mapped onto two subunits of the hTRPM4 atomic model. (C) Zoomed in view of the pore domain and its mutants.

reduction in surface expression. In contrast, T873I and L1075P mutations resulted in a significant increase of total and surface expressions of the channel (Liu et al., 2013).

It was reported that PKCδ activity influences the translocation of TRPM4 (Crnich et al., 2010; Garcia et al., 2011). When PKCδ was down-regulated or inhibited, TRPM4 was translocated from the plasma membrane to the cytosol. It was proposed that in myocytes, PKCδ activity could support cell excitability by maintaining TRPM4 at the plasma membrane. Since it was shown that the truncated form TRPM4a is mostly expressed in intracellular compartments and the full-length TRPM4b is mostly expressed on the plasma membrane, the N-terminal segment was used to identify interacting proteins that may be involved in trafficking. It was observed that 14-3-3γ, a trafficking chaperone, interacted with the N-terminal fragment of TRPM4. In addition, mutation of S88 at the N-terminus resulted in impaired trafficking to the plasma membrane (Cho et al., 2014). More recently, it was observed that TRPM4 interacts with EB proteins, which are microtubule plus-end tracking proteins (Blanco et al., 2019). Mutations to the putative EB-binding site lead to reduced TRPM4 expression at the plasma membrane. In addition to protein interactions that affect trafficking, it has been shown that glycosylation and SUMOylation may have effects on TRPM4 stabilization (Woo et al., 2013). It was shown that mutations N992Q and R164W affected current density but not the number of expressed channels at the plasma membrane (Syam et al., 2014). Other post-translational modifications, such as phosphorylation, also seem to be important means to modulate TRPM4 activity. PKC-dependent phosphorylation of TRPM4 was shown to increase the sensitivity to Ca^{2+} activation and decrease desensitization. When two residues, S1145 and S1152, were mutated, TRPM4 could no longer be activated, suggesting that phosphorylation is essential for channel activation (Nilius et al., 2005).

More recently, TRPM4 has been observed to play an important role in the regulation of skin physiology. Two mutations (I1033M and I1040T) were discovered in human TRPM4 and linked to progressive symmetric erythrokeratodermia (PSEK) (Wang et al., 2019). PSEK is characterized by early-onset, slow progression, symmetrical, erythematous skin plaques predominantly localized on the extremities. These two mutated residues are located on the S6 transmembrane segment within the ion conduction pathway of the channel, with I1040 positioning at the narrowest part. Given this, it is expected that these residues may play a role in the gating of the channel. Indeed, it was shown that the channel displayed increased basal activity and increased sensitivity to intracellular Ca^{2+}, and its expression leads to an elevated resting membrane potential. In addition, the cellular consequences of these gain-of-function mutants include enhanced proliferation and differentiation of keratinocytes (Wang et al., 2019).

Much less is known about the potential involvement of TRPM5 in human diseases. However, given its role in chemosensation for taste and gastrointestinal functions, malfunctions in TRPM5 could lead to various health issues. Indeed, genetic variations of TRPM5 are suggested to link to prediabetic phenotypes (Ketterer et al., 2011). The underlying mechanism remains to be explored.

6.11 CONCLUSION

Since the discovery of TRPM4 and TRRPM5, our understanding of these channels, from their biophysical and molecular activation mechanisms to their potential role in disease, has significantly advanced. An uncanny resemblance in biophysical characteristics to the earlier known CAN channels is noted. TRPM4 and TRPM5 are recognized as important effectors for Ca^{2+} signaling in many cell types. In addition, while the immediate consequence of their activation is to translate an increase in intracellular Ca^{2+} concentration into membrane depolarization, it is realized that their activities can feedback to regulate Ca^{2+} signaling (e.g., see (Launay et al., 2004)). This can be achieved in several ways; for example, membrane depolarization alters the electrical driving force for Ca^{2+} entry through various Ca^{2+} permeant pathways. Given their unique physiological roles in cellular signaling, TRPM4 and TRRPM5 are attractive targets for pharmacological regulation of cellular functions for clinical purposes.

ACKNOWLEDGMENTS

We thank Wei Lu (Van Andel Institute) and members of the Zheng lab for insightful discussions of the manuscript. This work is funded by NIH R01NS128180 to J.Z.

REFERENCES

Adler, M., Wong, B. S., Sabol, S. L., Busis, N., Jackson, M. B., Weight, F. F. Action potentials and membrane ion channels in clonal anterior pituitary cells. *Proc Natl Acad Sci U S A*, 1983, 80(7): 2086–2090.

Amarouch, M. Y., Syam, N., Abriel, H. Biochemical, single-channel, whole-cell patch clamp, and pharmacological analyses of endogenous TRPM4 channels in HEK293 cells. *Neurosci Lett*, 2013, 541: 105–110.

Autzen, H. E., Myasnikov, A. G., Campbell, M. G., Asarnow, D., Julius, D., Cheng, Y. Structure of the human TRPM4 ion channel in a lipid nanodisc. *Science*, 2018, 359(6372): 228–232.

Barbet, G., Demion, M., Moura, I. C., Serafini, N., Leger, T., Vrtovsnik, F., Monteiro, R. C., Guinamard, R., Kinet, J. P., Launay, P. The calcium-activated nonselective cation channel TRPM4 is essential for the migration but not the maturation of dendritic cells. *Nat Immunol*, 2008, 9(10): 1148–1156.

Benham, C. D., Gunthorpe, M. J., Davis, J. B. TRPV channels as temperature sensors. *Cell Calcium*, 2003, 33(5-6): 479–487.

Bevan, S., Gray, P. T., Ritchie, J. M. A calcium-activated cation-selective channel in rat cultured Schwann cells. *Proc R Soc Lond B Biol Sci*, 1984, 222(1228): 349–355.

Bezencon, C., le Coutre, J., Damak, S. Taste-signaling proteins are coexpressed in solitary intestinal epithelial cells. *Chem Senses*, 2007, 32(1): 41–49.

Bianchi, B., Ozhathil, L. C., Medeiros-Domingo, A., Gollob, M. H., Abriel, H. Four TRPM4 cation channel mutations found in cardiac conduction diseases lead to altered protein stability. *Front Physiol*, 2018, 9: 177.

Blanco, C., Morales, D., Mogollones, I., Vergara-Jaque, A., Vargas, C., Alvarez, A., Riquelme, D., Leiva-Salcedo, E., Gonzalez, W., Morales, D., Maureira, D., Aldunate, I., Caceres, M., Varela, D., Cerda, O. EB1- and EB2-dependent anterograde trafficking of TRPM4 regulates focal adhesion turnover and cell invasion. *FASEB J*, 2019, 33(8): 9434–9452.

Bousova, K., Barvik, I., Herman, P., Hofbauerova, K., Monincova, L., Majer, P., Zouharova, M., Vetyskova, V., Postulkova, K., Vondrasek, J. Mapping of CaM, S100A1 and PIP2-binding epitopes in the intracellular N- and C-Termini of TRPM4. *Int J Mol Sci*, 2020, 21(12): 4323.

Bousova, K., Herman, P., Vecer, J., Bednarova, L., Monincova, L., Majer, P., Vyklicky, L., Vondrasek, J., Teisinger, J. Shared CaM- and S100A1-binding epitopes in the distal TRPM4 N terminus. *FEBS J*, 2018, 285(3): 599–613.

Bousova, K., Jirku, M., Bumba, L., Bednarova, L., Sulc, M., Franek, M., Vyklicky, L., Vondrasek, J., Teisinger, J. PIP2 and PIP3 interact with N-terminus region of TRPM4 channel. *Biophys Chem*, 2015, 205: 24–32.

Brauchi, S., Orio, P., Latorre, R. Clues to understanding cold sensation: Thermodynamics and electrophysiological analysis of the cold receptor TRPM8. *Proc Natl Acad Sci U S A*, 2004, 101(43): 15494–15499.

Brixel, L. R., Monteilh-Zoller, M. K., Ingenbrandt, C. S., Fleig, A., Penner, R., Enklaar, T., Zabel, B. U., Prawitt, D. TRPM5 regulates glucose-stimulated insulin secretion. *Pflugers Arch*, 2010, 460(1): 69–76.

Cho, C. H., Kim, E., Lee, Y. S., Yarishkin, O., Yoo, J. C., Park, J. Y., Hong, S. G., Hwang, E. M. Depletion of 14-3-3gamma reduces the surface expression of transient receptor potential melastatin 4b (TRPM4b) channels and attenuates TRPM4b-mediated glutamate-induced neuronal cell death. *Mol Brain*, 2014, 7: 52.

Clapham, D. E. TRP channels as cellular sensors. *Nature*, 2003, 426(6966): 517–524.

Clemencon, B., Fine, M., Luscher, B., Baumann, M. U., Surbek, D. V., Abriel, H., Hediger, M. A. Expression, purification, and projection structure by single particle electron microscopy of functional human TRPM4 heterologously expressed in Xenopus laevis oocytes. *Protein Expr Purif*, 2014, 95: 169–176.

Colquhoun, D., Neher, E., Reuter, H., Stevens, C. F. Inward current channels activated by intracellular Ca in cultured cardiac cells. *Nature*, 1981, 294(5843): 752–754.

Colsoul, B., Schraenen, A., Lemaire, K., Quintens, R., Van Lommel, L., Segal, A., Owsianik, G., Talavera, K., Voets, T., Margolskee, R. F., Kokrashvili, Z., Gilon, P., Nilius, B., Schuit, F. C., Vennekens, R. "Loss of high-frequency glucose-induced Ca^{2+} oscillations in pancreatic islets correlates with impaired glucose tolerance in Trpm5-/- mice. *Proc Natl Acad Sci U S A*, 2010, 107(11): 5208–5213.

Constantine, M., Liew, C. K., Lo, V., Macmillan, A., Cranfield, C. G., Sunde, M., Whan, R., Graham, R. M., Martinac, B. Heterologously-expressed and liposome-reconstituted human transient receptor potential melastatin 4 channel (TRPM4) is a functional tetramer. *Sci Rep*, 2016, 6: 19352.

Crnich, R., Amberg, G. C., Leo, M. D., Gonzales, A. L., Tamkun, M. M., Jaggar, J. H., Earley, S. Vasoconstriction resulting from dynamic membrane trafficking of TRPM4 in vascular smooth muscle cells. *Am J Physiol Cell Physiol*, 2010, 299(3): C682–C694.

Damak, S., Rong, M., Yasumatsu, K., Kokrashvili, Z., Perez, C. A., Shigemura, N., Yoshida, R., Mosinger, B. Jr., Glendinning, J. I., Ninomiya, Y., Margolskee, R. F. Trpm5 null mice respond to bitter, sweet, and umami compounds. *Chem Senses*, 2006, 31(3): 253–264.

Dedkova, E. N., Blatter, L. A. Mitochondrial Ca^{2+} and the heart. *Cell Calcium*, 2008, 44(1): 77–91.

Demion, M., Bois, P., Launay, P., Guinamard, R. TRPM4, a Ca^{2+}-activated nonselective cation channel in mouse sino-atrial node cells. *Cardiovasc Res*, 2007, 73(3): 531–538.

Dorn, G. W. 2nd, Maack, C. SR and mitochondria: Calcium cross-talk between kissing cousins. *J Mol Cell Cardiol*, 2013, 55: 42–49.

Duan, J., Li, Z., Li, J., Santa-Cruz, A., Sanchez-Martinez, S., Zhang, J., Clapham, D. E. Structure of full-length human TRPM4. *Proc Natl Acad Sci U S A*, 2018, 115(10): 2377–2382.

Duthoit, G., Fressart, V., Hidden-Lucet, F., Simon, F., Kattygnarath, D., Charron, P., Himbert, C., Aouate, P., Guicheney, P., Lecarpentier, Y., Frank, R., Hebert, J. L. Brugada ECG pattern: A physiopathological prospective study based on clinical, electrophysiological, angiographic, and genetic findings. *Front Physiol*, 2012, 3: 474.

Earley, S. TRPM4 channels in smooth muscle function. *Pflugers Arch*, 2013, 465(9): 1223–1231.

Earley, S., Straub, S. V., Brayden, J. E. Protein kinase C regulates vascular myogenic tone through activation of TRPM4. *Am J Physiol Heart Circ Physiol*, 2007, 292(6): H2613–H2622.

Enklaar, T., Esswein, M., Oswald, M., Hilbert, K., Winterpacht, A., Higgins, M., Zabel, B., Prawitt, D. Mtrl, a novel biallelically expressed gene in the center of the mouse distal chromosome 7 imprinting cluster, is a member of the Trp gene family. *Genomics*, 2000, 67(2): 179–187.

Feng, Z. P., Clark, R. B., Berthiaume, Y. Identification of nonselective cation channels in cultured adult rat alveolar type II cells. *Am J Respir Cell Mol Biol*, 1993, 9(3): 248–254.

Ferrier, G. R., Moe, G. K. Effect of calcium on acetylstrophanthidin-induced transient depolarizations in canine Purkinje tissue. *Circ Res*, 1973, 33(5): 508–515.

Fitz, J. G., Sostman, A. H., Middleton, J. P. Regulation of cation channels in liver cells by intracellular calcium and protein kinase C. *Am J Physiol*, 1994, 266(4 Pt 1): G677–G684.

Fonfria, E., Murdock, P. R., Cusdin, F. S., Benham, C. D., Kelsell, R. E., McNulty, S. Tissue distribution profiles of the human TRPM cation channel family. *J Recept Signal Transduct Res*, 2006, 26(3): 159–178.

Galietta, L. J., Barone, V., De Luca, M., Romeo, G. Characterization of chloride and cation channels in cultured human keratinocytes. *Pflügers Archiv*, 1991, 418(1–2): 18–25.

Gallacher, D. V., Maruyama, Y., Petersen, O. H. Patch-clamp study of rubidium and potassium conductances in single cation channels from mammalian exocrine acini. *Pflugers Arch*, 1984, 401(4): 361–367.

Garcia, Z. I., Bruhl, A., Gonzales, A. L., Earley, S. Basal protein kinase Cdelta activity is required for membrane localization and activity of TRPM4 channels in cerebral artery smooth muscle cells. *Channels (Austin)*, 2011, 5(3): 210–214.

Garland, C. J., Smirnov, S. V., Bagher, P., Lim, C. S., Huang, C. Y., Mitchell, R., Stanley, C., Pinkney, A., Dora, K. A. TRPM4 inhibitor 9-phenanthrol activates endothelial cell intermediate conductance calcium-activated potassium channels in rat isolated mesenteric artery. *Br J Pharmacol*, 2015, 172(4): 1114–1123.

Gonzales, A. L., Amberg, G. C., Earley, S. Ca^{2+} release from the sarcoplasmic reticulum is required for sustained TRPM4 activity in cerebral artery smooth muscle cells. *Am J Physiol Cell Physiol*, 2010, 299(2): C279–C288.

Gonzales, A. L., Earley, S. Endogenous cytosolic Ca^{2+} buffering is necessary for TRPM4 activity in cerebral artery smooth muscle cells. *Cell Calcium*, 2012, 51(1): 82–93.

Grand, T., Demion, M., Norez, C., Mettey, Y., Launay, P., Becq, F., Bois, P., Guinamard, R. 9-phenanthrol inhibits human TRPM4 but not TRPM5 cationic channels. *Br J Pharmacol*, 2008, 153(8): 1697–1705.

Gray, M. A., Argent, B. E. Non-selective cation channel on pancreatic duct cells. *Biochim Biophys Acta*, 1990, 1029(1): 33–42.

Guinamard, R., Bouvagnet, P., Hof, T., Liu, H., Simard, C., Salle, L. TRPM4 in cardiac electrical activity. *Cardiovasc Res*, 2015, 108(1): 21–30.

Guinamard, R., Demion, M., Magaud, C., Potreau, D., Bois, P. Functional expression of the TRPM4 cationic current in ventricular cardiomyocytes from spontaneously hypertensive rats. *Hypertension*, 2006, 48(4): 587–594.

Guo, J., Ono, K., Noma, A. Monovalent cation conductance of the sustained inward current in rabbit sinoatrial node cells. *Pflugers Arch*, 1996, 433(1–2): 209–211.

Guo, J., She, J., Zeng, W., Chen, Q., Bai, X. C., Jiang, Y. Structures of the calcium-activated, non-selective cation channel TRPM4. *Nature*, 2017, 552(7684): 205–209.

Hamill, O. P., Marty, A., Neher, E., Sakmann, B., Sigworth, F. J. Improved patch-clamp techniques for high-resolution current recording from cells and cell-free membrane patches. *Pflugers Arch*, 1981, 391(2): 85–100.

Hille, B. (2001). *Ion Channels of Excitable Membranes*. Sunderland, MA, Sinauer Associates, Inc.

Hof, T., Simard, C., Rouet, R., Salle, L., Guinamard, R. Implication of the TRPM4 nonselective cation channel in mammalian sinus rhythm. *Heart Rhythm*, 2013, 10(11): 1683–1689.

Hofmann, T., Chubanov, V., Gudermann, T., Montell, C. TRPM5 is a voltage-modulated and Ca^{2+}-activated monovalent selective cation channel. *Curr Biol*, 2003, 13(13): 1153–1158.

Holzer, P. Transient receptor potential (TRP) channels as drug targets for diseases of the digestive system. *Pharmacol Ther*, 2011, 131(1): 142–170.

Huang, Y., Roth, B., Lu, W., Du, J. Ligand recognition and gating mechanism through three ligand-binding sites of human TRPM2 channel. *Elife*, 2019, 8: e50175.

Kaske, S., Krasteva, G., Konig, P., Kummer, W., Hofmann, T., Gudermann, T., Chubanov, V. TRPM5, a taste-signaling transient receptor potential ion-channel, is a ubiquitous signaling component in chemosensory cells. *BMC Neurosci*, 2007, 8: 49.

Kass, R. S., Lederer, W. J., Tsien, R. W., Weingart, R. Role of calcium ions in transient inward currents and aftercontractions induced by strophanthidin in cardiac Purkinje fibres. *J Physiol*, 1978a, 281: 187–208.

Kass, R. S., Tsien, R. W., Weingart, R. Ionic basis of transient inward current induced by strophanthidin in cardiac Purkinje fibres. *J Physiol*, 1978b, 281: 209–226.

Ketterer, C., Mussig, K., Heni, M., Dudziak, K., Randrianarisoa, E., Wagner, R., Machicao, F., Stefan, N., Holst, J. J., Fritsche, A., Haring, H. U., Staiger, H. Genetic variation within the TRPM5 locus associates with prediabetic phenotypes in subjects at increased risk for type 2 diabetes. *Metabolism*, 2011, 60(9): 1325–1333.

Kim, Y. S., Kang, E., Makino, Y., Park, S., Shin, J. H., Song, H., Launay, P., Linden, D. J. Characterizing the conductance underlying depolarization-induced slow current in cerebellar Purkinje cells. *J Neurophysiol*, 2013, 109(4): 1174–1181.

Kruse, M., Schulze-Bahr, E., Corfield, V., Beckmann, A., Stallmeyer, B., Kurtbay, G., Ohmert, I., Schulze-Bahr, E., Brink, P., Pongs, O. Impaired endocytosis of the ion channel TRPM4 is associated with human progressive familial heart block type I. *J Clin Invest*, 2009, 119(9): 2737–2744.

Launay, P., Cheng, H., Srivatsan, S., Penner, R., Fleig, A., Kinet, J. P. TRPM4 regulates calcium oscillations after T cell activation. *Science*, 2004, 306(5700): 1374–1377.

Launay, P., Fleig, A., Perraud, A. L., Scharenberg, A. M., Penner, R., Kinet, J. P. TRPM4 is a Ca^{2+}-activated nonselective cation channel mediating cell membrane depolarization. *Cell*, 2002, 109(3): 397–407.

Lederer, W. J., Tsien, R. W. Proceedings: Transient inward current underlying strophanthidin's enhancement of pace-maker activity in Purkinje fibres. *J Physiol*, 1975, 249(1): 40P–41P.

Lederer, W. J., Tsien, R. W. Transient inward current underlying arrhythmogenic effects of cardiotonic steroids in Purkinje fibres. *J Physiol*, 1976, 263(2): 73–100.

Liman, E. R. Trpm5. *Handb Exp Pharmacol*, 2014, 222: 489–502.

Lin, W., Margolskee, R., Donnert, G., Hell, S. W., Restrepo, D. Olfactory neurons expressing transient receptor potential channel M5 (TRPM5) are involved in sensing semiochemicals. *Proc Natl Acad Sci U S A*, 2007, 104(7): 2471–2476.

Lindau, M., Fernandez, J. M. A patch-clamp study of histamine-secreting cells. *J Gen Physiol*, 1986, 88(3): 349–368.

Lipton, S. A. Antibody activates cationic channels via second messenger Ca^{2+}. *Biochim Biophys Acta*, 1986, 856(1): 59–67.

Liu, B., Qin, F. Functional control of cold- and menthol-sensitive TRPM8 ion channels by phosphatidylinositol 4,5-bisphosphate. *J Neurosci*, 2005, 25(7): 1674–1681.

Liu, D., Liman, E. R. Intracellular Ca^{2+} and the phospholipid PIP2 regulate the taste transduction ion channel TRPM5. *Proc Natl Acad Sci U S A*, 2003, 100(25): 15160–15165.

Liu, D., Zhang, Z., Liman, E. R. Extracellular acid block and acid-enhanced inactivation of the Ca^{2+}-activated cation channel TRPM5 involve residues in the S3-S4 and S5-S6 extracellular domains. *J Biol Chem*, 2005, 280(21): 20691–20699.

Liu, H., Chatel, S., Simard, C., Syam, N., Salle, L., Probst, V., Morel, J., Millat, G., Lopez, M., Abriel, H., Schott, J. J., Guinamard, R., Bouvagnet, P. Molecular genetics and functional anomalies in a series of 248 Brugada cases with 11 mutations in the TRPM4 channel. *PLoS One*, 2013, 8(1): e54131.

Liu, H., El Zein, L., Kruse, M., Guinamard, R., Beckmann, A., Bozio, A., Kurtbay, G., Megarbane, A., Ohmert, I., Blaysat, G., Villain, E., Pongs, O., Bouvagnet, P. Gain-of-function mutations in TRPM4 cause autosomal dominant isolated cardiac conduction disease. *Circ Cardiovasc Genet*, 2010, 3(4): 374–385.

Lopez, F., Delgado, R., Lopez, R., Bacigalupo, J., Restrepo, D. Transduction for pheromones in the main olfactory epithelium is mediated by the Ca^{2+}-activated channel TRPM5. *J Neurosci*, 2014, 34(9): 3268–3278.

Marty, A., Tan, Y. P., Trautmann, A. Three types of calcium-dependent channel in rat lacrimal glands. *J Physiol*, 1984, 357: 293–325.

Maruyama, Y., Moore, D., Petersen, O. H. Calcium-activated cation channel in rat thyroid follicular cells. *Biochim Biophys Acta*, 1985, 821(2): 229–232.

Maruyama, Y., Petersen, O. H. Single calcium-dependent cation channels in mouse pancreatic acinar cells. *J Membr Biol*, 1984, 81(1): 83–87.

Maruyama, Y., Peterson, O. H. Single-channel currents in isolated patches of plasma membrane from basal surface of pancreatic acini. *Nature*, 1982, 299(5879): 159–161.

Mathar, I., Jacobs, G., Kecskes, M., Menigoz, A., Philippaert, K., Vennekens, R. Trpm4. *Handb Exp Pharmacol*, 2014, 222: 461–487.

Matsunaga, H., Yamashita, N., Miyajima, Y., Okuda, T., Chang, H., Ogata, E., Kurokawa, K. Ion channel activities of cultured rat mesangial cells. *Am J Physiol*, 1991, 261(5 Pt 2): F808–F814.

McKemy, D. (2015). TRPM channels. *Handbook of Ion Channels*. J. Zheng and M. Trudeau. Boca Raton, Fl, CRC press.

Mironov, S. L. Metabotropic glutamate receptors activate dendritic calcium waves and TRPM channels which drive rhythmic respiratory patterns in mice. *J Physiol*, 2008, 586(9): 2277–2291.

Mironov, S. L. Calmodulin and calmodulin kinase II mediate emergent bursting activity in the brainstem respiratory network (preBotzinger complex). *J Physiol*, 2013, 591(7): 1613–1630.

Morita, H., Honda, A., Inoue, R., Ito, Y., Abe, K., Nelson, M. T., Brayden, J. E. Membrane stretch-induced activation of a TRPM4-like nonselective cation channel in cerebral artery myocytes. *J Pharmacol Sci*, 2007, 103(4): 417–426.

Nilius, B., Mahieu, F., Prenen, J., Janssens, A., Owsianik, G., Vennekens, R., Voets, T. The Ca^{2+}-activated cation channel TRPM4 is regulated by phosphatidylinositol 4,5-biphosphate. *EMBO J*, 2006, 25(3): 467–478.

Nilius, B., Prenen, J., Droogmans, G., Voets, T., Vennekens, R., Freichel, M., Wissenbach, U., Flockerzi, V. Voltage dependence of the Ca^{2+}-activated cation channel TRPM4. *J Biol Chem*, 2003, 278(33): 30813–30820.

Nilius, B., Prenen, J., Janssens, A., Voets, T., Droogmans, G. Decavanadate modulates gating of TRPM4 cation channels. *J Physiol*, 2004, 560(Pt 3): 753–765.

Nilius, B., Prenen, J., Tang, J., Wang, C., Owsianik, G., Janssens, A., Voets, T., Zhu, M. X. Regulation of the Ca^{2+} sensitivity of the nonselective cation channel TRPM4. *J Biol Chem*, 2005, 280(8): 6423–6433.

Nilius, B., Prenen, J., Voets, T., Droogmans, G. Intracellular nucleotides and polyamines inhibit the Ca^{2+}-activated cation channel TRPM4b. *Pflugers Arch*, 2004, 448(1): 70–75.

Nilius, B., Talavera, K., Owsianik, G., Prenen, J., Droogmans, G., Voets, T. Gating of TRP channels: A voltage connection?. *J Physiol* 2005, 567(Pt 1): 35–44.

Nonaka, T., Matsuzaki, K., Kawahara, K., Suzuki, K., Hoshino, M. Monovalent cation selective channel in the apical membrane of rat inner medullary collecting duct cells in primary culture. *Biochim Biophys Acta*, 1995, 1233(2): 163-174.

Oike, H., Wakamori, M., Mori, Y., Nakanishi, H., Taguchi, R., Misaka, T., Matsumoto, I., Abe, K. Arachidonic acid can function as a signaling modulator by activating the TRPM5 cation channel in taste receptor cells. *Biochim Biophys Acta*, 2006, 1761(9): 1078–1084.

Omasits, U., Ahrens, C. H., Muller, S., Wollscheid, B. Protter: Interactive protein feature visualization and integration with experimental proteomic data. *Bioinformatics*, 2014, 30(6): 884–886.

Ono, S., Mougouris, T., DuBose, T. D. Jr, Sansom, S. C. ATP and calcium modulation of nonselective cation channels in IMCD cells. *Am J Physiol*, 1994, 267(4 Pt 2): F558-F565.

Ozhathil, L. C., Delalande, C., Bianchi, B., Nemeth, G., Kappel, S., Thomet, U., Ross-Kaschitza, D., Simonin, C., Rubin, M., Gertsch, J., Lochner, M., Peinelt, C., Reymond, J. L., Abriel, H. Identification of potent and selective small molecule inhibitors of the cation channel TRPM4. *Br J Pharmacol*, 2018, 175(12): 2504–2519.

Partridge, L. D., Swandulla, D. Single Ca-activated cation channels in bursting neurons of helix. *Pflugers Arch*, 1987, 410(6): 627–631.

Partridge, L. D., Swandulla, D. Calcium-activated non-specific cation channels. *Trends Neurosci*, 1988, 11(2): 69–72.

Perez, C. A., Huang, L., Rong, M., Kozak, J. A., Preuss, A. K., Zhang, H., Max, M., Margolskee, R. F. A transient receptor potential channel expressed in taste receptor cells. *Nat Neurosci*, 2002, 5(11): 1169–1176.

Petersen, O. H., Maruyama, Y. What is the mechanism of the calcium influx to pancreatic acinar cells evoked by secretagogues?. *Pflugers Arch* 1983, 396(1): 82–84.

Pezza, R. J., Villarreal, M. A., Montich, G. G., Argarana, C. E. Vanadate inhibits the ATPase activity and DNA binding capability of bacterial MutS. a structural model for the vanadate-MutS interaction at the walker a motif. *Nucleic Acids Res*, 2002, 30(21): 4700–4708.

Piao, H., Takahashi, K., Yamaguchi, Y., Wang, C., Liu, K., Naruse, K. Transient receptor potential melastatin-4 is involved in hypoxia-reoxygenation injury in the cardiomyocytes. *PLoS One*, 2015, 10(4): e0121703.

Prawitt, D., Enklaar, T., Klemm, G., Gartner, B., Spangenberg, C., Winterpacht, A., Higgins, M., Pelletier, J., Zabel, B. Identification and characterization of MTR1, a novel gene with homology to melastatin (MLSN1) and the trp gene family located in the BWS-WT2 critical region on chromosome 11p15.5 and showing allele-specific expression. *Hum Mol Genet*, 2000, 9(2): 203–216.

Prawitt, D., Monteilh-Zoller, M. K., Brixel, L., Spangenberg, C., Zabel, B., Fleig, A., Penner, R. TRPM5 is a transient Ca^{2+}-activated cation channel responding to rapid changes in $[Ca^{2+}]i$. *Proc Natl Acad Sci U S A*, 2003, 100(25): 15166–15171.

Rae, J. L., Levis, R. A. Patch clamp recordings from the epithelium of the Lens obtained using glasses selected for low noise and improved sealing properties. *Biophys J*, 1984, 45(1): 144–146.

Reuter, H. Ion channels in cardiac cell membranes. *Annu Rev Physiol*, 1984, 46: 473–484.

Riquelme, D., Silva, I., Philp, A. M., Huidobro-Toro, J. P., Cerda, O., Trimmer, J. S., Leiva-Salcedo, E. Subcellular localization and activity of TRPM4 in medial prefrontal cortex layer 2/3. *Front Cell Neurosci*, 2018, 12: 12.

Rohacs, T., Lopes, C. M., Michailidis, I., Logothetis, D. E. PI(4,5)P2 regulates the activation and desensitization of TRPM8 channels through the TRP domain. *Nat Neurosci*, 2005, 8(5): 626–634.

Ruan, Z., Haley, E., Orozco, I. J., Sabat, M., Myers, R., Roth, R., Du, J., Lu, W. Structures of the TRPM5 channel elucidate mechanisms of activation and inhibition. *Nat Struct Mol Biol*, 2021, 28(7): 604–613.

Runnels, L. W., Yue, L., Clapham, D. E. The TRPM7 channel is inactivated by PIP(2) hydrolysis. *Nat Cell Biol*, 2002, 4(5): 329–336.

Sala-Rabanal, M., Wang, S., Nichols, C. G. On potential interactions between non-selective cation channel TRPM4 and sulfonylurea receptor SUR1. *J Biol Chem*, 2012, 287(12): 8746–8756.

Salkoff, L. B., Wyman, R. J. Ion currents in Drosophila flight muscles. *J Physiol*, 1983, 337: 687–709.

Sasse, P., Zhang, J., Cleemann, L., Morad, M., Hescheler, J., Fleischmann, B. K. Intracellular $Ca2^{+}$ oscillations, a potential pacemaking mechanism in early embryonic heart cells. *J Gen Physiol*, 2007, 130(2): 133–144.

Sheu, S. S., Blaustein, M. P. Effects of calcium on membrane potential and sodium influx in barnacle muscle fibers. *Am J Physiol*, 1983, 244(3): C297–C302.

Shin, J. H., Kim, Y. S., Linden, D. J. Dendritic glutamate release produces autocrine activation of mGluR1 in cerebellar Purkinje cells. *Proc Natl Acad Sci U S A*, 2008, 105(2): 746–750.

Siemen, D., Reuhl, T. Non-selective cationic channel in primary cultured cells of brown adipose tissue. *Pflugers Arch*, 1987, 408(5): 534–536.

Sigworth, F. J. Voltage gating of ion channels. *Q Rev Biophys*, 1994, 27(1): 1–40.

Simard, C., Hof, T., Keddache, Z., Launay, P., Guinamard, R. The TRPM4 non-selective cation channel contributes to the mammalian atrial action potential. *J Mol Cell Cardiol*, 2013, 59: 11–19.

Simon, F., Leiva-Salcedo, E., Armisen, R., Riveros, A., Cerda, O., Varela, D., Eguiguren, A. L., Olivero, P., Stutzin, A. Hydrogen peroxide removes TRPM4 current desensitization conferring increased vulnerability to necrotic cell death. *J Biol Chem*, 2010, 285(48): 37150–37158.

Simonneau, M., Distasi, C., Tauc, L., Barbin, G. Potassium channels in mouse neonate dorsal root ganglion cells: A patch-clamp study. *Brain Res*, 1987, 412(2): 224–232.

Sipido, K. R., Callewaert, G., Porciatti, F., Vereecke, J., Carmeliet, E. $[Ca^{2+}]$i-dependent membrane currents in guinea-pig ventricular cells in the absence of Na/Ca exchange. *Pflugers Arch*, 1995, 430(5): 871–878.

Stallmeyer, B., Zumhagen, S., Denjoy, I., Duthoit, G., Hebert, J. L., Ferrer, X., Maugenre, S., Schmitz, W., Kirchhefer, U., Schulze-Bahr, E., Guicheney, P., Schulze-Bahr, E. Mutational spectrum in the Ca^{2+}–activated cation channel Gene TRPM4 in patients with cardiac conductance disturbances. *Hum Mutat*, 2012, 33(1): 109–117.

Sturgess, N. C., Hales, C. N., Ashford, M. L. Inhibition of a calcium-activated, non-selective cation channel, in a rat insulinoma cell line, by adenine derivatives. *FEBS Lett*, 1986, 208(2): 397–400.

Swandulla, D., Lux, H. D. Activation of a nonspecific cation conductance by intracellular Ca^{2+} elevation in bursting pacemaker neurons of Helix pomatia. *J Neurophysiol*, 1985, 54(6): 1430–1443.

Syam, N., Rougier, J. S., Abriel, H. Glycosylation of TRPM4 and TRPM5 channels: Molecular determinants and functional aspects. *Front Cell Neurosci*, 2014, 8: 52.

Szalai, G., Csordas, G., Hantash, B. M., Thomas, A. P., Hajnoczky, G. Calcium signal transmission between ryanodine receptors and mitochondria. *J Biol Chem*, 2000, 275(20): 15305–15313.

Takezawa, R., Cheng, H., Beck, A., Ishikawa, J., Launay, P., Kubota, H., Kinet, J. P., Fleig, A., Yamada, T., Penner, R. A pyrazole derivative potently inhibits lymphocyte Ca^{2+} influx and cytokine production by facilitating transient receptor potential melastatin 4 channel activity. *Mol Pharmacol*, 2006, 69(4): 1413–1420.

Talavera, K., Nilius, B., Voets, T. Neuronal TRP channels: Thermometers, pathfinders and life-savers. *Trends Neurosci*, 2008, 31(6): 287–295.

Talavera, K., Yasumatsu, K., Voets, T., Droogmans, G., Shigemura, N., Ninomiya, Y., Margolskee, R. F., Nilius, B. Heat activation of TRPM5 underlies thermal sensitivity of sweet taste. *Nature*, 2005, 438(7070): 1022–1025.

Teruyama, R., Sakuraba, M., Kurotaki, H., Armstrong, W. E. Transient receptor potential channel m4 and m5 in magnocellular cells in rat supraoptic and paraventricular nuclei. *J Neuroendocrinol*, 2011, 23(12): 1204–1213.

Teulon, J., Paulais, M., Bouthier, M. A Ca2+-activated cation-selective channel in the basolateral membrane of the cortical thick ascending limb of Henle's loop of the mouse. *Biochim Biophys Acta*, 1987, 905(1): 125–132.

Thorn, P., Petersen, O. H. Activation of nonselective cation channels by physiological cholecystokinin concentrations in mouse pancreatic acinar cells. *J Gen Physiol*, 1992, 100(1): 11–25.

Toyoshima, C., Nakasako, M., Nomura, H., Ogawa, H. Crystal structure of the calcium pump of sarcoplasmic reticulum at 2.6 A resolution. *Nature*, 2000, 405(6787): 647–655.

Uchida, K., Tominaga, M. Extracellular zinc ion regulates transient receptor potential melastatin 5 (TRPM5) channel activation through its interaction with a pore loop domain. *J Biol Chem*, 2013, 288(36): 25950–25955.

Uhlen, M., Fagerberg, L., Hallstrom, B. M., Lindskog, C., Oksvold, P., Mardinoglu, A., Sivertsson, A., Kampf, C., Sjostedt, E., Asplund, A., Olsson, I., Edlund, K., Lundberg, E., Navani, S., Szigyarto, C. A., Odeberg, J., Djureinovic, D., Takanen, J. O., Hober, S., Alm, T., Edqvist, P. H., Berling, H., Tegel, H., Mulder, J., Rockberg, J., Nilsson, P., Schwenk, J. M., Hamsten, M., von Feilitzen, K., Forsberg, M., Persson, L.,

Johansson, F., Zwahlen, M., von Heijne, G., Nielsen, J., Ponten, F. Proteomics. Tissue-based map of the human proteome. *Science*, 2015, 347(6220): 1260419.

Ullrich, N. D., Voets, T., Prenen, J., Vennekens, R., Talavera, K., Droogmans, G., Nilius, B. Comparison of functional properties of the Ca2+-activated cation channels TRPM4 and TRPM5 from mice. *Cell Calcium*, 2005, 37(3): 267–278.

Vennekens, R., Nilius, B. (2007). Insights into TRPM4 function, regulation and physiological role." *Handb Exp Pharmacol*, 2007, 179: 269285.

Voets, T., Owsianik, G., Janssens, A., Talavera, K., Nilius, B. TRPM8 voltage sensor mutants reveal a mechanism for integrating thermal and chemical stimuli. *Nat Chem Biol*, 2007, 3(3): 174–182.

von Tscharner, V., Prod'hom, B., Baggiolini, M., Reuter, H. Ion channels in human neutrophils activated by a rise in free cytosolic calcium concentration. *Nature*, 1986, 324(6095): 369–372.

Wang, H., Xu, Z., Lee, B. H., Vu, S., Hu, L., Lee, M., Bu, D., Cao, X., Hwang, S., Yang, Y., Zheng, J., Lin, Z. Gain-of-function mutations in TRPM4 activation gate cause progressive symmetric erythrokeratodermia. *J Invest Dermatol*, 2019, 139(5): 1089–1097.

Winkler, P. A., Huang, Y., Sun, W., Du, J., Lu, W. Electron cryo-microscopy structure of a human TRPM4 channel. *Nature*, 2017, 552(7684): 200–204.

Woo, S. K., Kwon, M. S., Ivanov, A., Geng, Z., Gerzanich, V., Simard, J. M. Complex N-glycosylation stabilizes surface expression of transient receptor potential melastatin 4b protein. *J Biol Chem*, 2013, 288(51): 36409–36417.

Xu, X. Z., Moebius, F., Gill, D. L., Montell, C. Regulation of melastatin, a TRP-related protein, through interaction with a cytoplasmic isoform. *Proc Natl Acad Sci U S A*, 2001, 98(19): 10692–10697.

Yamaguchi, S., Tanimoto, A., Iwasa, S., Otsuguro, K. I. TRPM4 and TRPM5 channels share crucial amino acid residues for Ca^{2+} sensitivity but not significance of PI(4,5)P2. *Int J Mol Sci*, 2019, 20(8): 2012.

Yamaguchi, S., Tanimoto, A., Otsuguro, K., Hibino, H., Ito, S. "Negatively charged amino acids near and in transient receptor potential (TRP) domain of TRPM4 channel are one determinant of its $Ca2^{+}$ sensitivity. *J Biol Chem*, 2014, 289(51): 35265–35282.

Yan, J., Bengtson, C. P., Buchthal, B., Hagenston, A. M., Bading, H. Coupling of NMDA receptors and TRPM4 guides discovery of unconventional neuroprotectants. *Science*, 2020, 370(6513): eaay3302.

Yang, F., Vu, S., Yarov-Yarovoy, V., Zheng, J. Rational design and validation of a vanilloid-sensitive TRPV2 ion channel. *Proc Natl Acad Sci U S A*, 2016, 113(26): E3657–E3666.

Yang, F., Xu, L., Lee, B. H., Xiao, X., Yarov-Yarovoy, V., Zheng, J. An unorthodox mechanism underlying voltage sensitivity of TRPV1 ion channel. *Adv Sci (Weinh)*, 2020, 7(20): 2000575.

Yang, F., Zheng, J. High temperature sensitivity is intrinsic to voltage-gated potassium channels. *Elife*, 2014, 3: e03255.

Yarishkin, O. V., Hwang, E. M., Park, J. Y., Kang, D., Han, J., Hong, S. G. Endogenous TRPM4-like channel in Chinese hamster ovary (CHO) cells. *Biochem Biophys Res Commun*, 2008, 369(2): 712–717.

Yellen, G. Single Ca^{2+}-activated nonselective cation channels in neuroblastoma. *Nature*, 1982, 296(5855): 357–359.

Yin, Y., Le, S. C., Hsu, A. L., Borgnia, M. J., Yang, H., Lee, S. Y. Structural basis of cooling agent and lipid sensing by the cold-activated TRPM8 channel. *Science*, 2019, 363(6430): eaav9334.

Yoo, J. C., Yarishkin, O. V., Hwang, E. M., Kim, E., Kim, D. G., Park, N., Hong, S. G., Park, J. Y. Cloning and characterization of rat transient receptor potential-melastatin 4 (TRPM4). *Biochem Biophys Res Commun*, 2010, 391(1): 806–811.

Zhang, F., Swartz, K. J., Jara-Oseguera, A. Conserved allosteric pathways for activation of TRPV3 revealed through engineering vanilloid-sensitivity. *Elife*, 2019 8:e42756.

Zhang, Y., Hoon, M. A., Chandrashekar, J., Mueller, K. L., Cook, B., Wu, D., Zuker, C. S., Ryba, N. J. Coding of sweet, bitter, and umami tastes: Different receptor cells sharing similar signaling pathways. *Cell*, 2003, 112(3): 293–301.

Zhang, Z., Okawa, H., Wang, Y., Liman, E. R. Phosphatidylinositol 4,5-bisphosphate rescues TRPM4 channels from desensitization. *J Biol Chem*, 2005, 280(47): 39185–39192.

Zheng, J. Molecular mechanism of TRP channels. *Compr Physiol*, 2013, 3(1): 221–242.

Zheng, J., Trudeau, M. C., ed. (2023). *Textbook of Ion Channels: Three Volume Set*. Boca Raton, FL, CRC Press.

Zitt, C., Strauss, B., Schwarz, E. C., Spaeth, N., Rast, G., Hatzelmann, A., Hoth, M. Potent inhibition of Ca^{2+} release-activated Ca^{2+} channels and T-lymphocyte activation by the pyrazole derivative BTP2. *J Biol Chem*, 2004, 279(13): 12427–12437.

7 Physiological and Pathological Functions of TRPML1

Mengnan Xu, Rose Yang Dong, Pingping Wang, Elisa Dai, and Xian-Ping Dong

7.1 INTRODUCTION

The lysosome is a single-membrane organelle that contains an acidic lumen and 60 or so hydrolytic enzymes. It has been considered to be the waste disposal and recycling center of a cell since its discovery by Christian de Duve in 1955 [1–3]. The lysosome not only degrades materials taken up from outside the cell through the endocytic membrane transport pathway but also digests obsolete components of the cell itself through the autophagic pathway [4–6]. Depending on the cell type and conditions, one mammalian cell could have several hundred lysosomes, making up as a high as 5% of the cellular volume. These lysosomes display great heterogeneity in size, morphology, and function [7–9].

In addition to the role of degradation, the lysosome is also considered a multifunctional signaling hub [10–13]. For example, the cytosolic surface of the limiting membrane of the lysosome serves as a platform for the activation of the mammalian target of rapamycin complex 1 (mTORC1), a serine threonine protein kinase present in all eukaryotic organisms. By phosphorylating various cell growth-related substrates, such as ribosomal S6 kinase 1 (S6K1) [14], eukaryotic translation initiation factor 4E (eIF4E)-binding protein 1 (4E-BP1) [15, 16], and the transcription factor EB (TFEB), it acts as the master nutrient sensor and plays a crucial role in cell proliferation, growth, and metabolism in response to a diverse set of environmental inputs [17, 18]. The lysosome also dynamically interacts with other organelles and membrane compartments such as the endoplasmic reticulum (ER) [19], mitochondria [20], and the plasma membrane (PM) [21] to regulate cellular homeostasis.

The membrane of lysosomes contains a variety of ion channels and transporters. Their activation allows the movement of ions across its membrane, thereby regulating the ionic composition within the compartment and the overall functionality and homeostasis of the lysosome [8, 22]. Transient receptor potential mucolipin 1 (TRPML1) is one of the major and most understood lysosomal ion channels. Dysfunction of TRPML1 has been correlated with various diseases, such as lysosomal storage disorders (LSDs) [23, 24], neurodegenerative disorders, and cancers [25, 26]. This chapter endeavors to summarize and discuss the physiological and pathological functions of TRPML1.

7.2 LYSOSOMAL ION HOMEOSTASIS

7.2.1 Lysosomal Ions

The lysosome serves as a reservoir for various ions, including hydrogen (H^+), calcium (Ca^{2+}), sodium (Na^+), potassium (K^+), iron (Fe^{2+}), zinc (Zn^{2+}), and chloride (Cl^-) ions [8]. As the end point of the endocytic, phagocytic, and autophagic pathways, dynamic membrane transport in these pathways is accompanied by changes in ion concentrations within the lysosomal lumen. For example, the concentration of H^+ continuously increases during endosome maturation and late endosome (LE)-lysosome fusion, establishing an essential acidic environment (pH of ~4.5–5.5) for hydrolytic enzyme function [8], in stark contrast to the neutral environment (pH ~7.2) of the cytosol. As a ubiquitous second messenger, lysosomal Ca^{2+} has received much attention in the past years. The lysosome has been suggested to be an important Ca^{2+} store, with its free Ca^{2+} concentration estimated to be ~0.5 mM [27, 28], which is almost 5000 times higher than that in the

DOI: 10.1201/9781003402732-7

cytosol [29]. This significant Ca^{2+} gradient is thought to be established by an as-yet-unidentified Ca^{2+} transporter or Ca^{2+}/H^+ exchanger [8, 22, 30–33]. Lysosomal Ca^{2+} can be released via various Ca^{2+} channels to regulate a wide range of cellular processes, such as membrane trafficking and gene expression [8]. In addition, lysosomal Na^+ and K^+ concentrations have been estimated to be ~100 mM and ~10 mM, respectively, which are much different from their cytosolic concentrations (~12 mM for Na^+ and ~150 mM for K^+) [34, 35]. The lysosome also stores a variety of trace metals, including millimolar concentrations of Zn^{2+} and Fe^{2+} [36–39]. Cl^- is the major anion identified in the lysosome, with a concentration of > 80 mM, which is ~10 fold higher than Cl^- concentration in the cytosol (5–10 mM) [40–43].

7.2.2 Lysosomal Ion Channels and Transporters

Lysosomal ion homeostasis is maintained by lysosomal ion channels and transporters. The study of lysosomal ion channels and transporters has been relatively limited compared to those on the PM due to the inaccessibility of intracellular membranes. In recent years, advancements in biomedical methodologies, especially lysosome-patch-clamp technology and genetic approaches, have led to the identification and characterization of several ion channels and transporters in the lysosome. These include the ATP-driven H^+ pump (V-ATPase) [37, 44], TRPMLs [39, 45–47], Two-pore Channels (TPCs) [48–54], P2X4 purinoceptor [55, 56], Transient Receptor Potential Melastatin 2 (TRPM2) [57, 58], Ca^{2+}-activated big-conductance K^+ channels (BK) [59, 60], H^+/K^+ permeable channel Transmembrane Protein 175 (TMEM175) [61–66], Cl^- channel CLN7 [67], Cl^-/H^+ exchanger CLC7 [68], and lysosomal volume-regulated anion channel (Lyso-VRAC) or Leucine-rich repeat-containing protein 8A (LRRC8) [69] (Fig. 7.1). These ion channels are not only important in maintaining

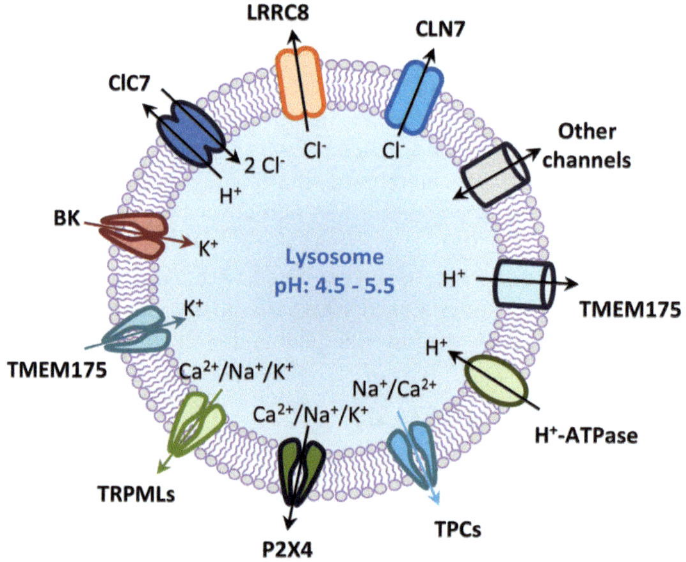

FIGURE 7.1 Lysosomal ion channels and transporters. Numerous ion channels and transporters have been identified in the limiting membrane of lysosomes. The luminal acidic environment of lysosomes is mainly established by V-ATPase. The "H^+ leak" channel TMEM175 prevents over acidification of the lumen. The homeostasis of lysosomal pH and ions is also coordinated by the activities of many other ion channels in lysosomes. These channels include nonselective cation channels such as TRPMLs and P2X4, Na^+/Ca^{2+}-selective channels such as TPCs, K^+-selective channel such as BK and TMEM175, Cl^-/H^+ exchanger ClC7, Cl^--selective channel CLN7, and volume-regulated anion channel LRRC8. Other channels such as TRPM2 and voltage-gated Ca^{2+} channels have also been suggested to be expressed in lysosomes.

lysosomal ion homeostasis at rest but also essential for cells to adapt to environmental changes by regulating ion flux upon different stimulations. Bear in mind that while these ion channels and transporters have been found in the lysosome, some of them, like TRPM2 and P2X4, are also localized in the PM and other subcellular compartments, suggesting a diverse function of these channels in the cell.

7.3 TRPML1

7.3.1 The TRPML Family

The TRPML family (TRPMLs, i.e., TRPML1-3) is a group of nonselective cation channels that are mainly expressed and function in endolysosomes [8, 22, 70, 71]. Although TRPMLs share ~75% similarity in amino acid sequence [72] and have similar structures [73], their physiological properties are different, which allows them to regulate different cellular processes. In mammals, TRPML1 is ubiquitously expressed, whereas the expressions of TRPML2 and TRPML3 are restricted to certain types of cells and tissues [70, 71, 74–78]. At the cellular level, TRPML1 predominantly resides in late endosomes (LEs) and lysosomes, which have a more acidic lumen [79–81]. On the other hand, TRPML2 and TRPML3 are also enriched in recycling endosomes (REs) [82–84] and early endosomes (EEs) [83, 85, 86], respectively, both of which have a relatively less acidic lumen [39, 79, 82]. All three TRPMLs are inwardly rectifying, conducting cation release from endolysosomes. In addition to Ca^{2+} and Na^{+} release [39, 87, 88], TRPML1 and TRPML2 also conduct the release of heavy trace metals, including Fe^{2+} and Zn^{2+} [39, 89]. Mutations in TRPML1 have been associated with the neurodegenerative Mucolipidosis type IV (ML-IV) disease [39, 70]. However, TRPML2 and TRPML3 have not been linked with any human diseases.

7.3.2 Characterization of TRPML1

The human disease relevance renders TRPML1 the subject of particularly extensive research within the realm of lysosomal ion channels. By using artificially enlarged lysosomes, *Dong et al.* first directly performed patch-clamp recording of TRPML1 [39]. Since then, lysosomal patch clamp technology has significantly facilitated more detailed studies on lysosomal ion channels [39]. It is now believed that TRPML1 is the major lysosomal channel conducting Na^{+}, K^{+}, Ca^{2+}, Fe^{2+}, and Zn^{2+} release from the lysosome to the cytosol [37] (Fig. 7.2).

In accordance with its subcellular localization, TRPML1 is activated by a low lysosomal luminal pH [90, 91] and phosphatidylinositol 3,5-bisphosphate [PI(3,5)P2] [46, 92], the main form of phosphoinositides localized in LEs and lysosomes. Recent studies have also suggested that reactive oxygen species (ROS) can activate TRPML1 [93]. On the other hand, TRPML1 is inhibited by sphingomyelin (SM) [94], phosphoinositides PI(4,5)P2, PI(3,4,5)P3, and PI(3,4)P2 [95], and intraluminal adenosine [96]. In addition, TRPML1 can be phosphorylated and inhibited by mTORC1 kinase. Nutrient starvation and other stress conditions when mTORC1 activity becomes low increase TRPML1 activity [97]. Moreover, TRPML1 undergoes proteolytic cleavage mediated by cathepsin B, resulting in its inactivation [98]. The palmitoylation at the C-terminal region of TRPML1 may also regulate TRPML1 internalization and its subsequent levels in the lysosomes [99]. These studies suggest that TRPML1 is regulated by several mechanisms under different cellular conditions (Table 7.1).

In addition to the endogenous factors that regulate TRPML1 activity, TRPML1 is also regulated by some synthetic small molecules as presented in Table 7.1. The application of these synthetic small molecules has significantly facilitated the study of the physiological and pathological roles of TRPML1.

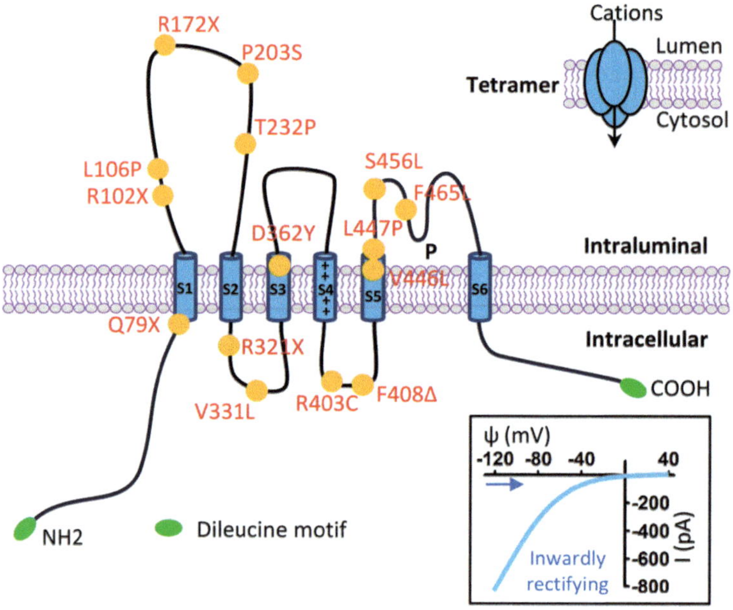

FIGURE 7.2 Topological structure and channel activity of TRPML1. Topology of TRPML1 pore forming subunit with the ML-IV causing mutations. Its lysosomal localization is determined by two dileucine motifs. Four subunits form a tetrameric channel in the lysosomal membrane, allowing cation release from the lysosomal lumen to the cytosol. TRPML1 currents are strongly inwardly rectifying.

TABLE 7.1
Endogenous Factors and Synthetic Small Molecules That Can Regulate TRPML1 Activity

Name	Activator or Inhibitor	Endogenous or Synthetic	Reference(s)
PI(3,5)P2	activator	endogenous	[46, 92]
PI(4,5)P2	inhibitor	endogenous	[95]
PI(3,4,5)P3	inhibitor	endogenous	[95]
PI(3,4)P2	inhibitor	endogenous	[95]
Lysosomal lumen SM	inhibitor	endogenous	[94]
Lysosomal adenosine	inhibitor	endogenous	[96]
ROS	activator	endogenous	[93]
Lysosomal luminal low pH	activator	endogenous	[90, 91]
mTORC1	inhibitor	endogenous	[97]
LAMTOR1	inhibitor	endogenous	[100]
Cathepsin B	inhibitor	endogenous	[98]
ML-SAs	activator	synthetic	[92, 94, 101–104]
ML1-SA1	activator	synthetic	[105]
MK6-83	activator	synthetic	[106, 107]
Rapamycin	activator	synthetic	[92, 108]
Temsirolimus	activator	synthetic	[92]
ML-SIs	inhibitor	synthetic	[102–104, 109–111]

7.3.3 Physiological Functions of TRPML1

Recent studies have suggested that TRPML1 plays a crucial role in various physiological processes such as cellular signaling, membrane trafficking, autophagy, lysosomal pH regulation, and lysosome crosstalk with the ER and mitochondria [38, 39, 101, 112, 113] (Fig. 7.3).

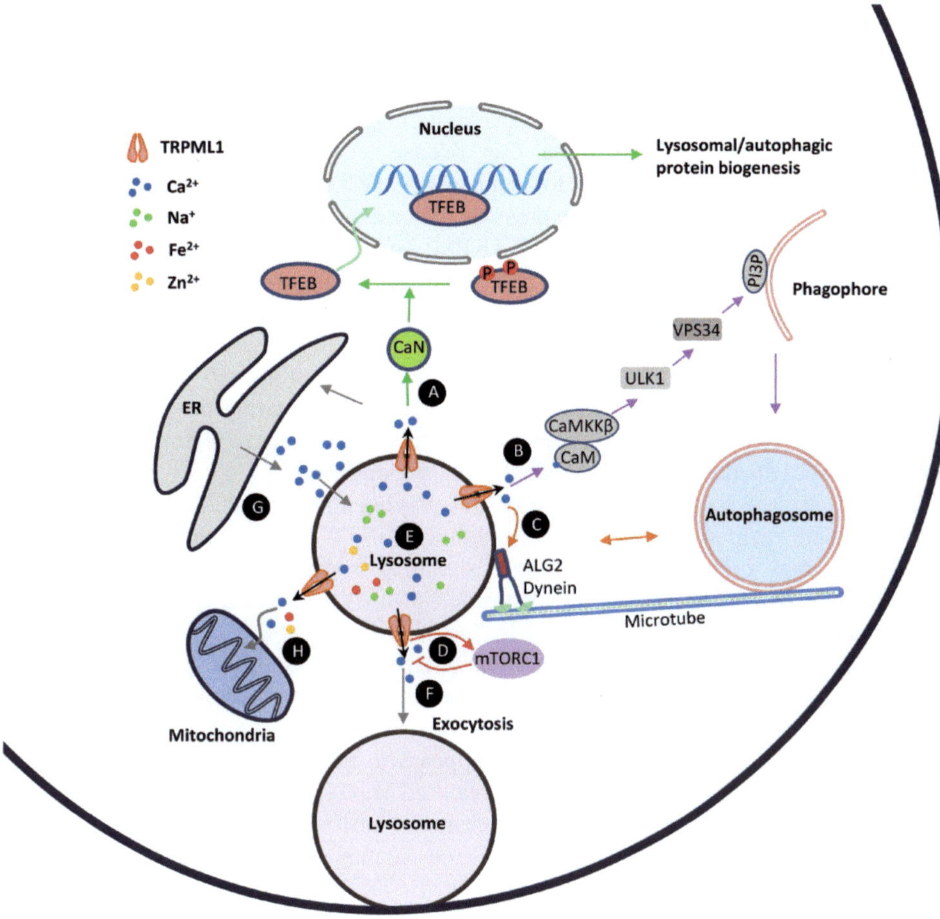

FIGURE 7.3 Physiological functions of TRPML1. (A) TRPML1 modulates autophagy by inducing TFEB nuclear translocation. TRPML1-mediated Ca^{2+} release promotes TFEB dephosphorylation by activating phosphatase calcineurin (CaN), resulting in TFEB nuclear translocation and consequent transcriptional activation of lysosomal and autophagic genes. (B) TRPML1 modulates autophagy by rapidly inducing autophagosome biogenesis. TRPML1-induced Ca^{2+} release triggers the activation of CaMKKβ signaling pathway. This activation further orchestrates the activation of two pivotal protein complexes, namely ULK1 and VPS34, to initiate autophagy. This cascade of events culminates in the production of PI3P and the subsequent recruitment of indispensable PI3P-binding proteins to the emerging phagophore. (C) TRPML1 modulates autophagy by regulating autophagosome-lysosome fusion. TRPML1 promotes Ca^{2+}-dependent centripetal movement of lysosomes toward the perinuclear region to fuse with autophagosomes by ALG-2. (D) TRPML1 and mTORC1 form a negative feedback loop to sustain mTORC1 activity during starvation. This is essential for autophagic lysosome reformation and cell survival. (E) TRPML1 activation promotes lysosomal degradation, potentially by facilitating lysosomal acidification. (F) TRPML1 activation promotes the release of lysosomal contents such as cellular waste, toxic materials, and signaling molecules via lysosomal exocytosis. (G) Ca^{2+} release from lysosomes via TRPML1 induces ER Ca^{2+} release, resulting in a global Ca^{2+} increase. (H) TRPML1 facilitates Ca^{2+} transfer from lysosomes to mitochondria at lysosome-mitochondria contacts. Lysosomal TRPML1-mediated Zn^{2+}/Fe^{2+} release is also involved in many cellular events.

7.3.3.1 Cellular Signaling

In recent years, the lysosome has garnered significant attention as a major intracellular Ca^{2+} store. Accordingly, TRPML1 has been suggested to act as a predominant Ca^{2+} release channel in lysosomes, thereby playing a pivotal role in both local and global Ca^{2+}-dependent signaling pathways. By releasing Ca^{2+} from lysosomes, TRPML1 contributes to the regulation of local Ca^{2+}-dependent signaling pathways, which are often mediated by the Ca^{2+} sensing protein calmodulin (CaM). This TRPML1-Ca^{2+}/CaM signaling pathway is particularly important for autophagy, a conserved degradation and recycling process of the cell that removes unnecessary or dysfunctional components through a lysosome-dependent regulated mechanism. Depending on the stages of autophagy, the downstream targets of the TRPML1-Ca^{2+}/CaM signaling pathway are different. At the transcriptional level, TRPML1-mediated lysosomal Ca^{2+} release activates the only known CaM-activated protein phosphatase, calcineurin (CaN), to dephosphorylate TFEB, leading to nuclear translocation and activation of TFEB and subsequent expression of numerous autolysosomal proteins [114]. At the post-transcriptional level, CaM activation by TRPML1-mediated lysosomal Ca^{2+} release promotes autophagy induction by evoking the CaM-dependent protein kinase kinase β *(*CaMKKβ*)* - AMP-activated protein kinase (AMPK) signaling pathway [106]. By stimulating the mTORC1 signaling pathway, CaM activation by TRPML1-mediated lysosomal Ca^{2+} release also promotes autophagic lysosome reformation, an important process for the cell to continuously generate lysosomes for the removal of newly formed autophagosomes [97, 115]. By regulating these signaling pathways, TRPML1 helps the cell adapt to conditions of nutrient scarcity. Please refer to Section 7.3.3.4. for the role of TRPML1 in global Ca^{2+}-dependent signaling pathways.

TRPML1 also functions as a senser of ROS [93], playing an important role in mitigating cellular oxidative stress by regulating the autophagy pathway [109]. TRPML1 activation also facilitates lysosomal exocytosis, promoting the release of signaling molecules that are stored in lysosomes, such as ATP (~1 mM) [56, 116, 117] and adenosine (~1 mM) [96], to the extracellular environment to activate related signaling pathways.

Zinc, one of the most abundant essential trace elements in mammals, is also reserved in lysosomes. By using a newly engineered GZnP3 sensor with sub-nanomolar sensitivity, a recent study suggested a direct role of TRPML1 in facilitating Zn^{2+} release from lysosomes to the cytosol [118]. Interestingly, TRPML1-mediated Zn^{2+} release appears to be unique in neurons [118]. Similarly, lysosomes store high levels of Fe^{2+} that can be released via TRPML1 to regulate intracellular Fe^{2+} homeostasis [39]. Given that Fe^{2+} acts as an essential transition metal for virtually all biological processes, disruptions in TRPML1-mediated Fe^{2+} release may give rise to a diverse range of human diseases [119], including iron deficiency anemia in ML-IV disease [39, 120, 121]. It remains to be determined if iron and zinc can function as signaling molecules.

7.3.3.2 Membrane Trafficking

It is believed that TRPML1 is one of the major lysosomal channels that functions as a regulator for both the fusion and fission of lysosomes. First, as a degradation center, the lysosome needs to frequently fuse with other cellular organelles that contain cargo [7, 8]. These organelles include endosomes that contain small materials from the extracellular environment, phagosomes that contain solid materials from the extracellular environment, such as bacteria, and autophagosomes that contain harmful intracellular materials engulfed from the cytosol, such as damaged mitochondria. After completing the digestion process, lysosomes undergo fission processes, which are key for their reformation and homeostasis [7, 122]. All these processes can be regulated by TRPML1-mediated Ca^{2+} signaling with the assistance of different Ca^{2+} sensors [33, 123–130]. For example, TRPML1-mediated lysosomal Ca^{2+} release activates the Ca^{2+}-binding protein apoptosis-linked gene 2 (ALG-2) to regulate lysosomal positioning, autophagosome-lysosome fusion, and lysosome tubulation and reformation through dynein [113].

As mentioned, the lysosome frequently exports its intraluminal cargos through fusing with the PM, a process termed lysosomal exocytosis. TRPML1 is believed to be a key regulator of lysosomal exocytosis [21]. By facilitating lysosomal exocytosis, TRPML1 has been shown to play an important role in multiple physiological processes, such as removing indigestible or toxic materials by exporting them to extracellular places [132, 133], delivering lysosomal membranes to phagocytic cups on the PM to help the ingestion of large particles [134, 135], and delivering intracellular vesicles to the sites of injury for membrane repair in skeletal muscles [136].

7.3.3.3 Autophagy Regulation

Autophagy is a lysosome-dependent "self-eating" process in which non-essential or harmful cytoplasmic components, such as damaged organelles and bacteria, are engulfed into autophagosomes [137, 138]. Autophagosomes then fuse with lysosomes, where the cargo is degraded by lysosomal hydrolases into small molecules that are then reused by the cells or exported outside of the cells. Autophagy plays a key role in the survival of the cell during challenging conditions such as nutrient deficiency, bacterial infection, oxidative stress, and some disease conditions. Basal autophagy is also important for a variety of physiological processes such as cell growth [139] and differentiation [140].

The autophagy pathway can be divided into four stages, including autophagosome biogenesis, autophagosome-lysosome fusion, autolysosomal degradation, and autophagic lysosome reformation [137, 138]. All four steps contribute to "autophagic flux", an essential concept in autophagy research emphasizing the continuous and overall flow of materials through the autophagy pathway [137, 138]. Accumulative evidence has suggested that TRPML1 appears to be an important regulator of all four stages of autophagy. By using genetic and pharmacological approaches to modulate TRPML1 activity, a recent study demonstrated that TRPML1-mediated lysosomal Ca^{2+} release activates CaMKKβ and AMPK. This subsequently activates the Unc-51-like autophagy activating kinase 1 (ULK1) and vacuolar protein sorting 34 (VPS34) protein complexes, two essential protein complexes for autophagy initiation [141], to facilitate autophagosome formation [106]. In the same study, the authors suggested that acute TRPML1 activation increases the fusion between autophagosomes and lysosomes [106]. This is in agreement with a previous report showing that TRPML1-mediated lysosomal Ca^{2+} release activates ALG-2, an EF-hand-containing protein in lysosomes, to promote centripetal movement of lysosomes towards the perinuclear region and subsequent fusion of lysosomes and autophagosomes [113]. A role of TRPML1 in autophagic degradation was suggested by an early study showing that TRPML1 is required for proteolytic degradation in autolysosomes, thus regulating autophagic flux [142]. Lastly, TRPML1 activation upon starvation helps to retain mTORC1 in the lysosome by interacting with CaM, where luminal nutrients activate mTORC1 to prevent excess reduction of mTORC1 activity, thereby assisting autophagic lysosome reformation [97].

In addition to these post-transcriptional effects of TRPML1 on autophagy, TRPML1 also sustains autophagic flux at a transcriptional level. For this to occur, lysosomal Ca^{2+} release through TRPML1 activates CaN to dephosphorylate TFEB. This induces TFEB nuclear translocation and consequent transcriptional activation of lysosomal and autophagic genes, resulting in increased autophagic flux by augmenting lysosome biogenesis [114, 144]. Because TRPML1 is also upregulated by TFEB during starvation, TRPML1 and TFEB form a feedback loop to accelerate the expression of lysosomal and autophagic proteins, greatly facilitating autophagic flux in response to nutrient depletion [114, 145].

In summary, to maintain a high autophagic flux in response to acute or long-term nutrient starvation, TRPML1 plays a multifaceted role in autophagy to help cells adapt to adverse environments and survive extreme conditions. These include CaMKKβ-dependent autophagy induction, ALG2-dependent autophagosome-lysosome fusion, autolysosome degradation, autophagic lysosome reformation, and TFEB-dependent autophagy-related gene expression.

7.3.3.4 Lysosomal Acidification

Lysosomal H^+ accumulation is mainly maintained by V-ATPase, which pumps H^+ into the lysosome to promote its lumen acidification [44, 146, 147]. Because unidirectional pumping of H^+ into the lysosomal lumen across the membrane by the V-ATPase can establish an inside positive lysosomal membrane potential that suppresses the V-ATPase activity, a compensatory influx of anions such as Cl^- or efflux of cations such as Ca^{2+}, Na^+, and K^+ is needed to neutralize such membrane potential [5, 148, 149]. It was first suggested that the lysosomal $2Cl^-/H^+$ exchanger ClC7 mediated Cl^- influx into lysosomes to dissipate the membrane potential caused by V-ATPase [150]. However, loss of ClC7 does not cause an apparent change in lysosomal pH [151–154]. A later study suggested that cation efflux instead of anion influx is essential for lysosome acidification [149]. In line with this, both TPC2 [51] and TRPML1 have been suggested to play an important role in lysosomal acidification [155–157] and degradation [142, 158–160], likely under stress conditions when mTORC1 is suppressed because both channels are phosphorylated and inhibited by mTORC1. Recently, TMEM175 is suggested to function as a lysosomal "H^+ leak" channel to prevent over-acidification when lysosomal pH is below the normal range [62]. Interestingly, lysosomal-associated membrane proteins (LAMP) can bind to and inhibit the TMEM175 channel, regulating lysosomal pH [161]. Thus, lysosomal pH optimum is maintained by the collaboration of V-ATPase, TMEM175 and LAMP proteins. Upon stress, other ion channels such as TPC2 and TRPML1 are activated to regulate lysosomal pH, helping the cell adapt to environmental changes.

7.3.3.5 Lysosomal Cross Talk with the ER and Mitochondria

In addition to the role of TRPML1 in controlling local Ca^{2+} signaling, TRPML1-mediated lysosomal Ca^{2+} release can evoke global Ca^{2+} signals [162]. This is likely to occur at lysosome-ER membrane contact sites where TRPML1 releases Ca^{2+} from lysosomes to induce ER Ca^{2+} release using a mechanism similar to Ca^{2+}-induced Ca^{2+} release (CICR) [163]. This results in Ca^{2+} depletion in the ER, further promoting extracellular Ca^{2+} influx via store-operated Ca^{2+} entry (SOCE) to refill the ER with Ca^{2+} [164]. In agreement with this, TRPML1 channels are critical for the generation of RyR2-mediated Ca^{2+} sparks in bladder smooth muscles, bladder smooth muscle contractility, and normal micturition [165].

Emerging evidence also suggests physical contact between lysosomes and mitochondria [166]. Interestingly, *Peng et al.* recently suggested that TRPML1 regulates the transfer of Ca^{2+} from lysosomes to mitochondria at lysosome-mitochondrion contact sites [112]. Because lysosomal Fe^{2+} can be translocated to mitochondria [167–170] and TRPML1 functions as a lysosomal Fe^{2+} release channel [39], TRPML1 may also assist in the Fe^{2+} transfer between lysosomes and mitochondria.

7.3.4 Pathological Functions of TRPML1

Because TRPML1 is involved in a variety of important cellular processes, it is not surprising that TRPML1 dysregulation is involved in many pathological conditions, such as neurodegeneration, cancer, and infection.

7.3.4.1 Neurodegeneration

Neurodegenerative disorders represent a group of debilitating conditions characterized by the progressive loss of structure and function of neurons in the central nervous system, resulting in the loss of both motor and cognitive functions [171, 172]. These disorders affect millions of people worldwide, posing a serious burden on society and healthcare systems, especially within aging populations due to increasing incidence rates. Alzheimer's disease (AD) stands as the most prevalent type of neurodegenerative disorder, followed by Parkinson's disease (PD), Huntington's disease (HD), amyotrophic lateral sclerosis (ALS), and multiple sclerosis (MS) [173–177]. Many neurodegenerative disorders are associated with the accumulation of protein aggregates, such as intraneuronal hyperphosphorylated Tau (in AD) or α-Synuclein (in PD).

Although neurodegenerative diseases are prevalent in the elderly, in rare cases, they can also affect children. Lysosomal storage disorders (LSDs) are the most common neurodegenerative disorders in children [178]. They encompass a group of inherited metabolic disorders that are characterized by an abnormal build-up of various toxic materials in the cell as a result of deficiency in specific proteins integral to lysosomal function, such as enzymes and lysosomal membrane proteins. For example, ML-IV is caused by mutations in the TRPML1 channel and is characterized by the storage of mucopolysaccharides and lipids; Niemann-Pick disease type C1 (NPC1) is caused by mutations in the cholesterol transporter NPC1 in the lysosomal membrane and is characterized by the storage of cholesterol and sphingolipids; Fabry disease is caused by the deficiency of α-galactosidase-A and is characterized by the storage of globotriaosylceramide [179]. Most LSDs are autosomal recessively inherited, but a few are X-linked recessively inherited [180]. In addition to the nervous system, LSDs may affect many other parts of the body, such as the skeleton, skin, skeletal muscle, and heart [178, 180, 181]. To date, more than 70 LSDs have been identified [178]. Although individually rare, collectively they affect about 1 in 5,000–7,500 live births, making them a relatively common morbidity [181].

During the past decades, overwhelming evidence has suggested that abnormal membrane trafficking, autophagy, and lysosomal pH contribute to neurodegenerative disorders. Since these biological processes are closely regulated by TRPML1, the role of TRPML1 in neurodegenerative disorders has received considerable attention.

As mentioned above, mutations in TRPML1 are the direct cause of ML-IV [39, 70, 182], an LSD with neurodegeneration. These mutations include missense mutations, small deletions, small insertions, gross deletions/insertions, or aberrant splicing mutations [182]. While many of the mutations lead to the loss of TRPML1, some mutations produce TRPML1 with reduced function, causing mild symptoms. Most ML-IV patients develop severe motor defects, visual impairment, intellectual disability, limited speech, iron deficiency, and impaired production of stomach acids. At the subcellular level, excessive lysosomal storage is commonly observed in patients [183, 184], suggesting a defect in TRPML1-mediated lysosomal trafficking and lysosomal degradation function as described in Section 7.3.3.3. In addition, several studies have established a connection between TRPML1 function and various other LSDs. For example, *Shen et al.* revealed that abnormal lipid accumulation in the lysosomes of NPC1 patients disrupts TRPML1-dependent lysosomal trafficking, leading to secondary lysosomal storage and impeding the clearance of lysosomal materials [23]. Encouragingly, activation of TRPML1 using small molecule agonists ameliorates cellular phenotypes in NPC1 mutant cells [23], suggesting potential therapeutic benefits of TRPML1 agonists for NPC1. Likewise, a compromised TRPML1-mediated Ca^{2+} release has also been suggested in Niemann-Pick type A (NPA), Niemann-Pick type B (NPB), and Fabry disease [185]. Because TRPML1 channels are strongly inwardly rectifying, their activation could collapse lysosomal membrane potential, preventing further Ca^{2+}/Na^{+} release. *Cao et al.* [60] suggested that BK function as a lysosomal K^{+} channel to provide the needed counter cation influx to facilitate continued Ca^{2+}/Na^{+} release via activated TRPML1 channels, and in turn lysosomal trafficking and function. Consistently, BK agonists can effectively restore impaired TRPML1-mediated Ca^{2+} release and rescue abnormal lysosomal storage in these LSDs [185].

In addition to the rare neurodegenerative LSDs, TRPML1 has also been implicated in some common neurodegenerative diseases, including AD, PD, and ALS. AD is the most common neurodegenerative disease associated with a gradual decline in memory, thinking, behavior, and social skills. Histologically, brain tissues are characterized by senile plaques, neurofibrillary tangles (NFTs), and neuron loss. Senile plaques are mainly composed of the amyloid-β (Aβ) peptide, and neurofibrillary tangles comprise hyperphosphorylated aggregates of the microtubule-associated protein Tau. They are considered to be the pathological hallmarks of AD. Interestingly, Aβ deposition precedes NFT formation [186], and as the level of Aβ increases, a rapid spread of Tau hyperphosphorylation occurs in the brain [187]. Eventually, a complex interplay among abnormal Aβ, Tau, and several other factors causes AD [173]. Therefore, it is hypothesized that the accumulation of Aβ peptide is the causative component of familial AD pathogenesis.

Familial AD is associated with the mutations of three genes, i.e., Presenilin 1, Presenilin 2, and Amyloid Precursor Protein (APP). Presenilin 1 and Presenilin 2 are the catalytic components of the γ-secretase that cut APP to form Aβ peptides [188, 189]. Although extracellular Aβ deposition is classically viewed as a distinguishing characteristic of AD, its biogenesis occurs in the cytoplasm [189]. Recently, Aβ, Presenilins, and APP have been found in lysosomes and autophagic compartments of neurons (52-63), and defective autophagy-lysosome pathways such as autophagosome accumulation (52-58) have been associated with AD pathogenesis [190–196]. Because of its role in regulating the autophagy-lysosome pathway, TRPML1 has been implicated in AD progression. On the one hand, TRPML1 agonists facilitate the clearance of accumulated Aβ peptides in the lysosomal and autophagic compartments of neurons by promoting lysosomal Ca^{2+} efflux and lysosomal acidification [190]. By increasing Ca^{2+}-dependent lysosomal exocytosis, TRPML1 also facilitates the clearance of mutant Tau that is accumulated in lysosomes [197]. On the other hand, AD-linked Presenilin 1 mutations disrupt lysosomal acidification [192], facilitating Ca^{2+} efflux from lysosomes mediated by TRPML1 [198]. This further causes deficits in the retrograde transport of lysosomes in axons [199]. These findings appear to contradict previous studies that suggested TRPML1 being activated by low pH [39]. However, because elevated lysosomal pH can inhibit mTORC1 [200,201], which normally suppresses TRPML1 [97], Presenilin 1 deficiency may increase TRPML1 activity by inhibiting mTORC1. Paradoxically, hyperactive mTORC1 is often detected in AD [202, 203]. Therefore, more work needs to be done to determine the activation mechanism and the role of TRPML1 in AD. In addition to the genetic variants that cause AD, some genes can also increase AD risk. One well-known gene that influences AD risk is the apolipoprotein E (APOE) gene that is involved in the metabolism of cholesterol and other types of fat in the body. Interestingly, new evidence indicates that induced pluripotent stem cell (iPSC)-derived human cortical neurons expressing APOE ε4, a variant with increased risk of AD, have significantly diminished TRPML1-mediated lysosomal Ca^{2+} release and impaired autophagy-lysosome pathway, and the AD-like neuronal defects in autophagy-lysosome pathway in the cells were rescued by a TRPML1 agonist, ML-SA1 [204]. Overall, existing evidence favors the development of disease-modifying therapies by targeting TRPML1.

PD is another prevalent neurodegenerative disorder in the elderly population, impacting approximately 1% of individuals aged 60 and above [205, 206]. Clinically, PD is characterized by a triad of motor symptoms, including resting tremor, rigidity, and bradykinesia, alongside nonmotor symptoms and neuropsychiatric changes [207–209]. Aging is the greatest risk factor for the development of PD. However, contemporary molecular genetic investigations have unveiled the pivotal involvement of genetic determinants in the pathogenesis of this disorder. Approximately 20 distinct causative genes have been ascertained for familial PD, while an extensive 200 genes have been associated with sporadic cases of the condition [210]. The primary neuropathological hallmark of PD involves the progressive degeneration of dopaminergic neurons [211], with Lewy bodies predominantly composed of the α-Synuclein protein in affected brain regions of PD patients [212]. In a PD model bearing the common G2019S-LRRK2 mutation, TRPML1-mediated lysosomal release of Fe^{2+} contributes to Fe^{2+}-mediated cytotoxicity, and TFEB expression reverts the Fe^{2+} overload-associated cytotoxicity by increasing TRPML1-dependent lysosomal exocytosis [213]. In a PD model associated with mutations in ATP13A2/PARK9, impaired lysosomal exocytosis is found to contribute to α-Synuclein accumulation in dopaminergic neurons. Activation of the lysosomal TRPML1 is sufficient to upregulate lysosomal exocytosis and prevent α-Synuclein accumulation [214]. These studies suggest TRPML1 as a potential therapeutic target in PD.

TRPML1 also appears to be implicated in ALS, a neurodegenerative ailment predominantly affecting motor neurons, resulting in a gradual loss of muscular strength and eventual respiratory failure within a span of 2 to 5 years from the onset of symptoms [215]. For example, chronic exposure to β-N-methylamino-L-alanine (L-BMAA), an environmental risk factor of the ALS/parkinsonism-dementia complex (ALS/PDC) [216, 217], reduces TRPML1 protein expression and produces an impairment of both lysosomal and ER Ca^{2+} homeostasis in primary motor neurons [218]. Pharmacological stimulation of TRPML1 rescues motor neurons from L-BMAA-induced toxicity [218]. TRPML1 may also be implicated in ALS caused by mutations in FIG4, a phosphoinositide phosphatase that regulates the biogenesis of PI(3,5)P2, the endogenous agonist of TRPML1 [219, 220]. In alignment with this, lysosomal Ca^{2+} accumulation in FIG4-deficient cells is corrected by a TRPML synthetic agonist [219].

In summary, TRPML1 has been implicated in a variety of LSDs and common neurodegenerative disorders. In most cases, TRPML1 is inhibited due to lysosome storage. Impaired TRPML1 in turn leads to defects in membrane trafficking, autophagy, lysosomal degradation, and lysosomal exocytosis, causing further storage accumulation in the lysosome. This positive feedback mechanism could dramatically increase the accumulation of toxic materials in lysosomes to aggravate the conditions. In this scenario, boosting TRPML1 to speed up membrane trafficking and autophagic degradation may break the vicious cycle and ameliorate phenotypes of LSDs and neurodegenerative disorders.

7.3.4.2 Cancer

One of the major differences between cancer cells and normal cells is that cancer cells grow uncontrollably. To meet the needs for growth, invasion, and drug resistance [221], cancer cells normally enhance their lysosome function by increasing the expression of lysosomal hydrolases [222–226] and lysosomal membrane proteins [227–233]. This helps lysosomes degrade intracellular materials through autophagy and extracellular materials gained through endocytosis, macropinocytosis, and phagocytosis, generating nutrients to support cancer cell survival and growth [221, 234, 235]. For example, an increase in V-ATPase [227–233] and aspartic protease cathepsin D (CTSD) [226] has been reported in several cancers. On the other hand, cancer cells have increased lysosomal exocytosis, resulting in the change of the tumor microenvironment to favor cancer cell migration and invasion [26, 224, 225, 236, 237]. For example, cancer cells secrete cathepsins to promote their migration and invasion by degrading the extracellular matrix (ECM) [224, 236]. In agreement with this, lysosomes in cancer cells often localize to the periphery area [236], suggesting more frequent lysosomal exocytosis in cancer cells. Moreover, lysosomal exocytosis may help cancer cells compartmentalize anticancer drugs away from the cellular targets [238–240], resulting in the insensitivity of the cancer cells to treatment.

Changes in expression and function of lysosomal TRPML1 have been frequently observed in a variety of cancer types, indicating the important role of this channel in cancer. This aligns with TRPML1's role in promoting autolysosomal degradation, nutrient regeneration, and lysosomal exocytosis. For example, TRPML1 expression is increased in cells bearing oncogenic *HRAS* mutations [25, 241], melanoma cells [242], nonsmall-cell lung cancer (NSCLC) [243], pancreatic ductal adenocarcinoma [244], and Triple-Negative Breast Cancer (TNBC) [26]. Importantly, elevated TRPML1 expression is often correlated with the poor prognosis of cancer patients [25, 243–245]. However, some studies suggest that TRPML1 can negatively regulate cancer progression. For example, reduction of TRPML1 mRNA expression strongly correlates with short survival in glioblastoma patients [245]. Thus, the role of TRPML1 may differ in different cancer types.

Likewise, the mechanism by which TRPML1 regulates cancer progression appears to be complex and sometimes very different among different cancers. Given that cancer cells constantly face stressful conditions, such as nutrient starvation and enhanced oxidative stress, cancer cells often adopt autophagy to ensure the high demand for nutrients for survival and growth. Because TRPML1 controls multiple steps of autophagy, including autophagic initiation [246], autophagosome-lysosome fusion [8, 113], autolysosome degradation [142, 158, 247], and autophagic lysosome reformation (ALR) [97, 247], TRPML1 may help cancer cells survive the tumor microenvironment by upregulating autophagy. In NSCLC cells, suppressing TRPML1 impairs autophagy and reduces cancer cell proliferation, migration, and invasion [243]. Cancer cells also acquire nutrients from extracellular space via macropinocytosis [248–250]. By negatively regulating MAPK and mTOR signaling, TRPML1 promotes macropinocytosis to acquire nutrients in melanoma cells, increasing melanoma cell survival and proliferation [242, 251]. TRPML1 may also promote cancer progression by facilitating lysosomal exocytosis [25, 26, 252]. For example, by promoting lysosomal exocytosis to release lysosomal ATP to the extracellular space, TRPML1 facilitates TNBC cell invasion and metastasis [26]. TRPML1 mediated lysosomal exocytosis may also promote the proliferation of cancer cells bearing oncogenic *HRAS* mutations [25]. In addition, TRPML1 has also been suggested to increase the chemoresistance of cancer cells by promoting lysosomal exocytosis [238].

7.3.4.3 Infection

TRPML1 is associated with immune responses in several cell types. In macrophages, large particle insults cause an increase in lysosomal PI(3,5)P2 to activate TRPML1, which further results in the rapid delivery of lysosomal membranes to nascent phagosomes via lysosomal exocytosis, facilitating phagosome formation [135, 253]. TRPML1 is also required for phagosome maturation in macrophages through promoting phagosome-lysosome fusion [254, 255]. In addition, TRPML1 can stimulate the TFEB signaling pathway to boost the expression of lysosome genes in macrophages, thereby enhancing lysosome-based proteolysis and killing of phagocytosed pathogens [256].

TRPML1 is also suggested to play a key role in reversing the induction of lipid-laden foamy macrophages, a cellular hallmark of tuberculosis (TB) disease caused by Mycobacterium tuberculosis [257]. In this study, the authors suggested that 1-tuberculosinyladenosine (1-TbAd), a terpenyl nucleoside shed from Mycobacterium tuberculosis, caused lysosomal maturation arrest and autophagy blockade, leading to lipid storage in M1 macrophages. This further led to the transformation of infected phagolysosomes from a site of killing into a nutrient-rich replicative niche. Interestingly, TRPML1 activation rescued the M. tuberculosis-induced lysosomal failure, reversing lipid storage in cells and promoting survival of macrophages [257].

In Helicobacter pylori (H. pylori)-infected human gastric epithelium cells, vacuolating cytotoxin A (VacA), a major virulence factor secreted by H. pylori, inhibits TRPML1 activity to disrupt lysosomal and autophagy pathways. This leads to the generation of a protected reservoir for H. pylori, protecting the bacteria from antibiotic treatment and leading to infection recrudescence after therapy. TRPML1 agonists reverse the toxic effects of VacA on lysosomal and autophagy pathways, promoting the clearance of intracellular bacteria. These results suggest that TRPML1 may represent a therapeutic target for chronic H. pylori infection [258].

In dendritic cells (DCs), TRPML1 positively regulates the responses of Toll-like receptor 7 (TLR7), which recognizes nucleic acids from pathogens [259–261], by facilitating RNA transportation to lysosomes [262]. In addition, upon bacterial infection, TRPML1 promotes directional DC migration and DC chemotaxis to the lymph nodes for antigen presentation to T cells [263].

TRPML1 also regulates the remodeling of secretory lysosomes during natural killer (NK) cell education, suggesting a possibility of boosting NK cell functionality through targeted manipulation of TRPML1 [264].

Additionally, TRPML1 plays a significant role in viral infection. The transactivator of transcription (Tat) protein is pivotal for HIV-1 replication and is implicated in the pathogenesis of HIV-related neurological complications. Upon secretion from infected cells, Tat undergoes uptake via receptor-mediated endocytosis. Post-endocytosis, Tat necessitates release into the cytoplasm to trigger the HIV-1 TLR promoter, thereby facilitating viral replication within the nucleus. Activation of TRPML1 serves to augment Tat protein degradation through lysosomal acidification, effectively impeding Tat-mediated HIV-1 replication and mitigating disease progression [155, 157].

7.4 SUMMARY AND PERSPECTIVES

Acting as an important $Ca^{2+}/Na^{+}/Fe^{2+}/Zn^{2+}$ release channel in the lysosome, TRPML1 has been implicated in numerous crucial cellular processes such as biomolecule digestion and recycling, exocytosis of metabolites and storage, nutrient sensing and autophagy, and cell signaling. Particularly, its important role in orchestrating nutrient sensing and autophagy places it at the center of maintaining cellular homeostasis in response to a variety of cellular stress and diseases. The dysfunction of TRPML1 has been associated with multiple pathological conditions such as LSD, neurodegenerative disease, cancer, and infection. Considering that TRPML1 activity is significantly increased under stress and disease conditions, therapeutic strategies targeting TRPML1 could be more specific because healthy cells may be minimally affected. However, effective and specific agonists and inhibitors of TRPML1 have not yet been discovered. One of the future research directions should be to identify effective and specific agonists and inhibitors that can be eventually used in the clinics.

REFERENCES

1. De Duve, C., Pressman, B.C., Gianetto, R., Wattiaux, R., Appelmans, F. Tissue fractionation studies. 6. Intracellular distribution patterns of enzymes in rat-liver tissue. *Biochem J*, 1955, 60(4): 604–617.
2. Novikoff, A.B., Beaufay, H., De Duve, C. Electron microscopy of lysosomerich fractions from rat liver. *J Biophys Biochem Cytol*, 1956, 2(4 Suppl): 179–184.
3. Sabatini, D.D., Adesnik, M. Christian de Duve: Explorer of the cell who discovered new organelles by using a centrifuge. *Proc Natl Acad Sci U S A*, 2013, 110(33): 13234–13235.
4. Hu, Y.B., Dammer, E.B., Ren, R.J., Wang, G. The endosomal-lysosomal system: From acidification and cargo sorting to neurodegeneration. *Transl Neurodegener*, 2015, 4(1): 18.
5. Huotari, J., Helenius, A. Endosome maturation. *EMBO J*, 2011, 30(17): 3481–3500.
6. Stoorvogel, W., Strous, G.J., Geuze, H.J., Oorschot, V., Schwartz, A.L. Late endosomes derive from early endosomes by maturation. *Cell*, 1991, 65(3): 417–427.
7. Luzio, J.P., Pryor, P.R., Bright, N.A. Lysosomes: Fusion and function. *Nat Rev Mol Cell Biol*, 2007, 8(8): 622–632.
8. Xu, H., Ren, D. Lysosomal physiology. *Annu Rev Physiol*, 2015, 77: 57–80.
9. Bustin, S. Molecular biology of the cell, sixth edition; ISBN: 9780815344643; and molecular biology of the cell, sixth edition, the problems book*; ISBN 9780815344537. Int J Mol Sci*, 2015, 16(12): 28123–28125.
10. Patra, S., Patil, S., Klionsky, D.J., Bhutia, S.K. Lysosome signaling in cell survival and programmed cell death for cellular homeostasis. *J Cell Physiol*, 2023, 238(2): 287–305.
11. Inpanathan, S., Botelho, R.J. The lysosome signaling platform: Adapting with the times. *Front Cell Dev Biol*, 2019, 7: 113.
12. Savini, M., Zhao, Q., Wang, M.C. Lysosomes: Signaling hubs for metabolic sensing and longevity. *Trends Cell Biol*, 2019, 29(11): 876–887.
13. Lamming, D.W., Bar-Peled, L. Lysosome: The metabolic signaling hub. *Traffic*, 2019, 20(1): 27–38.
14. Laplante, M., Sabatini, D.M. mTOR signaling in growth control and disease. *Cell*, 2012, 149(2): 274–293.
15. Qin, X., Jiang, B., Zhang, Y. 4E-BP1, a multifactor regulated multifunctional protein. *Cell Cycle*, 2016, 15(6): 781–786.
16. Ma, X.M., Blenis, J. Molecular mechanisms of mTOR-mediated translational control. *Nature Rev Mol Cell Biol*, 2009, 10(5): 307–318.
17. Zhang, W., Li, X., Wang, S., Chen, Y., Liu, H. Regulation of TFEB activity and its potential as a therapeutic target against kidney diseases. *Cell Death Discov*, 2020, 6(1): 32.
18. Carosi, J.M., Fourrier, C., Bensalem, J., Sargeant, T.J. The mTOR-lysosome axis at the centre of ageing. *FEBS Open Bio*, 2022, 12(4): 739–757.
19. Raiborg, C., Wenzel, E.M., Stenmark, H. ER–endosome contact sites: Molecular compositions and functions. *EMBO J*, 2015, 34(14): 1848–1858.
20. Deus, C.M., Yambire, K.F., Oliveira, P.J., Raimundo, N. Mitochondria-Lysosome crosstalk: From physiology to neurodegeneration. *Trends Mol Med*, 2020, 26(1): 71–88.
21. Buratta, S., Tancini, B., Sagini, K., Delo, F., Chiaradia, E., Urbanelli, L., Emiliani, C. Lysosomal exocytosis, exosome release and secretory autophagy: The autophagic- and endo-lysosomal systems go extracellular. *Int J Mol Sci*, 2020, 21(7): 2576.
22. Li, P., Gu, M., Xu, H. Lysosomal ion channels as decoders of cellular signals. *Trends Biochem Sci*, 2019, 44(2): 110–124.
23. Shen, D., Wang, X., Li, X., Zhang, X., Yao, Z., Dibble, S., Dong, X.-p., Yu, T., Lieberman, A.P., Showalter, H.D., Xu, H. Lipid storage disorders block lysosomal trafficking by inhibiting a TRP channel and lysosomal calcium release. *Nat Commun*, 2012, 3(1): 731.
24. Weiss, N., Cross-talk between TRPML1 channel, lipids and lysosomal storage diseases. *Commun Integr Biol*, 2012, 5(2): 111–113.

25. Jung, J., Cho, K.J., Naji, A.K., Clemons, K.N., Wong, C.O., Villanueva, M., Gregory, S., Karagas, N.E., Tan, L., Liang, H., Rousseau, M.A., Tomasevich, K.M., Sikora, A.G., Levental, I., van der Hoeven, D., Zhou, Y., Hancock, J.F., Venkatachalam, K. HRAS-driven cancer cells are vulnerable to TRPML1 inhibition. *EMBO Rep*, 2019, 20(4): 20.
26. Xu, M., Almasi, S., Yang, Y., Yan, C., Sterea, A.M., Rizvi Syeda, A.K., Shen, B., Richard Derek, C., Huang, P., Gujar, S., Wang, J., Zong, W.X., Trebak, M., El Hiani, Y., Dong, X.P. The lysosomal TRPML1 channel regulates triple negative breast cancer development by promoting mTORC1 and purinergic signaling pathways. *Cell Calcium*, 2019, 79: 80–88.
27. Lloyd-Evans, E., Morgan, A.J., He, X., Smith, D.A., Elliot-Smith, E., Sillence, D.J., Churchill, G.C., Schuchman, E.H., Galione, A., Platt, F.M. Niemann-pick disease type C1 is a sphingosine storage disease that causes deregulation of lysosomal calcium. *Nat Med*, 2008, 14(11): 1247–1255.
28. Christensen, K.A., Myers, J.T., Swanson, J.A. pH-dependent regulation of lysosomal calcium in macrophages. *J Cell Sci*, 2002, 115(3): 599–607.
29. Raffaello, A., Mammucari, C., Gherardi, G., Rizzuto, R. Calcium at the center of cell signaling: Interplay between endoplasmic reticulum, mitochondria, and lysosomes. *Trends Biochem Sci*, 2016, 41(12): 1035–1049.
30. Narayanaswamy, N., Chakraborty, K., Saminathan, A., Zeichner, E., Leung, K., Devany, J., Krishnan, Y. A pH-correctable, DNA-based fluorescent reporter for organellar calcium. *Nat Methods*, 2019, 16(1): 95–102.
31. Melchionda, M., Pittman, J.K., Mayor, R., Patel, S. Ca^{2+}/H^+ exchange by acidic organelles regulates cell migration in vivo. *J Cell Biol*, 2016, 212(7): 803–813.
32. Wang, W., Zhang, X., Gao, Q., Lawas, M., Yu, L., Cheng, X., Gu, M., Sahoo, N., Li, X., Li, P., Ireland, S., Meredith, A., Xu, H. A voltage-dependent K^+ channel in the lysosome is required for refilling lysosomal Ca^{2+} stores. *J Cell Biol*, 2017, 216(6): 1715–1730.
33. Morgan, A.J., Platt, F.M., Lloyd-Evans, E., Galione, A. Molecular mechanisms of endolysosomal Ca^{2+} signalling in health and disease. *Biochem J*, 2011, 439(3): 349–374.
34. Steinberg, B.E., Huynh, K.K., Brodovitch, A., Jabs, S., Stauber, T., Jentsch, T.J., Grinstein, S. A cation counterflux supports lysosomal acidification. *J Cell Biol*, 2010, 189(7): 1171–1186.
35. Wang, X., Zhang, X., Dong, X.P., Samie, M., Li, X., Cheng, X., Goschka, A., Shen, D., Zhou, Y., Harlow, J., Zhu, M.X., Clapham, D.E., Ren, D., Xu, H. TPC proteins are phosphoinositide- activated sodium-selective ion channels in endosomes and lysosomes. *Cell*, 2012, 151(2): 372–383.
36. Rizzollo, F., More, S., Vangheluwe, P., Agostinis, P. The lysosome as a master regulator of iron metabolism. *Trends Biochem Sci*, 2021, 46(12): 960–975.
37. Xu, H., Ren, D. Lysosomal physiology. *Annu Rev Physiol*, 2015, 77(1): 57–80.
38. Minckley, T.F., Zhang, C., Fudge, D.H., Dischler, A.M., LeJeune, K.D., Xu, H., Qin, Y. Sub-nanomolar sensitive GZnP3 reveals TRPML1-mediated neuronal Zn^{2+} signals. *Nat Commun*, 2019, 10(1): 4806.
39. Dong, X.P., Cheng, X., Mills, E., Delling, M., Wang, F., Kurz, T., Xu, H. The type IV mucolipidosis-associated protein TRPML1 is an endolysosomal iron release channel. *Nature*, 2008, 455(7215): 992–996.
40. Saito, M., Hanson, P.I., Schlesinger, P. Luminal chloride-dependent activation of endosome calcium channels: Patch clamp study of enlarged endosomes. *J Biol Chem*, 2007, 282(37): 27327–27333.
41. Stauber, T., Jentsch, T.J. Chloride in vesicular trafficking and function. *Annu Rev Physiol*, 2013, 75: 453–477.
42. Mindell, J.A. Lysosomal acidification mechanisms. *Annu Rev Physiol*, 2012, 74: 69–86.
43. Chakraborty, K., Leung, K., Krishnan, Y. High lumenal chloride in the lysosome is critical for lysosome function. *Elife*, 2017, 25(6): 28862.
44. Mindell, J.A. Lysosomal acidification mechanisms. *Annu Rev Physiol*, 2012, 74(1): 69–86.
45. Cheng, X., Shen, D., Samie, M., Xu, H. Mucolipins: Intracellular TRPML1-3 channels. *FEBS Lett*, 2010, 584(10): 2013–2021.
46. Dong, X.P., Shen, D., Wang, X., Dawson, T., Li, X., Zhang, Q., Cheng, X., Zhang, Y., Weisman, L.S., Delling, M., Xu, H. PI(3,5)P2 controls membrane trafficking by direct activation of mucolipin Ca^{2+} release channels in the endolysosome. *Nat Commun*, 2010, 1(4): 38.
47. Shen, D., Wang, X., Xu, H. Pairing phosphoinositides with calcium ions in endolysosomal dynamics: Phosphoinositides control the direction and specificity of membrane trafficking by regulating the activity of calcium channels in the endolysosomes. *Bioessays*, 2011, 33(6): 448–457.
48. Brailoiu, E., Churamani, D., Cai, X., Schrlau, M.G., Brailoiu, G.C., Gao, X., Hooper, R., Boulware, M.J., Dun, N.J., Marchant, J.S., Patel, S. Essential requirement for two-pore channel 1 in NAADP-mediated calcium signaling. *J Cell Biol*, 2009, 186(2): 201–209.

49. Calcraft, P.J., Ruas, M., Pan, Z., Cheng, X., Arredouani, A., Hao, X., Tang, J., Rietdorf, K., Teboul, L., Chuang, K.T., Lin, P., Xiao, R., Wang, C., Zhu, Y., Lin, Y., Wyatt, C.N., Parrington, J., Ma, J., Evans, A.M., Galione, A., Zhu, M.X. NAADP mobilizes calcium from acidic organelles through two-pore channels. *Nature*, 2009, 459(7246): 596–600.
50. Ruas, M., Davis, L.C., Chen, C.C., Morgan, A.J., Chuang, K.T., Walseth, T.F., Grimm, C., Garnham, C., Powell, T., Platt, N., Platt, F.M., Biel, M., Wahl-Schott, C., Parrington, J., Galione, A. Expression of Ca^{2+}-permeable two-pore channels rescues NAADP signalling in TPC-deficient cells. *EMBO J*, 2015, 34(13): 1743–1758.
51. Cang, C., Zhou, Y., Navarro, B., Seo, Y.-j., Aranda, K., Shi, L., Battaglia-Hsu, S., Nissim, I., Clapham, David E., Ren, D. mTOR regulates lysosomal ATP-sensitive two-Pore Na^+ channels to adapt to metabolic state. *Cell*, 2013, 152(4): 778–790.
52. Cang, C., Bekele, B., Ren, D. The voltage-gated sodium channel TPC1 confers endolysosomal excitability. *Nat Chem Biol*, 2014, 10(6): 463–469.
53. Gerndt, S., Chen, C.C., Chao, Y.K., Yuan, Y., Burgstaller, S., Scotto Rosato, A., Krogsaeter, E., Urban, N., Jacob, K., Nguyen, O.N.P., Miller, M.T., Keller, M., Vollmar, A.M., Gudermann, T., Zierler, S., Schredelseker, J., Schaefer, M., Biel, M., Malli, R., Wahl-Schott, C., … Grimm, C. Agonist-mediated switching of ion selectivity in TPC2 differentially promotes lysosomal function. *ELife*, 2020, 9: e54712.
54. Zhang, X., Chen, W., Li, P., Calvo, R., Southall, N., Hu, X., Bryant-Genevier, M., Feng, X., Geng, Q., Gao, C., Yang, M., Tang, K., Ferrer, M., Marugan, J.J., Xu, H. Agonist-specific voltage-dependent gating of lysosomal two-pore Na^+ channels. *ELife*, 2019, 8: e51423.
55. Qureshi, O.S., Paramasivam, A., Yu, J.C., Murrell-Lagnado, R.D. Regulation of P2X4 receptors by lysosomal targeting, glycan protection and exocytosis. *J Cell Sci*, 2007, 120(Pt 21): 3838–3849.
56. Huang, P., Zou, Y., Zhong, X.Z., Cao, Q., Zhao, K., Zhu, M.X., Murrell-Lagnado, R., Dong, X.P. P2X4 forms functional ATP-activated cation channels on lysosomal membranes regulated by luminal pH. *J Biol Chem*, 2014, 289(25): 17658–17667.
57. Sumoza-Toledo, A., Lange, I., Cortado, H., Bhagat, H., Mori, Y., Fleig, A., Penner, R., Partida-Sánchez, S. Dendritic cell maturation and chemotaxis is regulated by TRPM2-mediated lysosomal Ca^{2+} release. *FASEB J*, 2011, 25(10): 3529–3542.
58. Lange, I., Yamamoto, S., Partida-Sanchez, S., Mori, Y., Fleig, A., Penner, R. TRPM2 functions as a lysosomal Ca^{2+}-release channel in beta cells. *Sci Signal*, 2009, 2(71): 2000278.
59. Wang, W., Zhang, X., Gao, Q., Lawas, M., Yu, L., Cheng, X., Gu, M., Sahoo, N., Li, X., Li, P., Ireland, S., Meredith, A., Xu, H. A voltage-dependent K^+ channel in the lysosome is required for refilling lysosomal Ca^{2+} stores. *J Cell Biol*, 2017, 216(6): 1715–1730.
60. Cao, Q., Zhong, X.Z., Zou, Y., Zhang, Z., Toro, L., Dong, X.P. BK channels alleviate lysosomal storage diseases by providing positive feedback regulation of lysosomal Ca^{2+} release. *Dev Cell*, 2015, 33(4): 427–441.
61. Chapel, A., Kieffer-Jaquinod, S., Sagné, C., Verdon, Q., Ivaldi, C., Mellal, M., Thirion, J., Jadot, M., Bruley, C., Garin, J., Gasnier, B., Journet, A. An extended proteome map of the lysosomal membrane reveals novel potential transporters. *Mol Cell Proteomics*, 2013, 12(6): 1572–1588.
62. Hu, M., Li, P., Wang, C., Feng, X., Geng, Q., Chen, W., Marthi, M., Zhang, W., Gao, C., Reid, W., Swanson, J., Du, W., Hume, R.I., Xu, H. Parkinson's disease-risk protein TMEM175 is a proton-activated proton channel in lysosomes. *Cell*, 2022, 185(13): 2292–2308.e20.
63. Cang, C., Aranda, K., Seo, Y.-j., Gasnier, B., Ren, D. TMEM175 is an organelle K^+ channel regulating lysosomal function. *Cell*, 2015, 162(5): 1101–1112.
64. Brunner, J.D., Jakob, R.P., Schulze, T., Neldner, Y., Moroni, A., Thiel, G., Maier, T., Schenck, S. Structural basis for ion selectivity in TMEM175 K^+ channels. *Elife*, 2020, 9: e53683.
65. Oh, S., Paknejad, N., Hite, R.K. Gating and selectivity mechanisms for the lysosomal K^+ channel TMEM175. *Elife*, 2020, 9: e53430.
66. Wie, J., Liu, Z., Song, H., Tropea, T.F., Yang, L., Wang, H., Liang, Y., Cang, C., Aranda, K., Lohmann, J. A growth-factor-activated lysosomal K^+ channel regulates Parkinson's pathology. *Nature*, 2021, 591(7850): 431–437.
67. Wang, Y., Zeng, W., Lin, B., Yao, Y., Li, C., Hu, W., Wu, H., Huang, J., Zhang, M., Xue, T., Ren, D., Qu, L., Cang, C. CLN7 is an organellar chloride channel regulating lysosomal function. *Sci Adv*, 2021, 7 (51): 15.
68. Jentsch, T.J., CLC chloride channels and transporters: From genes to protein structure, pathology and physiology. *Crit Rev Biochem Mol Biol*, 2008, 43(1): 3–36.
69. Li, P., Hu, M., Wang, C., Feng, X., Zhao, Z., Yang, Y., Sahoo, N., Gu, M., Yang, Y., Xiao, S., Sah, R., Cover, T.L., Chou, J., Geha, R., Benavides, F., Hume, R.I., Xu, H. LRRC8 family proteins within

lysosomes regulate cellular osmoregulation and enhance cell survival to multiple physiological stresses. *Proc Natl Acad Sci U S A*, 2020, 117(46): 29155–29165.
70. Cheng, X., Shen, D., Samie, M., Xu, H. Mucolipins: Intracellular TRPML1-3 channels. *FEBS Lett*, 2010, 584(10): 2013–2021.
71. Venkatachalam, K., Wong, C.-O., Zhu, M.X. The role of TRPMLs in endolysosomal trafficking and function. *Cell Calcium*, 2015, 58(1): 48–56.
72. Puertollano, R., Kiselyov, K. TRPMLs: In sickness and In health. *Am J Physiol Renal Physiol*, 2009, 296(6): 21.
73. Xu, M., Dong, X.-P. Endolysosomal TRPMLs in cancer. *Biomolecules*, 2021, 11(1): 65.
74. Sun, L., Hua, Y., Vergarajauregui, S., Diab, H.I., Puertollano, R. Novel role of TRPML2 in the regulation of the innate immune response. *J Immunol*, 2015, 195(10): 4922–4932.
75. Samie, M.A., Grimm, C., Evans, J.A., Curcio-Morelli, C., Heller, S., Slaugenhaupt, S.A., Cuajungco, M.P. The tissue-specific expression of TRPML2 (MCOLN-2) gene is influenced by the presence of TRPML1. *Pflugers Arch*, 2009, 459(1): 79–91.
76. Grimm, C., Jörs, S., Guo, Z., Obukhov, A.G., Heller, S. Constitutive activity of TRPML2 and TRPML3 channels versus activation by low extracellular sodium and small molecules. *J Biol Chem*, 2012, 287(27): 22701–22708.
77. Spix, B., Chao, Y.-K., Abrahamian, C., Chen, C.-C., Grimm, C. TRPML cation channels in inflammation and immunity. *Front Immunol*, 2020, 11: 225.
78. Xu, H., Delling, M., Li, L., Dong, X., Clapham, D.E. Activating mutation in a mucolipin transient receptor potential channel leads to melanocyte loss in varitint–waddler mice. *Proc Natl Acad Sci U S A*, 2007, 104(46): 18321–18326.
79. Li, M., Zhang, W.K., Benvin, N.M., Zhou, X., Su, D., Li, H., Wang, S., Michailidis, I. E., Tong, L., Li, X., Yang, J. Structural basis of dual Ca^{2+}/pH regulation of the endolysosomal TRPML1 channel. *Nat Struct Mol Biol*, 2017, 24(3): 205–213.
80. Chen, C.-C., Keller, M., Hess, M., Schiffmann, R., Urban, N., Wolfgardt, A., Schaefer, M., Bracher, F., Biel, M., Wahl-Schott, C., Grimm, C. A small molecule restores function to TRPML1 mutant isoforms responsible for mucolipidosis type IV. *Nat Commun*, 2014, 5(1): 4681.
81. Plesch, E., Chen, C.C., Butz, E., Scotto Rosato, A., Krogsaeter, E.K., Yinan, H., Bartel, K., Keller, M., Robaa, D., Teupser, D., Holdt, L.M., Vollmar, A.M., Sippl, W., Puertollano, R., Medina, D., Biel, M., Wahl-Schott, C., Bracher, F., Grimm, C. Selective agonist of TRPML2 reveals direct role in chemokine release from innate immune cells. *Elife*, 2018, 7: e39720.
82. Venkatachalam, K., Hofmann, T., Montell, C. Lysosomal localization of TRPML3 depends on TRPML2 and the mucolipidosis-associated protein TRPML1. *J Biol Chem*, 2006, 281(25): 17517–17527.
83. Karacsonyi, C., Miguel, A.S., Puertollano, R. Mucolipin-2 localizes to the Arf6-associated pathway and regulates recycling of GPI-APs. *Traffic*, 2007, 8(10): 1404–1414.
84. Kim, H.J., Soyombo, A.A., Tjon-Kon-Sang, S., So, I., Muallem, S. The Ca^{2+} channel TRPML3 regulates membrane trafficking and autophagy. *Traffic*, 2009, 10(8): 1157–1167.
85. Martina, J.A., Lelouvier, B., Puertollano, R. The calcium channel mucolipin-3 is a novel regulator of trafficking along the endosomal pathway. *Traffic*, 2009, 10(8): 1143–1156.
86. Dong, X.P., Shen, D., Wang, X., Dawson, T., Li, X., Zhang, Q., Cheng, X., Zhang, Y., Weisman, L.S., Delling, M., Xu, H. PI(3,5)P2 controls membrane trafficking by direct activation of mucolipin Ca^{2+} release channels in the endolysosome. *Nat Commun*, 2010, 1(4): 38.
87. Chen, C.-C., Butz, E.S., Chao, Y.-K., Grishchuk, Y., Becker, L., Heller, S., Slaugenhaupt, S.A., Biel, M., Wahl-Schott, C., Grimm, C. Small molecules for early endosome-specific patch clamping. *Cell Chem Biol*, 2017, 24(7): 907–916.e4.
88. Eichelsdoerfer, J.L., Evans, J.A., Slaugenhaupt, S.A., Cuajungco, M.P. Zinc dyshomeostasis is linked with the loss of mucolipidosis IV-associated TRPML1 ion channel. *J Biol Chem*, 2010, 285(45): 34304–34308.
89. Li, M., Zhang, W.K., Benvin, N.M., Zhou, X., Su, D., Li, H., Wang, S., Michailidis, I.E., Tong, L., Li, X., Yang, J. Structural basis of dual Ca^{2+}/pH regulation of the endolysosomal TRPML1 channel. *Nat Struct Mol Biol*, 2017, 24(3): 205–213.
90. Chen, C.C., Keller, M., Hess, M., Schiffmann, R., Urban, N., Wolfgardt, A., Schaefer, M., Bracher, F., Biel, M., Wahl-Schott, C., Grimm, C. A small molecule restores function to TRPML1 mutant isoforms responsible for mucolipidosis type IV. *Nat Commun*, 2014, 5: 4681.
91. Gan, N., Han, Y., Zeng, W., Wang, Y., Xue, J., Jiang, Y. Structural mechanism of allosteric activation of TRPML1 by PI(3,5)P2 and rapamycin. *Proc Natl Acad Sci U S A*, 2022, 119(7): e2120404119.

92. Zhang, X., Cheng, X., Yu, L., Yang, J., Calvo, R., Patnaik, S., Hu, X., Gao, Q., Yang, M., Lawas, M., Delling, M., Marugan, J., Ferrer, M., Xu, H. MCOLN1 is a ROS sensor in lysosomes that regulates autophagy. *Nat Commun*, 2016, **7**(1): 12109.
93. Shen, D., Wang, X., Li, X., Zhang, X., Yao, Z., Dibble, S., Dong, X.-P., Yu, T., Lieberman, A.P., Showalter, H.D. Lipid storage disorders block lysosomal trafficking by inhibiting a TRP channel and lysosomal calcium release. *Nat Commun*, 2012, 3: 731.
94. Zhang, X., Li, X., Xu, H. Phosphoinositide isoforms determine compartment-specific ion channel activity. *Proc Natl Acad Sci U S A*, 2012, 109(28): 11384–11389.
95. Zhong, X.Z., Zou, Y., Sun, X., Dong, G., Cao, Q., Pandey, A., Rainey, J.K., Zhu, X., Dong, X.P. Inhibition of transient receptor potential channel mucolipin-1 (TRPML1) by lysosomal adenosine involved in severe combined immunodeficiency diseases. *J Biol Chem*, 2017, 292(8): 3445–3455.
96. Sun, X., Yang, Y., Zhong, X.Z., Cao, Q., Zhu, X.-H., Zhu, X., Dong, X.P. A negative feedback regulation of MTORC1 activity by the lysosomal Ca^{2+}channel MCOLN1 (mucolipin 1) using a CALM (calmodulin)-dependent mechanism. *Autophagy*, 2018, 14(1): 38–52.
97. Kiselyov, K., Chen, J., Rbaibi, Y., Oberdick, D., Tjon-Kon-Sang, S., Shcheynikov, N., Muallem, S., Soyombo, A. TRP-ML1 is a lysosomal monovalent cation channel that undergoes proteolytic cleavage. *J Biol Chem*, 2005, 280(52): 43218–43223.
98. Vergarajauregui, S., Puertollano, R. Two di-leucine motifs regulate trafficking of mucolipin-1 to lysosomes. *Traffic*, 2006, 7(3): 337–353.
99. Sun, J., Liu, Y., Hao, X., Lin, W., Su, W., Chiang, E., Baudry, M., Bi, X. LAMTOR1 inhibition of TRPML1-dependent lysosomal calcium release regulates dendritic lysosome trafficking and hippocampal neuronal function. *EMBO J*, 2022, 41(5): e108119.
100. Yu, L., Zhang, X., Yang, Y., Li, D., Tang, K., Zhao, Z., He, W., Wang, C., Sahoo, N., Converso-Baran, K., Davis, C.S., Brooks, S.V., Bigot, A., Calvo, R., Martinez, N.J., Southall, N., Hu, X., Marugan, J., Ferrer, M., Xu, H. Small-molecule activation of lysosomal TRP channels ameliorates Duchenne muscular dystrophy in mouse models. *Sci Adv*, 2020, 6(6): eaaz2736.
101. Sahoo, N., Gu, M., Zhang, X., Raval, N., Yang, J., Bekier, M., Calvo, R., Patnaik, S., Wang, W., King, G., Samie, M., Gao, Q., Sahoo, S., Sundaresan, S., Keeley, T.M., Wang, Y., Marugan, J., Ferrer, M., Samuelson, L.C., Merchant, J.L., … Xu, H. Gastric acid secretion from parietal cells is mediated by a Ca^{2+} efflux channel in the tubulovesicle. *Dev Cell*, 2017, 41(3): 262–273 e6.
102. Du, W., Gu, M., Hu, M., Pinchi, P., Chen, W., Ryan, M., Nold, T., Bannaga, A., Xu, H. Lysosomal Zn^{2+} release triggers rapid, mitochondria-mediated, non-apoptotic cell death in metastatic melanoma. *Cell Rep*, 2021, 37(3): 109848.
103. Siow, W.X., Kabiri, Y., Tang, R., Chao, Y.K., Plesch, E., Eberhagen, C., Flenkenthaler, F., Fröhlich, T., Bracher, F., Grimm, C., Biel, M., Zischka, H., Vollmar, A.M., Bartel, K. Lysosomal TRPML1 regulates mitochondrial function in hepatocellular carcinoma cells. *J Cell Sci*, 2022, 135(6): 21.
104. Scotto Rosato, A., Montefusco, S., Soldati, C., Di Paola, S., Capuozzo, A., Monfregola, J., Polishchuk, E., Amabile, A., Grimm, C., Lombardo, A., De Matteis, M.A., Ballabio, A., Medina, D.L. TRPML1 links lysosomal calcium to autophagosome biogenesis through the activation of the CaMKKβ/VPS34 pathway. *Nat Commun*, 2019, 10(1): 5630.
105. He, H., Liu, X., Lv, L., Liang, H., Leng, B., Zhao, D., Zhang, Y., Du, Z., Chen, X., Li, S., Lu, Y., Shan, H. Calcineurin suppresses AMPK-dependent cytoprotective autophagy in cardiomyocytes under oxidative stress. *Cell Death Dis*, 2014, 5: e997.
106. Zhang, X., Chen, W., Gao, Q., Yang, J., Yan, X., Zhao, H., Su, L., Yang, M., Gao, C., Yao, Y., Inoki, K., Li, D., Shao, R., Wang, S., Sahoo, N., Kudo, F., Eguchi, T., Ruan, B., Xu, H. Rapamycin directly activates lysosomal mucolipin TRP channels independent of mTOR. *PLOS Biol*, 2019, 17(5): e3000252.
107. Zhang, X., Cheng, X., Yu, L., Yang, J., Calvo, R., Patnaik, S., Hu, X., Gao, Q., Yang, M., Lawas, M., Delling, M., Marugan, J., Ferrer, M., Xu, H. MCOLN1 is a ROS sensor in lysosomes that regulates autophagy. *Nat Commun*, 2016, 7: 12109.
108. Schmiege, P., Fine, M., Blobel, G., Li, X. Human TRPML1 channel structures in open and closed conformations. *Nature*, 2017, 550(7676): 366–370.
109. Schmiege, P., Fine, M., Li, X. Atomic insights into ML-SI3 mediated human TRPML1 inhibition. *Structure*, 2021, 29(11): 1295–1302 e3.
110. Peng, W., Wong, Y.C., Krainc, D. Mitochondria-lysosome contacts regulate mitochondrial Ca^{2+} dynamics via lysosomal TRPML1. *Proc Natl Acad Sci U S A*, 2020, 117(32): 19266–19275.
111. Li, X., Rydzewski, N., Hider, A., Zhang, X., Yang, J., Wang, W., Gao, Q., Cheng, X., Xu, H. A molecular mechanism to regulate lysosome motility for lysosome positioning and tubulation. *Nat Cell Biol*, 2016, 18(4): 404–417.

112. Medina, D.L., Di Paola, S., Peluso, I., Armani, A., De Stefani, D., Venditti, R., Montefusco, S., Scotto-Rosato, A., Prezioso, C., Forrester, A., Settembre, C., Wang, W., Gao, Q., Xu, H., Sandri, M., Rizzuto, R., De Matteis, M.A., Ballabio, A. Lysosomal calcium signalling regulates autophagy through calcineurin and TFEB. *Nat Cell Biol*, 2015, 17(3): 288–299.
1135. Li, R.-J., Xu, J., Fu, C., Zhang, J., Zheng, Y.G., Jia, H., Liu, J.O. Regulation of mTORC1 by lysosomal calcium and calmodulin. *Elife*, 2016, 5: e19360.
114. Zhang, Z., Chen, G., Zhou, W., Song, A., Xu, T., Luo, Q., Wang, W., Gu, X.S., Duan, S. Regulated ATP release from astrocytes through lysosome exocytosis. *Nat Cell Biol*, 2007, 9(8): 945–953.
115. Cao, Q., Zhao, K., Zhong, X.Z., Zou, Y., Yu, H., Huang, P., Xu, T.L., Dong, X.P. SLC17A9 protein functions as a lysosomal ATP transporter and regulates cell viability. *J Biol Chem*, 2014, 289(33): 23189–23199.
116. Minckley, T.F., Zhang, C., Fudge, D.H., Dischler, A.M., LeJeune, K.D., Xu, H., Qin, Y. Sub-nanomolar sensitive GZnP3 reveals TRPML1-mediated neuronal Zn^{2+} signals. *Nat Commun*, 2019, 10(1): 019–12761.
117. Lane, D.J.R., Merlot, A.M., Huang, M.L.H., Bae, D.H., Jansson, P.J., Sahni, S., Kalinowski, D.S., Richardson, D.R. Cellular iron uptake, trafficking and metabolism: Key molecules and mechanisms and their roles in disease. *Biochim et Biophys Acta*, 2015, 1853(5): 1130–1144.
118. Altarescu, G., Sun, M., Moore, D.F., Smith, J.A., Wiggs, E.A., Solomon, B.I., Patronas, N.J., Frei, K.P., Gupta, S., Kaneski, C.R., Quarrell, O.W., Slaugenhaupt, S.A., Goldin, E., Schiffmann, R. The neurogenetics of mucolipidosis type IV. *Neurology*, 2002, 59(3): 306–313.
119. Schiffmann, R., Dwyer, N.K., Lubensky, I.A., Tsokos, M., Sutliff, V.E., Latimer, J.S., Frei, K.P., Brady, R.O., Barton, N.W., Blanchette-Mackie, E.J. Constitutive achlorhydria in mucolipidosis type IV. *Proc Natl Acad Sci U S A*, 1998, 95(3): 1207–1212.
120. Saftig, P., Klumperman, J. Lysosome biogenesis and lysosomal membrane proteins: Trafficking meets function. *Nat Rev Mol Cell Biol*, 2009, 10(9): 623–35.
121. Lloyd-Evans, E., Platt, F.M. Lysosomal Ca^{2+} homeostasis: Role in pathogenesis of lysosomal storage diseases. *Cell Calcium*, 2011, 50(2): 200–205.
122. Cao, Q., Yang, Y., Zhong, X.Z., Dong, X.P. The lysosomal Ca^{2+} release channel TRPML1 regulates lysosome size by activating calmodulin. *J Biol Chem*, 2017, 292(20): 8424–8435.
123. Pryor, P.R., Mullock, B.M., Bright, N.A., Gray, S.R., Luzio, J.P. The role of intraorganellar Ca^{2+} in late endosome-lysosome heterotypic fusion and in the reformation of lysosomes from hybrid organelles. *J Cell Biol*, 2000, 149(5): 1053–1062.
124. Treusch, S., Knuth, S., Slaugenhaupt, S.A., Goldin, E., Grant, B.D., Fares, H. Caenorhabditis elegans functional orthologue of human protein h-mucolipin-1 is required for lysosome biogenesis. *Proc Natl Acad Sci U S A*, 2004, 101(13): 4483–4488.
125. Peters, C., Mayer, A. Ca^{2+}/calmodulin signals the completion of docking and triggers a late step of vacuole fusion. *Nature*, 1998, 396(6711): 575–580.
126. Miller, A., Schafer, J., Upchurch, C., Spooner, E., Huynh, J., Hernandez, S., McLaughlin, B., Oden, L., Fares, H. Mucolipidosis type IV protein TRPML1-dependent lysosome formation. *Traffic*, 2015, 16(3): 284–297.
127. Campbell, E.M., Fares, H. Roles of CUP-5, the *Caenorhabditis elegans* orthologue of human TRPML1, in lysosome and gut granule biogenesis. *BMC Cell Biol*, 2010, 11(1): 40.
128. Piper, R.C., Luzio, J.P. CUPpling calcium to lysosomal biogenesis. *Trends Cell Biol*, 2004, 14(9): 471–473.
129. Zhong, D., Wang, R., Zhang, H., Wang, M., Zhang, X., Chen, H. Induction of lysosomal exocytosis and biogenesis via TRPML1 activation for the treatment of uranium-induced nephrotoxicity. *Nat Commun*, 2023, 14(1): 3997.
130. Samie, M.A., Xu, H. Lysosomal exocytosis and lipid storage disorders. *J Lipid Res*, 2014, 55(6): 995–1009.
131. Sun, X., Xu, M., Cao, Q., Huang, P., Zhu, X., Dong, X.P. A lysosomal K^+ channel regulates large particle phagocytosis by facilitating lysosome Ca^{2+} release. *Sci Rep*, 2020, 10(1): 1038.
132. Samie, M., Wang, X., Zhang, X., Goschka, A., Li, X., Cheng, X., Gregg, E., Azar, M., Zhuo, Y., Garrity, A.G., Gao, Q., Slaugenhaupt, S., Pickel, J., Zolov, S.N., Weisman, L.S., Lenk, G.M., Titus, S., Bryant-Genevier, M., Southall, N., Juan, M., … Xu, H. A TRP channel in the lysosome regulates large particle phagocytosis via focal exocytosis. *Dev Cell*, 2013, 26(5): 511–524.
133. Cheng, X., Zhang, X., Gao, Q., Ali Samie, M., Azar, M., Tsang, W.L., Dong, L., Sahoo, N., Li, X., Zhuo, Y., Garrity, A.G., Wang, X., Ferrer, M., Dowling, J., Xu, L., Han, R., Xu, H. The intracellular Ca^{2+} channel MCOLN1 is required for sarcolemma repair to prevent muscular dystrophy. *Nat Med*, 2014, 20(10): 1187–1192.

134. Finkbeiner, S. The autophagy lysosomal pathway and neurodegeneration. *Cold Spring Harb Perspect Biol*, 2020, 12(3): a033993.
135. Ryter, S.W., Bhatia, D., Choi, M.E. Autophagy: A lysosome-dependent process with implications in cellular redox homeostasis and human disease. *Antioxid Redox Signal*, 2019, 30(1): 138–159.
136. Wang, R.C., Levine, B. Autophagy in cellular growth control. *FEBS Lett*, 2010, 584(7): 1417–1426.
137. Mizushima, N., Levine, B. Autophagy in mammalian development and differentiation. *Nat Cell Biol*, 2010, 12(9): 823–830.
138. Russell, R.C., Tian, Y., Yuan, H., Park, H.W., Chang, Y.Y., Kim, J., Kim, H., Neufeld, T.P., Dillin, A., Guan, K.L. ULK1 induces autophagy by phosphorylating beclin-1 and activating VPS34 lipid kinase. *Nat Cell Biol*, 2013, 15(7): 741–750.
139. Sun, T., Wang, X., Lu, Q., Ren, H., Zhang, H. CUP-5, the C. elegans ortholog of the mammalian lysosomal channel protein MLN1/TRPML1, is required for proteolytic degradation in autolysosomes. *Autophagy*, 2011, 7(11): 1308–1315.
140. Medina, D.L., Ballabio, A. Lysosomal calcium regulates autophagy. *Autophagy*, 2015, 11(6): 970–971.
141. Settembre, C., Di Malta, C., Polito, V.A., Garcia Arencibia, M., Vetrini, F., Erdin, S., Erdin, S.U., Huynh, T., Medina, D., Colella, P., Sardiello, M., Rubinsztein, D.C., Ballabio, A. TFEB links autophagy to lysosomal biogenesis. *Science*, 2011, 332(6036): 1429–1433.
142. Forgac, M., Vacuolar ATPases: Rotary proton pumps in physiology and pathophysiology. *Nat Rev Mol Cell Biol*, 2007, 8(11): 917–929.
143. Toei, M., Saum, R., Forgac, M. Regulation and isoform function of the v-ATPases. *Biochemistry*, 2010, 49(23): 4715–4723.
144. Riederer, E., Cang, C., Ren, D. Lysosomal ion channels: What are they good for and are they druggable targets? *Annu Rev Pharmacol Toxicol*, 2023, 63(1): 19–41.
145. Steinberg, B.E., Huynh, K.K., Brodovitch, A., Jabs, S., Stauber, T., Jentsch, T.J., Grinstein, S. A cation counterflux supports lysosomal acidification. *J Cell Biol*, 2010, 189(7): 1171–1186.
146. Graves, A.R., Curran, P.K., Smith, C.L., Mindell, J.A. The cl^-/H^+ antiporter ClC-7 is the primary chloride permeation pathway in lysosomes. *Nature*, 2008, 453(7196): 788–792.
147. Weinert, S., Jabs, S., Supanchart, C., Schweizer, M., Gimber, N., Richter, M., Rademann, J., Stauber, T., Kornak, U., Jentsch, T.J. Lysosomal pathology and osteopetrosis upon loss of H^+-driven lysosomal Cl^- accumulation. *Science*, 2010, 328(5984): 1401–1403.
148. Kasper, D., Planells-Cases, R., Fuhrmann, J.C., Scheel, O., Zeitz, O., Ruether, K., Schmitt, A., Poët, M., Steinfeld, R., Schweizer, M., Kornak, U., Jentsch, T.J. Loss of the chloride channel ClC-7 leads to lysosomal storage disease and neurodegeneration. *EMBO J*, 2005, 24(5): 1079–1091.
149. Lange, P.F., Wartosch, L., Jentsch, T.J., Fuhrmann, J.C. ClC-7 requires Ostm1 as a β-subunit to support bone resorption and lysosomal function. *Nature*, 2006, 440(7081): 220–223.
150. Weinert, S., Jabs, S., Hohensee, S., Chan, W.L., Kornak, U., Jentsch, T.J. Transport activity and presence of ClC-7/Ostm1 complex account for different cellular functions. *EMBO Rep*, 2014, 15(7): 784–791.
151. Bae, M., Patel, N., Xu, H., Lee, M., Tominaga-Yamanaka, K., Nath, A., Geiger, J., Gorospe, M., Mattson, M.P., Haughey, N.J. Activation of TRPML1 clears intraneuronal $A\beta$ in preclinical models of HIV infection. *J Neurosci*, 2014, 34(34): 11485–11503.
152. Bach, G., Chen, C.S., Pagano, R.E. Elevated lysosomal pH in mucolipidosis type IV cells. *Clin Chim Acta*, 1999, 280(1-2): 173–179.
153. Khan, N., Lakpa, K.L., Halcrow, P.W., Afghah, Z., Miller, N.M., Geiger, J.D., Chen, X. BK channels regulate extracellular Tat-mediated HIV-1 LTR transactivation. *Sci Rep*, 2019, 9(1): 12285.
154. Venkatachalam, K., Long, A.A., Elsaesser, R., Nikolaeva, D., Broadie, K., Montell, C. Motor deficit in a Drosophila model of mucolipidosis type IV due to defective clearance of apoptotic cells. *Cell*, 2008, 135(5): 838–851.
155. Schaheen, L., Dang, H., Fares, H. Basis of lethality in C. elegans lacking CUP-5, the mucolipidosis type IV orthologue. *Dev Biol*, 2006, 293(2): 382–91.
156. Fares, H., Greenwald, I. Regulation of endocytosis by CUP-5, the Caenorhabditis elegans mucolipin-1 homolog. *Nat Genet*, 2001, 28(1): 64–68.
157. Zhang, J., Zeng, W., Han, Y., Lee, W.R., Liou, J., Jiang, Y. Lysosomal LAMP proteins regulate lysosomal pH by direct inhibition of the TMEM175 channel. *Mol Cell*, 2023, 83(14): 2524–2539 e7.
158. Kilpatrick, B.S., Yates, E., Grimm, C., Schapira, A.H., Patel, S. Endo-lysosomal TRP mucolipin-1 channels trigger global ER Ca^{2+} release and Ca^{2+} influx. *J Cell Sci*, 2016, 129(20): 3859–3867.
159. Endo, M., Calcium-induced calcium release in skeletal muscle. *Physiol Rev*, 2009, 89(4): 1153–1176.
160. Hogan, P.G., Rao, A. Store-operated calcium entry: Mechanisms and modulation. *Biochem Biophys Res Commun*, 2015, 460(1): 40–49.

161. Griffin, C.S., Alvarado, M.G., Yamasaki, E., Drumm, B.T., Krishnan, V., Ali, S., Nagle, E.M., Sanders, K.M., Earley, S. The intracellular Ca^{2+} release channel TRPML1 regulates lower urinary tract smooth muscle contractility. *Proc Natl Acad Sci U S A*, 2020, 117(48): 30775–30786.
162. Wong, Y.C., Ysselstein, D., Krainc, D. Mitochondria-lysosome contacts regulate mitochondrial fission via RAB7 GTP hydrolysis. *Nature*, 2018, 554(7692): 382–386.
163. Hung, H.I., Schwartz, J.M., Maldonado, E.N., Lemasters, J.J., Nieminen, A.L. Mitoferrin-2-dependent mitochondrial iron uptake sensitizes human head and neck squamous carcinoma cells to photodynamic therapy. *J Biol Chem*, 2013, 288(1): 677–686.
164. Hu, J., Kholmukhamedov, A., Lindsey, C.C., Beeson, C.C., Jaeschke, H., Lemasters, J.J. Translocation of iron from lysosomes to mitochondria during acetaminophen-induced hepatocellular injury: Protection by starch-desferal and minocycline. *Free Radic Biol Med*, 2016, 97: 418–426.
165. Uchiyama, A., Kim, J.S., Kon, K., Jaeschke, H., Ikejima, K., Watanabe, S., Lemasters, J.J. Translocation of iron from lysosomes into mitochondria is a key event during oxidative stress-induced hepatocellular injury. *Hepatology*, 2008, 48(5): 1644–1654.
166. Zhang, X., Lemasters, J.J. Translocation of iron from lysosomes to mitochondria during ischemia predisposes to injury after reperfusion in rat hepatocytes. *Free Radic Biol Med*, 2013, 63: 243–253.
167. Sweeney, P., Park, H., Baumann, M., Dunlop, J., Frydman, J., Kopito, R., McCampbell, A., Leblanc, G., Venkateswaran, A., Nurmi, A., Hodgson, R. Protein misfolding in neurodegenerative diseases: Implications and strategies. *Transl Neurodegener*, 2017, 6(6): 017–0077.
168. Wilson, D.M., III, Cookson, M.R., Van Den Bosch, L., Zetterberg, H., Holtzman, D.M., Dewachter, I. Hallmarks of neurodegenerative diseases. *Cell*, 2023, 186(4): 693–714.
169. DeTure, M.A., Dickson, D.W. The neuropathological diagnosis of Alzheimer's disease. *Mol Neurodegener*, 2019, 14(1): 32.
170. Garofalo, M., Pandini, C., Bordoni, M., Pansarasa, O., Rey, F., Costa, A., Minafra, B., Diamanti, L., Zucca, S., Carelli, S., Cereda, C., Gagliardi, S. Alzheimer's, Parkinson's disease and amyotrophic lateral sclerosis Gene expression patterns divergence reveals different grade of RNA metabolism involvement. *Int J Mol Sci*, 2020, 21(24): 9500.
171. Ahmad, K., Baig, M.H., Mushtaq, G., Kamal, M.A., Greig, N.H., Choi, I. Commonalities in biological pathways, genetics, and cellular mechanism between Alzheimer disease and other neurodegenerative diseases: An in silico-updated overview. *Curr Alzheimer Res*, 2017, 14(11): 1190–1197.
172. Krogsaeter, E., Rosato, A.S., Grimm, C. TRPMLs and TPCs: Targets for lysosomal storage and neurodegenerative disease therapy? *Cell Calcium*, 2022, 103: 102553
173. Hu, Y., Hu, K., Song, H., Pawitan, Y., Piehl, F., Fang, F. Infections among individuals with multiple sclerosis, Alzheimer's disease and Parkinson's disease. *Brain Commun*, 2023, 5(2): fcad065.
174. Parenti, G., Medina, D.L., Ballabio, A. The rapidly evolving view of lysosomal storage diseases. *EMBO Mol Med*, 2021, 13(2): 18.
175. Platt, F.M., Boland, B., van der Spoel, A.C. Lysosomal storage disorders: The cellular impact of lysosomal dysfunction. *J Cell Biol*, 2012, 199(5): 723–734.
176. Kloska, A., Tylki-Szymańska, A., Wegrzyn, G. Lysosomal storage diseases–an overview. *Postepy Biochem*, 2011, 57(2): 128–32.
177. Sheth, J., Nair, A., Jee, B. Lysosomal storage disorders: From biology to the clinic with reference to India. *Lancet Reg Health Southeast Asia*, 2022, 9: 100108.
178. Misko, A., Wood, L., Kiselyov, K., Slaugenhaupt, S., Grishchuk, Y. Progress in elucidating pathophysiology of mucolipidosis IV. *Neurosci Lett*, 2021, 755(135944): 11.
179. Micsenyi, M.C., Dobrenis, K., Stephney, G., Pickel, J., Vanier, M.T., Slaugenhaupt, S.A., Walkley, S.U. Neuropathology of the Mcoln1$^{-/-}$ knockout mouse model of mucolipidosis type IV. *J Neuropathol Exp Neurol*, 2009, 68(2): 125–135.
180. Raghavan, S., Leshinsky, E., Kolodny, E. G M2-ganglioside metabolism in situ in mucolipidosis IV fibroblasts. *Neurochem Res*, 1999, 24: 475–479.
181. Zhong, X.Z., Sun, X., Cao, Q., Dong, G., Schiffmann, R., Dong, X.P. BK channel agonist represents a potential therapeutic approach for lysosomal storage diseases. *Sci Rep*, 2016, 6(1): 33684.
182. Oddo, S., Caccamo, A., Kitazawa, M., Tseng, B.P., LaFerla, F.M. Amyloid deposition precedes tangle formation in a triple transgenic model of Alzheimer's disease. *Neurobiol Aging*, 2003, 24(8): 1063–1070.
183. Medeiros, R., Baglietto-Vargas, D., LaFerla, F.M. The role of tau in Alzheimer's disease and related disorders. *CNS Neurosci Ther*, 2011, 17(5): 514–524.
184. Bertram, L., Tanzi, R.E. Thirty years of Alzheimer's disease genetics: The implications of systematic meta-analyses. *Nat Rev Neurosci*, 2008, 9(10): 768–78.

185. LaFerla, F.M., Green, K.N., Oddo, S. Intracellular amyloid-beta in Alzheimer's disease. *Nat Rev Neurosci*, 2007, 8(7): 499–509.
186. Bae, M., Patel, N., Xu, H., Lee, M., Tominaga-Yamanaka, K., Nath, A., Geiger, J., Gorospe, M., Mattson, M.P., Haughey, N.J. Activation of TRPML1 clears intraneuronal Aβ in preclinical models of HIV infection. *J Neurosci*, 2014, 34(34): 11485–11503.
187. Schutzmann, M.P., Hasecke, F., Bachmann, S., Zielinski, M., Hansch, S., Schroder, G.F., Zempel, H., Hoyer, W. Endo-lysosomal Abeta concentration and pH trigger formation of Abeta oligomers that potently induce Tau missorting. *Nat Commun*, 2021, 12(1): 4634.
188. Lee, J.H., Yu, W.H., Kumar, A., Lee, S., Mohan, P.S., Peterhoff, C.M., Wolfe, D.M., Martinez-Vicente, M., Massey, A.C., Sovak, G., Uchiyama, Y., Westaway, D., Cuervo, A.M., Nixon, R.A. Lysosomal proteolysis and autophagy require presenilin 1 and are disrupted by Alzheimer-related PS1 mutations. *Cell*, 2010, 141(7): 1146–1158.
189. Nixon, R.A., Amyloid precursor protein and endosomal-lysosomal dysfunction in Alzheimer's disease: Inseparable partners in a multifactorial disease. *FASEB J*, 2017, 31(7): 2729–2743.
190. Tagawa, K., Yazaki, M., Kinouchi, T., Maruyama, K., Sorimachi, H., Tsuchiya, T., Suzuki, K., Ishiura, S. Amyloid precursor protein is found in lysosomes. *Gerontology*, 1993, 39 **(Suppl 1)**: 24–29.
191. Haass, C., Koo, E.H., Mellon, A., Hung, A.Y., Selkoe, D.J. Targeting of cell-surface beta-amyloid precursor protein to lysosomes: Alternative processing into amyloid-bearing fragments. *Nature*, 1992, 357(6378): 500–503.
192. Pasternak, S.H., Bagshaw, R.D., Guiral, M., Zhang, S., Ackerley, C.A., Pak, B.J., Callahan, J.W., Mahuran, D.J. Presenilin-1, nicastrin, amyloid precursor protein, and gamma-secretase activity are co-localized in the lysosomal membrane. *J Biol Chem*, 2003, 278(29): 26687–26694.
193. Xu, Y., Du, S., Marsh, J.A., Horie, K., Sato, C., Ballabio, A., Karch, C.M., Holtzman, D.M., Zheng, H. TFEB regulates lysosomal exocytosis of tau and its loss of function exacerbates tau pathology and spreading. *Mol Psychiatry*, 2021, 26(10): 5925–5939.
194. Lee, J.H., McBrayer, M.K., Wolfe, D.M., Haslett, L.J., Kumar, A., Sato, Y., Lie, P.P., Mohan, P., Coffey, E.E., Kompella, U., Mitchell, C.H., Lloyd-Evans, E., Nixon, R.A. Presenilin 1 maintains lysosomal Ca^{2+} homeostasis via TRPML1 by regulating vATPase-mediated lysosome acidification. *Cell Rep*, 2015, 12(9): 1430–1444.
195. Lie, P.P.Y., Yoo, L., Goulbourne, C.N., Berg, M.J., Stavrides, P., Huo, C., Lee, J.H., Nixon, R.A. Axonal transport of late endosomes and amphisomes is selectively modulated by local Ca^{2+} efflux and disrupted by PSEN1 loss of function. *Sci Adv*, 2022, 8(17): 29.
196. Zoncu, R., Bar-Peled, L., Efeyan, A., Wang, S., Sancak, Y., Sabatini, D.M. mTORC1 senses lysosomal amino acids through an inside-out mechanism that requires the vacuolar H^+-ATPase. *Science*, 2011, 334(6056): 678–83.
197. Fedele, A.O., Proud, C.G. Chloroquine and bafilomycin a mimic lysosomal storage disorders and impair mTORC1 signalling. *Biosci Rep*, 2020, 40(4): BSR20200905.
198. Talboom, J.S., Velazquez, R. Oddo, S. The mammalian target of rapamycin at the crossroad between cognitive aging and Alzheimer's disease. *NPJ Aging Mech Dis*, 2015, 1: 15008.
199. Oddo, S., The role of mTOR signaling in Alzheimer disease. *Front Biosci (Schol Ed)*, 2012, 4(3): 941–952.
200. Somogyi, A., Kirkham, E.D., Lloyd-Evans, E., Winston, J., Allen, N.D., Mackrill, J.J., Anderson, K.E., Hawkins, P.T., Gardiner, S.E., Waller-Evans, H., Sims, R., Boland, B., O'Neill, C. The synthetic TRPML1 agonist ML-SA1 rescues Alzheimer-related alterations of the endosomal-autophagic-lysosomal system. *J Cell Sci*, 2023, 136(6): 21.
201. Klein, A.D., Mazzulli, J.R. Is Parkinson's disease a lysosomal disorder? *Brain*, 2018, 141(8): 2255–2262.
202. Tysnes, O.B., Storstein, A. Epidemiology of Parkinson's disease. *J Neural Transm*, 2017, 124(8): 901–905.
203. Sassone, J., Reale, C., Dati, G., Regoni, M., Pellecchia, M.T., Garavaglia, B. The role of VPS35 in the pathobiology of Parkinson's disease. *Cell Mol Neurobiol*, 2021, 41(2): 199–227.
204. Arotcarena, M.L., Teil, M., Dehay, B. Autophagy in synucleinopathy: The overwhelmed and defective machinery. *Cells*, 2019, 8(6): 565.
205. Tysnes, O.-B., Storstein, A. Epidemiology of Parkinson's disease. *J Neural Transmi*, 2017, 124(8): 901–905.
206. Funayama, M., Nishioka, K., Li, Y., Hattori, N. Molecular genetics of Parkinson's disease: Contributions and global trends. *J Hum Genet*, 2023, 68(3): 125–130.
207. Braak, H., Del Tredici, K. Neuropathological staging of brain pathology in sporadic Parkinson's disease: Separating the wheat from the chaff. *J Parkinsons Dis*, 2017, 7(s1): S71–S85.
208. Gómez-Benito, M., Granado, N., García-Sanz, P., Michel, A., Dumoulin, M., Moratalla, R. Modeling Parkinson's disease with the alpha-synuclein protein. *Front Pharmacol*, 2020, 11: 356.

209. Fernández, B., Olmedo, P., Gil, F., Fdez, E., Naaldijk, Y., Rivero-Ríos, P., Bracher, F., Grimm, C., Churchill, G.C., Hilfiker, S. Iron-induced cytotoxicity mediated by endolysosomal TRPML1 channels is reverted by TFEB. *Cell Death Dis*, 2022, 13(12): 1047.
210. Tsunemi, T., Perez-Rosello, T., Ishiguro, Y., Yoroisaka, A., Jeon, S., Hamada, K., Rammonhan, M., Wong, Y.C., Xie, Z., Akamatsu, W., Mazzulli, J.R., Surmeier, D.J., Hattori, N., Krainc, D. Increased lysosomal exocytosis induced by lysosomal Ca^{2+} channel agonists protects human dopaminergic neurons from α-synuclein toxicity. *J Neurosci*, 2019, 39(29): 5760–5772.
211. Rowland, L.P., Shneider, N.A. Amyotrophic lateral sclerosis. *N Engl J Med*, 2001, 344(22): 1688–700.
212. Murch, S.J., Cox, P.A., Banack, S.A., Steele, J.C., Sacks, O.W. Occurrence of beta-methylamino-l-alanine (BMAA) in ALS/PDC patients from Guam. *Acta Neurol Scand*, 2004, 110(4): 267–269.
213. Blokhuis, A.M., Groen, E.J., Koppers, M., van den Berg, L.H., Pasterkamp, R.J. Protein aggregation in amyotrophic lateral sclerosis. *Acta Neuropathol*, 2013, 125(6): 777–794.
214. Tedeschi, V., Petrozziello, T., Sisalli, M.J., Boscia, F., Canzoniero, L.M.T., Secondo, A. The activation of mucolipin TRP channel 1 (TRPML1) protects motor neurons from l-BMAA neurotoxicity by promoting autophagic clearance. *Sci Rep*, 2019, 9(1): 10743.
215. Zou, J., Hu, B., Arpag, S., Yan, Q., Hamilton, A., Zeng, Y.-S., Vanoye, C.G., Li, J. Reactivation of lysosomal Ca^{2+} efflux rescues abnormal lysosomal storage in FIG4-deficient cells. *J Neurosci* 2015, 35(17): 6801–6812.
216. Chow, C.Y., Landers, J.E., Bergren, S.K., Sapp, P.C., Grant, A.E., Jones, J.M., Everett, L., Lenk, G.M., McKenna-Yasek, D.M., Weisman, L.S., Figlewicz, D., Brown, R.H., Meisler, M.H. Deleterious variants of FIG4, a phosphoinositide phosphatase, in patients with ALS. *Am J Hum Genet*, 2009, 84(1): 85–8.
217. Davidson, S.M., Vander Heiden, M.G. Critical functions of the lysosome in cancer biology. *Annu Rev Pharmacol Toxicol*, 2017, 57: 481–507.
218. Kuester, D., Lippert, H., Roessner, A., Krueger, S. The cathepsin family and their role in colorectal cancer. *Pathol Res Pract*, 2008, 204(7): 491–500.
219. Ibrahim, S.A., El-Ghonaimy, E.A., Hassan, H., Mahana, N., Mahmoud, M.A., El-Mamlouk, T., El-Shinawi, M., Mohamed, M.M. Hormonal-receptor positive breast cancer: IL-6 augments invasion and lymph node metastasis via stimulating cathepsin b expression. *J Adv Res*, 2016, 7(5): 661–70.
220. Mohamed, M.M., Sloane, B.F., Cysteine cathepsins: Multifunctional enzymes in cancer. *Nat Rev Cancer*, 2006, 6(10): 764–775.
221. Gocheva, V., Joyce, J.A. Cysteine cathepsins and the cutting edge of cancer invasion. *Cell Cycle*, 2007, 6(1): 60–64.
222. Basu, S., Cheriyamundath, S., Gavert, N., Brabletz, T., Haase, G., Ben-Ze'ev, A. Increased expression of cathepsin d is required for L1-mediated colon cancer progression. *Oncotarget*, 2019, 10(50): 5217–5228.
223. Hinton, A., Sennoune, S.R., Bond, S., Fang, M., Reuveni, M., Sahagian, G.G., Jay, D., Martinez-Zaguilan, R., Forgac, M. Function of a subunit isoforms of the v-ATPase in pH homeostasis and in vitro invasion of MDA-MB231 human breast cancer cells. *J Biol Chem*, 2009, 284(24): 16400–16408.
224. Pérez-Sayáns, M., Reboiras-López, M.D., Somoza-Martín, J.M., Barros-Angueira, F., Diz, P.G., Rey, J.M., García-García, A. Measurement of ATP6V1C1 expression in brush cytology samples as a diagnostic and prognostic marker in oral squamous cell carcinoma. *Cancer Biol Ther*, 2010, 9(12): 1057–1064.
225. Chung, C., Mader, C.C., Schmitz, J.C., Atladottir, J., Fitchev, P., Cornwell, M.L., Koleske, A.J., Crawford, S.E., Gorelick, F. The vacuolar-ATPase modulates matrix metalloproteinase isoforms in human pancreatic cancer. *Lab Invest*, 2011, 91(5): 732–743.
226. Nishisho, T., Hata, K., Nakanishi, M., Morita, Y., Sun-Wada, G.H., Wada, Y., Yasui, N., Yoneda, T. The a3 isoform vacuolar type H^+-ATPase promotes distant metastasis in the mouse B16 melanoma cells. *Mol Cancer Res*, 2011, 9(7): 845–855.
227. Huang, L., Lu, Q., Han, Y., Li, Z., Zhang, Z., Li, X. ABCG2/V-ATPase was associated with the drug resistance and tumor metastasis of esophageal squamous cancer cells. *Diagn Pathol*, 2012, 7(180): 1746–1596.
228. Xu, J., Xie, R., Liu, X., Wen, G., Jin, H., Yu, Z., Jiang, Y., Zhao, Z., Yang, Y., Ji, B., Dong, H., Tuo, B. Expression and functional role of vacuolar H^+-ATPase in human hepatocellular carcinoma. *Carcinogenesis*, 2012, 33(12): 2432–2440.
229. Ohta, T., Numata, M., Yagishita, H., Futagami, F., Tsukioka, Y., Kitagawa, H., Kayahara, M., Nagakawa, T., Miyazaki, I., Yamamoto, M., Iseki, S., Ohkuma, S. Expression of 16 kDa proteolipid of vacuolar-type H^+-ATPase in human pancreatic cancer. *Br J Cancer*, 1996, 73(12): 1511–1517.
230. Perera, R.M., Stoykova, S., Nicolay, B.N., Ross, K.N., Fitamant, J., Boukhali, M., Lengrand, J., Deshpande, V., Selig, M.K., Ferrone, C.R., Settleman, J., Stephanopoulos, G., Dyson, N.J., Zoncu, R., Ramaswamy, S., Haas, W., Bardeesy, N. Transcriptional control of autophagy-lysosome function drives pancreatic cancer metabolism. *Nature*, 2015, 524(7565): 361–365.

231. Kroemer, G., Jäättelä, M. Lysosomes and autophagy in cell death control. *Nat Rev Cancer*, 2005, 5(11): 886–897.
232. Hämälistö, S., Jäättelä, M. Lysosomes in cancer-living on the edge (of the cell). *Curr Opin Cell Biol*, 2016, 39: 69–76.
233. Kikuta, J., Ishii, M. Osteoclast migration, differentiation and function: Novel therapeutic targets for rheumatic diseases. *Rheumatology*, 2013, 52(2): 226–234.
234. Vyas, A., Gomez-Casal, R., Cruz-Rangel, S., Villanueva, H., Sikora, A.G., Rajagopalan, P., Basu, D., Pacheco, J., Hammond, G.R.V., Kiselyov, K., Duvvuri, U. Lysosomal inhibition sensitizes TMEM16A-expressing cancer cells to chemotherapy. *Proc Natl Acad Sci U S A*, 2022, 119(12): e2100670119.
235. Groth-Pedersen, L., Jäättelä, M. Combating apoptosis and multidrug resistant cancers by targeting lysosomes. *Cancer Lett*, 2013, 332(2): 265–274.
236. Yanes, R.E., Tarn, D., Hwang, A.A., Ferris, D.P., Sherman, S.P., Thomas, C.R., Lu, J., Pyle, A.D., Zink, J.I., Tamanoi, F. Involvement of lysosomal exocytosis in the excretion of mesoporous silica nanoparticles and enhancement of the drug delivery effect by exocytosis inhibition. *Small*, 2013, 9(5): 697–704.
237. Simanshu, D.K., Nissley, D.V., McCormick, F. RAS proteins and their regulators in human disease. *Cell*, 2017, 170(1): 17–33.
238. Kasitinon, S.Y., Eskiocak, U., Martin, M., Bezwada, D., Khivansara, V., Tasdogan, A., Zhao, Z., Mathews, T., Aurora, A.B., Morrison, S.J. TRPML1 promotes protein homeostasis in melanoma cells by negatively regulating MAPK and mTORC1 signaling. *Cell Rep*, 2019, 28(9): 2293–2305 e9.
239. Yin, C., Zhang, H., Liu, X., Zhang, Y., Bai, X., Wang, L., Li, H., Li, X., Zhang, S., Zhang, L. Downregulated MCOLN1 attenuates the progression of non-small-cell lung cancer by inhibiting lysosome-autophagy. *Cancer Manag Res*, 2019, 11: 8607–8617.
240. Hu, Z.D., Yan, J., Cao, K.Y., Yin, Z.Q., Xin, W.W., Zhang, M.F. MCOLN1 promotes proliferation and predicts poor survival of patients with pancreatic ductal adenocarcinoma. *Dis Markers*, 2019, 5: 9436047.
241. Morelli, M.B., Amantini, C., Tomassoni, D., Nabissi, M., Arcella, A., Santoni, G. Transient receptor potential mucolipin-1 channels in glioblastoma: Role in Patient's survival. *Cancers*, 2019, 11(4): 525.
242. Scotto Rosato, A., Montefusco, S., Soldati, C., Di Paola, S., Capuozzo, A., Monfregola, J., Polishchuk, E., Amabile, A., Grimm, C., Lombardo, A., De Matteis, M.A., Ballabio, A., Medina, D.L. TRPML1 links lysosomal calcium to autophagosome biogenesis through the activation of the CaMKKβ/VPS34 pathway. *Nat Commun*, 2019, 10(1): 5630.
243. Wang, W., Gao, Q., Yang, M., Zhang, X., Yu, L., Lawas, M., Li, X., Bryant-Genevier, M., Southall, N.T., Marugan, J., Ferrer, M., Xu, H. Up-regulation of lysosomal TRPML1 channels is essential for lysosomal adaptation to nutrient starvation. *Proc Natl Acad Sci U S A*, 2015, 112(11): E1373–E1381.
244. Commisso, C., Davidson, S.M., Soydaner-Azeloglu, R.G., Parker, S.J., Kamphorst, J.J., Hackett, S., Grabocka, E., Nofal, M., Drebin, J.A., Thompson, C.B., Rabinowitz, J.D., Metallo, C.M., Vander Heiden, M.G., Bar-Sagi, D. Macropinocytosis of protein is an amino acid supply route in Ras-transformed cells. *Nature*, 2013, 497(7451): 633–637.
245. Palm, W., Park, Y., Wright, K., Pavlova, N.N., Tuveson, D.A., Thompson, C.B. The utilization of extracellular proteins as nutrients is suppressed by mTORC1. *Cell*, 2015, 162(2): 259–270.
246. Kamphorst, J.J., Nofal, M., Commisso, C., Hackett, S.R., Lu, W., Grabocka, E., Vander Heiden, M.G., Miller, G., Drebin, J.A., Bar-Sagi, D., Thompson, C.B., Rabinowitz, J.D. Human pancreatic cancer tumors are nutrient poor and tumor cells actively scavenge extracellular protein. *Cancer Res*, 2015, 75(3): 544–553.
247. Krishna, S., Palm, W., Lee, Y., Yang, W., Bandyopadhyay, U., Xu, H., Florey, O., Thompson, C.B., Overholtzer, M. PIKfyve regulates vacuole maturation and nutrient recovery following engulfment. *Dev Cell*, 2016, 38(5): 536–547.
248. Lyu, L., Jin, X., Li, Z., Liu, S., Li, Y., Su, R., Su, H. TBBPA regulates calcium-mediated lysosomal exocytosis and thereby promotes invasion and migration in hepatocellular carcinoma. *Ecotoxicol Environ Saf*, 2020, 192: 110255.
249. Sun, X., Xu, M., Cao, Q., Huang, P., Zhu, X., Dong, X.P. A lysosomal K^+ channel regulates large particle phagocytosis by facilitating lysosome Ca^{2+} release. *Sci Rep*, 2020, 10(1): 1038.
250. Dayam, R.M., Saric, A., Shilliday, R.E., Botelho, R.J. The phosphoinositide-gated lysosomal Ca^{2+} channel, TRPML1, is required for phagosome maturation. *Traffic*, 2015, 16(9): 1010–1026.
251. Dayam, R.M., Sun, C.X., Choy, C.H., Mancuso, G., Glogauer, M., Botelho, R.J. The lipid kinase PIKfyve coordinates the neutrophil immune response through the activation of the Rac GTPase. *J Immunol*, 2017, 199(6): 2096–2105.

252. Gray, M.A., Choy, C.H., Dayam, R.M., Ospina-Escobar, E., Somerville, A., Xiao, X., Ferguson, S.M., Botelho, R.J. Phagocytosis enhances lysosomal and bactericidal properties by activating the transcription factor TFEB. *Curr Biol*, 2016, 26(15): 1955–1964.
253. Bedard, M., van der Niet, S., Bernard, E.M., Babunovic, G., Cheng, T.Y., Aylan, B., Grootemaat, A.E., Raman, S., Botella, L., Ishikawa, E., O'Sullivan, M.P., O'Leary, S., Mayfield, J.A., Buter, J., Minnaard, A.J., Fortune, S.M., Murphy, L.O., Ory, D.S., Keane, J., Yamasaki, S., … Moody, D.B. A terpene nucleoside from M. Tuberculosis induces lysosomal lipid storage in foamy macrophages. *J Clin Invest*, 2023, 133(6): e161944.
254. Capurro, M.I., Greenfield, L.K., Prashar, A., Xia, S., Abdullah, M., Wong, H., Zhong, X.Z., Bertaux-Skeirik, N., Chakrabarti, J., Siddiqui, I., O'Brien, C., Dong, X., Robinson, L., Peek, R.M., Jr, Philpott, D.J., Zavros, Y., Helmrath, M., Jones, N.L. VacA generates a protective intracellular reservoir for helicobacter pylori that is eliminated by activation of the lysosomal calcium channel TRPML1. *Nat Microbiol*, 2019, 4(8): 1411–1423.
255. Beutler, B., Jiang, Z., Georgel, P., Crozat, K., Croker, B., Rutschmann, S., Du, X., Hoebe, K. Genetic analysis of host resistance: Toll-like receptor signaling and immunity at large. *Annu Rev Immunol*, 2006, 24: 353–389.
256. Kaisho, T., Akira, S. Toll-like receptor function and signaling. *J Allergy Clin Immunol*, 2006, 117(5): 979–987.
257. Kawai, T., Akira, S. The role of pattern-recognition receptors in innate immunity: Update on toll-like receptors. *Nat Immunol*, 2010, 11(5): 373–384.
258. Li, X., Saitoh, S.-i., Shibata, T., Tanimura, N., Fukui, R., Miyake, K. Mucolipin 1 positively regulates TLR7 responses in dendritic cells by facilitating RNA transportation to lysosomes. *Int Immunol*, 2014, 27(2): 83–94.
259. Bretou, M., Sáez, P.J., Sanséau, D., Maurin, M., Lankar, D., Chabaud, M., Spampanato, C., Malbec, O., Barbier, L., Muallem, S., Maiuri, P., Ballabio, A., Helft, J., Piel, M., Vargas, P., Lennon-Duménil, A.M. Lysosome signaling controls the migration of dendritic cells. *Sci Immunol*, 2017, 2(16): eaak9573.
260. Goodridge, J.P., Jacobs, B., Saetersmoen, M.L., Clement, D., Hammer, Q., Clancy, T., Skarpen, E., Brech, A., Landskron, J., Grimm, C., Pfefferle, A., Meza-Zepeda, L., Lorenz, S., Wiiger, M.T., Louch, W.E., Ask, E.H., Liu, L.L., Oei, V.Y.S., Kjällquist, U., Linnarsson, S., … Malmberg, K.J. Remodeling of secretory lysosomes during education tunes functional potential in NK cells. *Nat Commun*, 2019, 10(1): 514.

8 TRP Channels in Cardiac Diseases

Jun Zhang and Xiaoqiang Yao

8.1 INTRODUCTION

TRP channels are activated by diverse physical (e.g., mechanical stretch) and chemical stimuli (e.g., hormones, neurotransmitters, cytokines). The channels are permeable to a wide range of cations, including Ca^{2+}, Na^{+}, and K^{+}. They are extensively expressed in cardiac tissues as detected by RT-PCR, in situ hybridization, Western blot, Northern blot, immunostaining, and patch clamp recordings (Yue et al., 2015). However, expression patterns of different TRP isoforms in the heart are heterogeneous across different species, different cardiac regions (sinoatrial node, atrial and ventricular myocardium, and Purkinje fibers), and different cell types (cardiomyocytes, cardiac fibroblasts, and endothelial cells) (Earley and Brayden, 2015; Hof et al., 2019). In the heart, the activity of TRP channels contributes to the regulation of Ca^{2+} homeostasis, contractility, pacemaking, and conduction and modulation of action potentials. These effects are achieved by mediating changes in intracellular Ca^{2+} concentrations via Ca^{2+} entry and/or changes in membrane potentials via cation influx through receptor- or store-operated mechanisms. Dysfunction of TRP channels may contribute to the pathogenesis of several cardiac diseases, most notably cardiac hypertrophy, fibrosis, arrhythmia, ischemia and reperfusion-induced myocardial injury. In this chapter, we summarize the current knowledge regarding the contribution of TRP channels to the development of cardiac hypertrophy, arrhythmia, and ischemia/reperfusion injury, based on the accumulating experimental evidence, particularly those using TRP channel-specific inhibitors or cardiac-specific transgenic or knockout animal models. These studies suggest that TRP isoforms could be potential therapeutic targets for cardiac diseases.

8.2 TRP CHANNELS IN CARDIAC HYPERTROPHY

Cardiac hypertrophy, an increase in heart muscle mass, reflects remodeling of the myocardium in response to mechanical stress and neurohumoral stimuli. Cardiac hypertrophy is initially compensatory as a result of adaptive physiological responses and is reversible. However, persistence of hypertrophy for longer periods leads to maladaptive pathological hypertrophy, characterized by impaired Ca^{2+} handling, myocardial cell death, tissue fibrosis, contractile dysfunction, and heart failure (Nakamura and Sadoshima, 2018).

Cardiac hypertrophy is a highly complicated process with the involvement of numerous signaling pathways. In physiological hypertrophy, the most well-known signaling pathway is the insulin-like growth factor 1-PI3K-Akt pathway (Nakamura and Sadoshima, 2018). In contrast, pathological cardiac hypertrophy may result from dysregulation in neurohumoral factors and mechanical stress (Nakamura and Sadoshima, 2018). The former is often mediated by G-protein coupled receptors, such as α-adrenergic receptors, angiotensin II (Ang II) receptors and endothelin receptors. Besides Ca^{2+} signaling, small G proteins, mitogen-activated protein kinases, and histone deacetylases also participate in the development of pathological hypertrophy (Bernardo et al., 2010). Because TRP channels induce cytosolic Ca^{2+} changes in cardiomyocytes and cardiac fibroblasts, dysregulation of these channels may contribute to the development of pathological cardiac hypertrophy. Of note, the Ca^{2+}-dependent transcriptional pathway of calcineurin (CaN)/nuclear factor of the activated T cell (NFAT) is a key signaling pathway in pathological cardiac hypertrophy (Houser and Molkentin,

DOI: 10.1201/9781003402732-8

2008). CaN dephosphorylates transcription factors of the NFAT family and induces translocation of NFAT into the nucleus, resulting in the activation of hypertrophic genes.

Cardiac hypertrophy is often studied in animal models and in cultured cardiac cells. The commonly used animal models of cardiac hypertrophy include: surgical transverse aortic constriction (TAC) to create pressure overload either by using sutures or by application of metallic clips (Camacho et al., 2016); renal artery occlusion; and infusion of hypertrophic neurohumoral factors, such as catecholamines, Ang II or endothelin-1 (ET-1).

8.2.1 TRPC SUBFAMILY

Among the TRP channels, TRPC isoforms are the best-studied in cardiac hypertrophy (Eder and Molkentin, 2011). In contrast to the Ca^{2+} entry mediated by voltage-gated Ca^{2+} channels (VGCCs) and the subsequent release of Ca^{2+} from the sarcoplasmic reticulum (SR), which causes transient, high-amplitude cytosolic Ca^{2+} alterations, TRPC-mediated Ca^{2+} entry in response to catecholamines, Ang II and ET-1 through the receptor-operated or store-operated mechanism may elicit a more sustained increase in cytosolic Ca^{2+} (Dolmetsch et al., 1997), which subsequently induces NFAT nuclear translocation to promote cardiac hypertrophy. In a store-operated mechanism, Ca^{2+} depletion in endoplasmic reticulum (ER) Ca^{2+} stores activates STIM1, which is an ER Ca^{2+} sensor. STIM1 in turn activates Orai1 and TRPC to induce store-operated Ca^{2+} entry (SOCE) (Lopez et al., 2020). The association between TRPC channels and the CaN/NFAT activation is well recognized, as TRPC genes harbor conserved NFAT consensus sites in the promoter. Once activated, NFAT can also stimulate the expression of TRPC channels through a positive feedback mechanism (Watanabe et al., 2008). Based on sequence homology, TRPC can be subdivided into several subfamilies: TRPC3/6/7, TRPC1, and TRPC4/5. Moreover, TRPC1 can co-assemble with TRPC4 or TRPC5 to form TRPC1/4 or TRPC1/5 heteromeric channels.

8.2.1.1 TRPC3/6

The expression of TRPC3 is increased in cultured hypertrophic cardiomyocytes and in hypertrophic left ventricles of mice subjected to TAC (Bush et al., 2006). Overexpression of TRPC3 stimulates CaN/NFAT (Bush et al., 2006; Nakayama et al., 2006), induces hypertrophy in cultured cardiomyocytes, and promotes cardiac hypertrophy in mice *in vivo* (Nakayama et al., 2006). In contrast, knockout of the TRPC3 gene alleviates high salt-induced cardiac hypertrophy (Ma et al., 2019). Mechanistically, in addition to the TRPC3-CaN/NFAT signaling axis (Bush et al., 2006; Nakayama et al., 2006), TRPC3 may couple with NADPH oxidase (NOX2) to form a protein complex to stabilize NOX2 (Kato et al., 2022; Shimauchi et al., 2017), which functions as a positive regulator of reactive oxygen species (ROS), eventually leading to the development of cardiac hypertrophy (Kitajima et al., 2016). A recent study also suggested that TRPC3 may act through soluble epoxide hydrolase to mediate cardiac hypertrophy in hyperhomocysteinemic rats (Zhou et al., 2023).

Similar to TRPC3, the expression of TRPC6 is also upregulated in the TAC-induced mouse hypertrophic heart and failing human hearts. Furthermore, mice with cardiac-specific TRPC6 overexpression have an increased expression of hypertrophic markers, whereas knockdown of TRPC6 attenuates hypertrophic signaling induced by hypertrophic agonists phenylephrine (PE) and ET-1 (Kuwahara et al., 2006). Intriguingly, a report shows that knockout of TRPC3 or TRPC6 alone fails to protect against pressure overload-induced hypertrophy, whereas combined deletion of both is cardioprotective, supporting a role for both channels in cardiac hypertrophy (Seo et al., 2014b).

We and others have shown that TRPC1/3/6 and Orai1 can be inhibited by protein kinase G (PKG)-mediated phosphorylation through guanylate cyclase-dependent cyclic guanosine 3',5'-monophosphate (cGMP) production (Koitabashi et al., 2010; Kwan et al., 2004; 2006). The cGMP/PKG inhibition on TRPC3/6 and Orai1 negatively regulates CaN/NFAT, thereby exerting antihypertrophic effects. This mechanism underlies the antihypertrophic effects of cardiac natriuretic peptides, guanylyl cyclase, and phosphodiesterase type-5 (Kinoshita et al., 2010). Therefore, it is tempting that targeting the guanylyl cyclase/cGMP/PKG/TRPC axis may serve as a novel strategy to alleviate cardiac hypertrophy.

8.2.1.2 TRPC1/4/5

TRPC1 expression is upregulated in hypertrophic hearts in mice (Ohba et al., 2007) and in induced pluripotent stem cell (iPSC)-derived cardiomyocytes derived from hypertrophic patients (Ohba et al., 2007). Deletion of the TRPC1 gene diminishes the hypertrophic response of the heart in mice (Seth et al., 2009) and also reduces cardiomyocyte hypertrophy in iPSC-derived cardiomyocytes derived from hypertrophic patients (Tang et al., 2019). A dominant-negative construct of TRPC4 also attenuates cardiac hypertrophy (Wu et al., 2010). Furthermore, TRPC1 and TRPC4 may form a heteromeric complex to mediate background Ca^{2+} entry in cardiomyocytes, consequently contributing to cardiac hypertrophy (Camacho Londono et al., 2015). Intriguingly, Camacho Londono et al. showed that only the double knockout of both TRPC1 and TRPC4, but not the single knockout of either TRPC1 or TRPC4, can attenuate cardiac hypertrophy (Camacho Londono et al., 2015). The study also showed that TRPC3/6 is not involved in neurohormonal and mechanical stimulation-induced cardiac hypertrophy (Camacho Londono et al., 2015), which contradicts other reports (Kuwahara et al., 2006; Ma et al., 2019; Seo et al., 2014b). Such discrepancy could be due to the different genetic backgrounds of different mice and/or compensatory changes in the expression of other TRPCs when one TRPC isoform is knocked out.

Unlike other TRPCs that have pro-hypertrophic roles, current evidence suggests that TRPC5 may exert an anti-hypertrophic role. ATP stimulation of TRPC5 increases the production of nitric oxide (NO), which is a protective agent against cardiac hypertrophy. Herein, TRPC5 acts through the NO-cGMP-PKG axis to suppress TRPC3/6-associated NFAT activation, consequently alleviating cardiac hypertrophy (Sunggip et al., 2018).

8.2.2 TRPM4

TRPM4 is a Ca^{2+}-activated voltage-dependent nonselective cation channel permeable to Na^{+} and K^{+}, but impermeable to Ca^{2+} (Owsianik et al., 2006). The expression level of TRPM4 is relatively high in the heart compared to other tissues, and the channel is present in different cardiac cell types, including atrial and ventricular cardiomyocytes, sinoatrial node cells, Purkinje cells, and fibroblasts (Amarouch and El Hilaly, 2020).

TRPM4 could be anti-hypertrophic or pro-hypertrophic under different conditions. TRPM4 has an anti-hypertrophic role in cardiac hypertrophy induced by abnormal neurohormonal stimuli. TRPM4 global knockout in mice results in left ventricle eccentric hypertrophy, with an increase in wall thickness and chamber size (Demion et al., 2014). These mice also display high blood pressure and high plasma catecholamine level, which are risk factors for cardiac hypertrophy (Mathar et al., 2010). Furthermore, cardiac-specific knockout of TRPM4 ($TRPM4^{cKO}$) in mice also causes cardiac hypertrophy, excessive expression of hypertrophic genes in response to chronic Ang II challenge, and increased store-operated Ca^{2+} current and CaN phosphatase activity in cardiomyocytes (Kecskes et al., 2015). However, a recent study reported contradictory results that suggest a pro-hypertrophic role of TRPM4 in TAC-induced cardiac hypertrophy (Guo et al., 2021). In this study, cardiac-specific TRPM4 knockout was found to reduce the cardiac hypertrophy in response to TAC by 50%, accompanied by attenuated activation of hypertrophic markers (Guo et al., 2021). Furthermore, Gueffier et al. also showed that TRPM4 knockout may reduce physiological cardiac hypertrophy under exercise (Gueffier et al., 2017). Together, it is likely that TRPM4 is anti-hypertrophic in cardiac hypertrophy induced by abnormal neurohormonal stimuli whereas it is pro-hypertrophic in pressure overload-induced cardiac hypertrophy.

8.2.3 TRPV1/2/4

There are controversies as to whether TRPV1 is protective or detrimental in cardiac hypertrophy. Several groups reported that TRPV1 protects against TAC- or high salt-induced cardiac hypertrophy by promoting formation of mitochondria-associated endoplasmic reticulum membranes and improving mitochondrial function via the AMP-activated protein kinase/mitofusin 2 pathway (Lang et al., 2015; Wang et al., 2022), upregulating peroxisome proliferation-activated receptor δ (Gao et al., 2014), or reducing inflammation (Zhong et al., 2018). However, several other groups suggested that TRPV1 is pro-hypertrophic because TRPV1 activators enlarge the size of cardiomyocytes and increase the expression of hypertrophic genes, probably via CaMKIIδ and the MAPK signaling pathway (Chen et al., 2016).

TRPV2 and V4 are stretch-activated channels that play pro-hypertrophic roles. TAC challenge causes the accumulation of TRPV2 in the sarcolemmal membrane, which is the common pathological feature observed in dilated cardiomyopathy (DCM) patients. TRPV2 knockout significantly attenuates the TAC-induced cardiac hypertrophy but does not affect the response to β-adrenergic agonists or Ang II, suggesting that TRPV2 only regulates cardiac hypertrophy due to increased afterload (Koch et al., 2017). TRPV4 expression is also increased in TAC-induced mouse hypertrophic hearts. Genetic deletion or pharmacological blockade of TRPV4 ameliorates TAC-induced cardiac hypertrophy, fibrosis, and inflammation (Zou et al., 2022). Furthermore, stretch-induced Ca^{2+} response, which is much higher in human hiPSC-derived cardiomyocytes from DCM patients than those from normal individuals, is substantially attenuated by TRPV4 inhibitors and partially inhibited by TRPV2 inhibitors (Lu et al., 2017). These data suggest that inhibiting TRPV2 and V4 can be potential therapeutic strategies for treating cardiac hypertrophy and/or DCM (Iwata et al., 2013; Iwata et al., 2020).

Figure 8.1 summarizes the role of TRP channels in cardiac hypertrophy.

8.3 TRP CHANNELS IN CARDIAC ARRHYTHMIA

Cardiac arrhythmias are a major cause of morbidity and sudden cardiac death and are estimated to account for 40–50% of all cardiovascular disease-related deaths (Srinivasan and Schilling, 2018). Abnormal rhythm can be too slow (bradycardia), too fast (tachycardia), too early (premature) or irregular (flutter or fibrillation). Arrhythmias may arise via two major mechanisms (Keefe et al., 2023). The first mechanism is abnormal impulse generation, which includes abnormal automaticity such as increased sinoatrial node activity and abnormal triggered activity such as early after-depolarization (EAD) and delayed after-depolarization (DAD). EAD and DAD occur when there is an elevated resting membrane potential or when there are changes in action potential duration (APD) (Keefe et al., 2023). The second major mechanism is related to conduction disturbances, which include slowed local conduction and/or abnormal refractory periods, resulting in reentry events.

TRP channels are nonselective cation channels, the activity of which could modulate the depolarization and repolarization phases of action potentials in cardiomyocytes. These channels may alter membrane depolarization or repolarization either directly via cation current or indirectly via cytosolic Ca^{2+} level changes. High intracellular Ca^{2+} levels can stimulate Na^{+}-Ca^{2+} exchangers (NCX) to induce membrane depolarization (Shattock et al., 2015). Therefore, abnormalities in TRP channels, either decrease or increase in expression/activity, are expected to alter membrane depolarization/repolarization to cause abnormal impulse generation, conduction disturbance, and/or abnormal refractory period, leading to cardiac arrhythmias. Indeed, TRP channels have been reported to underlie different types of cardiac arrhythmias, including abnormal automaticity, EAD, DAD and conduction disturbance (Pironet et al., 2023).

Mechanical stress is a frequent cause of cardiac arrhythmia (Bode et al., 2001). It has long been known that excitatory currents carried by stretch-activated channels can generate fast arrhythmia,

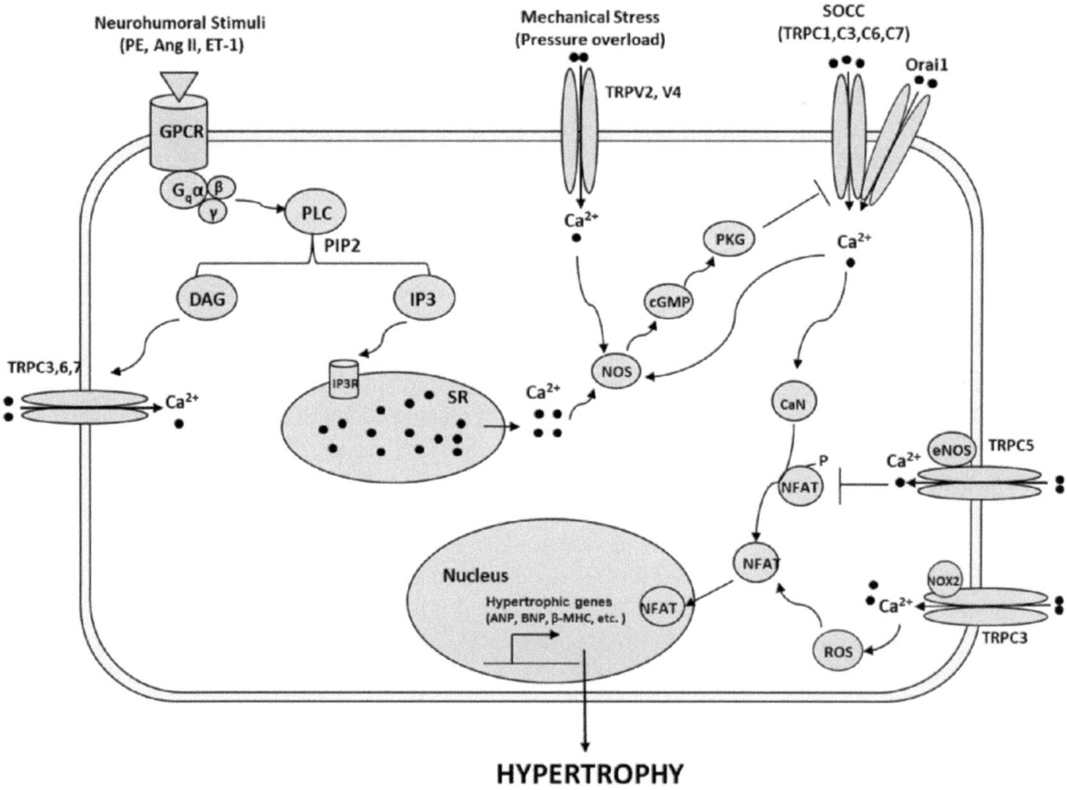

FIGURE 8.1 Role of TRP channels in cardiac hypertrophy. Neurohumoral stimuli, including endothelin-1 (ET-1), angiotensin II (Ang II) and phenylephrine (PE), stimulate Gq-protein-coupled receptors (GPCR) to deplete sarcoplasmic reticulum (SR) Ca^{2+} storage through the phosphatidylinositol (4,5) bisphosphate (PIP_2)-inositol trisphosphate (IP_3) pathway. SR Ca^{2+} depletion induces store-operated Ca^{2+} entry (SOCE), which involves Orai1 and TRPC1, C3, C6 and C7 channels. Diacylglycerol (DAG) can stimulate TRPC3, C6 and C7 channels directly. Mechanical stress stimulates mechanosensitive TRP channels, such as TRPV2 and V4, to allow Ca^{2+} entry. Ca^{2+} entry via TRP channels activates calcineurin (CaN), which in turn dephosphorylates nuclear factor of activated T cells (NFAT) to promote its translocation to the nucleus and subsequently activates the transcription of hypertrophic genes, such as atrial natriuretic peptide (ANP), brain natriuretic peptide (BNP) and beta major histocompatibility complex (β-MHC), etc. The TRPC3 channel interacts with and stabilizes NADPH oxidase 2 (NOX2) and promotes Ca^{2+} entry, leading to the generation of ROS, which are proarrhythmic factors (Jeong et al., 2012). Different from other members of the TRPC family, TRPC5 assembles with endothelial nitric oxide synthase (eNOS) to exert a protective role against cardiac hypertrophy. Note that an increased cytosolic Ca^{2+} level activates nitric oxide synthase (NOS) to produce NO. NO stimulates the production of cGMP, which activates PKG. PKG can phosphorylate multiple Ca^{2+}-permeable channels, including TRPC1, C3, C6, C7, P2 and Orai1, to inhibit these channels, forming a negative feedback loop to prevent excessive Ca^{2+} rise as a protection.

while blocking the stretch-activated channels with GsMTx4 inhibits the occurrence of atrial fibrillation (Bode et al., 2001). Many TRP channels can be activated by mechanical stimuli either directly or indirectly (Liu and Montell, 2015). These include but are not limited to TRPC1, 5, 6; TRPV1, 2, 4; TRPA1; TRPP2. Therefore, mechanical stress may act through TRP channels to induce cardiac arrhythmia. Ischemia is another cause of cardiac arrhythmia. Ischemia-induced ATP/UTP release may act through TRP channels to elicit cardiac arrhythmia (Alvarez et al., 2008). Ischemia may cause excessive production of ROS, which is also a risk factor for arrhythmia (Jeong et al., 2012).

8.3.1 TRPC3/6/7

Atrial fibrillation (AF) is the most common type of treated arrhythmia and is tightly linked to atrial fibrotic pathways because atrial fibrosis can alter the electrical conduction and excitability in the atrium and provide a substrate for the maintenance of AF (Han and Li, 2018). Excessive TRPC3 activity is pro-arrhythmic. In cardiac fibroblasts, the expression of TRPC3 is upregulated in patients with AF and in animal models of AF. Furthermore, TRPC3-mediated Ca^{2+} entry in fibroblasts activates extracellular signal-regulated kinase (ERK) to facilitate the proliferation and differentiation of the cells into myofibroblasts. In contrast, pharmacological inhibition or genetic deletion of TRPC3 suppresses the fibroblast activation *in vitro* and attenuates the progression of AF in animal models *in vivo* (Harada et al., 2012; Ju et al., 2015).

TRPC3 in cardiomyocytes may also contribute to cardiac arrhythmia. TRPC3 agonists trigger arrhythmia in *ex vivo* hearts from TRPC3-overexpressing mice via a mechanism that involves TRPC3/NCX1 interaction (Doleschal et al., 2015). The role of TRPC3/NCX1 in cardiac rhythm has been validated in stem cell-derived cardiomyocytes (Qi et al., 2016). Furthermore, TRPC3 interacts with and stabilizes NOX2 to enhance the generation of ROS in cardiomyocytes to induce arrhythmia (Jeong et al., 2012; Kitajima et al., 2016). In addition, ischemia-induced ATP/UTP release can activate TRPC3 channels in ventricular cardiomyocytes to elicit cardiac arrhythmia (Alvarez et al., 2008). Similar to TRPC3, excessive activity of TRPC6 is pro-arrhythmic. In Duchenne muscular dystrophy myocytes, mechanical stress-induced arrhythmia is reversed by TRPC6 gene deletion (Seo et al., 2014a). Besides TRPC3/6, Ca^{2+} entry via other TRPCs may also cause intracellular Ca^{2+} mishandling, EAD, DAD, and iron overload-induced arrhythmia (Wen et al., 2018). Interestingly, mechanical stress-induced arrhythmia can be reversed by a phosphodiesterase type-5 inhibitor via the cGMP-PKG-TRPC6 signaling pathway (Seo et al., 2014a). Therefore, the cGMP/PKG/TRPC pathway could be a target for an anti-arrhythmic strategy.

TRPC3 is also implicated in the dysfunction of the sinoatrial node (Yanni et al., 2011) and atrioventricular conduction block (Sabourin et al., 2012), both of which could induce arrhythmia.

8.3.2 TRPM4

TRPM4 activity may contribute to ventricular arrhythmia. In ventricular cardiomyocytes, spontaneous Ca^{2+} release from the sarcoplasmic reticulum may trigger the activation of an arrhythmogenic transient inward current (I_{ti}), causing EAD or DAD. A substantial amount of evidence suggests that TRPM4, together with NCX, mediates this arrhythmogenic I_{ti} in ventricular cardiomyocytes (Guinamard et al., 2004; Pironet et al., 2023). In agreement, TRPM4 inhibitor, 9-phenanthrol, markedly reduces the number of spontaneous Ca^{2+} increases and its associated arrhythmias in ventricular cardiomyocytes (Hedon et al., 2021) and also suppresses the hypoxia and re-oxygenation-induced ventricular arrhythmias (Simard et al., 2012). In addition to its contribution in I_{ti}, TRPM4 is also suggested to mediate a long-lasting Ca^{2+} overload-induced background current in ventricular cardiomyocytes, serving as another major mechanism of increasing arrhythmic vulnerability (Vandewiele et al., 2022).

TRPM4 activity may also contribute to atrial arrhythmia. Herein, TRPM4 participates in the repolarization phase of atrial action potentials (Hof et al., 2013; Simard et al., 2013). Mechanistically, in atrial cardiomyocytes, TRPM4 interacts with CaMKII, and this interaction is enhanced by Ang II, eventually inducing EAD. Such arrhythmic changes can be abolished by the TRPM4 inhibitor, 9-phenanthrol (Hu et al., 2021).

However, note that lack of TRPM4 may also cause cardiac arrhythmia. Atrial cardiomyocytes from *Trpm4*$^{-/-}$ mice exhibit a 20% shorter APD than that of wild-type mice, and furthermore, ventricular myocytes isolated from the *Trpm4*$^{-/-}$ mice show shortened APD_{50} and APD_{90} (Simard et al., 2013). TRPM4 also influences cardiac conduction. *Trpm4*$^{-/-}$ mice exhibit PR and QRS interval lengthening in the electrocardiogram and atrioventricular blocks (Demion et al., 2014). Intriguingly, mutations in TRPM4 are known to be associated with a variety of pro-arrhythmogenic human cardiac conduction disorders, such as progressive familial heart block type 1, Brugada syndrome (BrS) and long QT syndrome (Pironet et al., 2023). Some of these mutations are gain-of-function mutations, whereas others are loss-of-function mutations (Pironet et al., 2023).

8.3.3 TRPM6/7

In atrial cardiomyocytes and tissues from AF patients, the expression levels of TRPM6 and TRPM7 are increased, and such an increase contributes to atrial fibrosis, suggesting the contribution of TRPM6/7 in AF by promoting fibrogenesis (Zhang et al., 2012; Zhang et al., 2015). However, lack of TRPM7 may also cause arrhythmia. A loss-of-function mutation of TRPM7 is associated with long QT syndrome, which is a risk factor for cardiac arrhythmia (Arking et al., 2014). In addition, TRPM7 knockout mice show abnormal heart development (Jin et al., 2012).

8.3.4 TRPV4

TRPV4 is pro-arrhythmic. TRPV4 is upregulated in the atria of rats and AF patients. TRPV4 agonist GK1016790A increases both APD and cytosolic Ca^{2+} levels in atrial cardiomyocytes, whereas its antagonist GSK2193874 reverses the abnormal APD prolongation and decreases the vulnerability to AF (Liao et al., 2020). Furthermore, ventricular cardiomyocytes from aged mice show increased pro-arrhythmic diastolic Ca^{2+} signaling and greater incidence of arrhythmia upon ischemia-reperfusion stimulation, which can be attenuated by the perfusion of the TRPV4 antagonist HC067047 (Peana et al., 2022).

8.4 TRP CHANNELS IN ISCHEMIA/REPERFUSION (I/R) INJURY AND MYOCARDIAL INFARCTION

Myocardial infarction occurs when the heart muscle is severely damaged due to a lack of blood flow or ischemia. One treatment strategy is myocardial reperfusion, which can reduce the size of myocardial infarct. Although reperfusion is supposed to restore the original cardiac function, it is often followed by exacerbated damage because of impaired Ca^{2+} handling, mitochondrial dysfunction and excessive ROS (Hof et al., 2019; Liu et al., 2023). Compelling evidence shows that dysregulation of TRP channels contributes to the Ca^{2+} mishandling and excessive ROS production under I/R, leading to myocardial injury (Falcon et al., 2020).

8.4.1 TRPC

Myocardial I/R is characterized by the production of excessive ROS, which can cause tissue damage (Carden and Granger, 2000). Myocardial I/R and excessive ROS elevate the expression of multiple TRPCs (He et al., 2017; Wen et al., 2020) and stimulate the activity of TRPC3/6 channels (Wen et al., 2020). Furthermore, TRPC3/6 activity also enhances ROS production, forming a positive feedback loop in myocardial infarction and heart failure (Abramowitz and Birnbaumer, 2009). Indeed, transgenic overexpression of TRPC3 in myocardial cells sensitizes the myocardial injuries after I/R via calpain activation and consequent apoptosis (Shan et al., 2008), whereas inhibition of TRPCs by SKF96365 or genetic ablation of TRPC3/6/7 substantially reduces the I/R-induced myocardial injury (He et al., 2017).

8.4.2 TRPM2/4/8

TRPM2 is a ROS-sensitive channel. ROS stimulate TRPM2 to induce a sustained Ca^{2+} influx. However, the role of TRPM2 in myocardial I/R is controversial. Early studies suggested that TRPM2 is deletrious as its activation by ROS induces cardiomyocyte death, while inhibiting TRPM2 prevents the cell death (Roberge et al., 2014; Yang et al., 2006). Furthermore, TRPM2 global knockout mice (*Trpm2*^{gKO}) show reduced myocardial infarct and improved cardiac contractile function following I/R compared with wild-type mice (Hiroi et al., 2013). However, in contrast, Miller's and Chung's groups showed that TRPM2-mediated Ca^{2+} influx in cardiomyocytes protects against I/R injury by reducing mitochondrial oxidants. With the use of both global TRPM2 knockout mice (*Trpm2*^{gKO}) and cardiac-specific TRPM2 knockout mice (*Trpm2*^{cKO}), they found that myocardial I/R-challenged *Trpm2* knockout mice have exacerbated aberrant mitochondrial bioenergetics and increased ROS levels compared with wild-type mice (Hoffman et al., 2015; Miller et al., 2013; Miller et al., 2014).

TRPM4 is deletrious in I/R-induced myocardial infarction. Excessive activity of TRPM4 exacerbates I/R-induced myocardial infarction, whereas inhibition or gene deletion of TRPM4 reduces myocardial infarction (Boukenna et al., 2023; Wang et al., 2021). *Trpm4*$^{-/-}$ mice with ligation-induced myocardial infarction show lower mortality and higher cardiac performance during β-adrenergic stimulation than wild-type mice (Jacobs et al., 2015). Furthermore, TRPM4 gene silencing or TRPM4 inhibition with 9-phenanthrol attenuates the hypoxia-induced injury in rat cardiac myoblast cells H9c2 (Piao et al., 2015) and protects rat hearts from I/R injury (Simard et al., 2012; Wang et al., 2013).

TRPM8 is a cold-sensitive channel that has a cardiac protective role (Cheng et al., 2019; Pertusa et al., 2018). During I/R, hypothermic reperfusion activates TRPM8 to exert cardioprotective effect, whereas a TRPM8 antagonist, BCTC, diminishes the protective effect of hypothermic perfusion (Cheng et al., 2019).

8.4.3 TRPV1/4

TRPV1 has a protective role in myocardial I/R. Activation of TRPV1 by capsaicin attenuates I/R-induced myocardial injury, while genetic deletion or pharmacological inhibition of TRPV1 aggravates the injury (Jiang et al., 2018; Zhong et al., 2020). The underlying mechanisms for the cardioprotective role of TRPV1 may be related to an increased release of calcitonin gene-related peptide (CGRP) and substance P (Wang et al., 2012), activation of the NO signaling pathway and reduction of ROS production (Castrejon-Tellez et al., 2022), upregulation of the PI3K/Akt signaling pathway (Jiang et al., 2018), or inhibition of glycogen synthase kinase-3 to enhance gap junction coupling (Randhawa and Jaggi, 2017).

TRPV4 plays a detrimental role in I/R-induced cardiac injury. TRPV4 has high expression in the heart, and the expression level will further increase during myocardial I/R (Wu et al., 2017). Activation of TRPV4 by GSK1016790A increases myocardial infarct size and apoptotic cells (Zhang et al., 2021), whereas gene deletion or pharmacological inhibition of TRPV4 markedly reduces myocardial infarct size, decreases troponin T levels and improves cardiac function after I/R injury in mice (Wu et al., 2019; Wu et al., 2023). The mechanisms may be related to TRPV4-associated ROS production, a decrease in mitochondrial membrane potential and the opening of the mitochondrial permeability transition pore (Wu et al., 2017).

8.4.4 TRPML1

Lysosomal cation channel TRPML1 also contributes to I/R-induced myocardial injury. Mechanistically, I/R-induced ROS overproduction activates lysosomal TRPML1 in cardiomyocytes to release lysosomal zinc into the cytosol, which inhibits autophagic flux, leading to the accumulation of damaged mitochondria and consequent cardiomyocyte death (Xing et al., 2022). Therefore, targeting TRPML1 channels could be a potential strategy to restore the inhibited autophagic flux and protect against I/R-induced myocardial injury (Sui et al., 2022).

Figure 8.2 summarizes the role of TRP channels in myocardial ischemia/reperfusion injury.

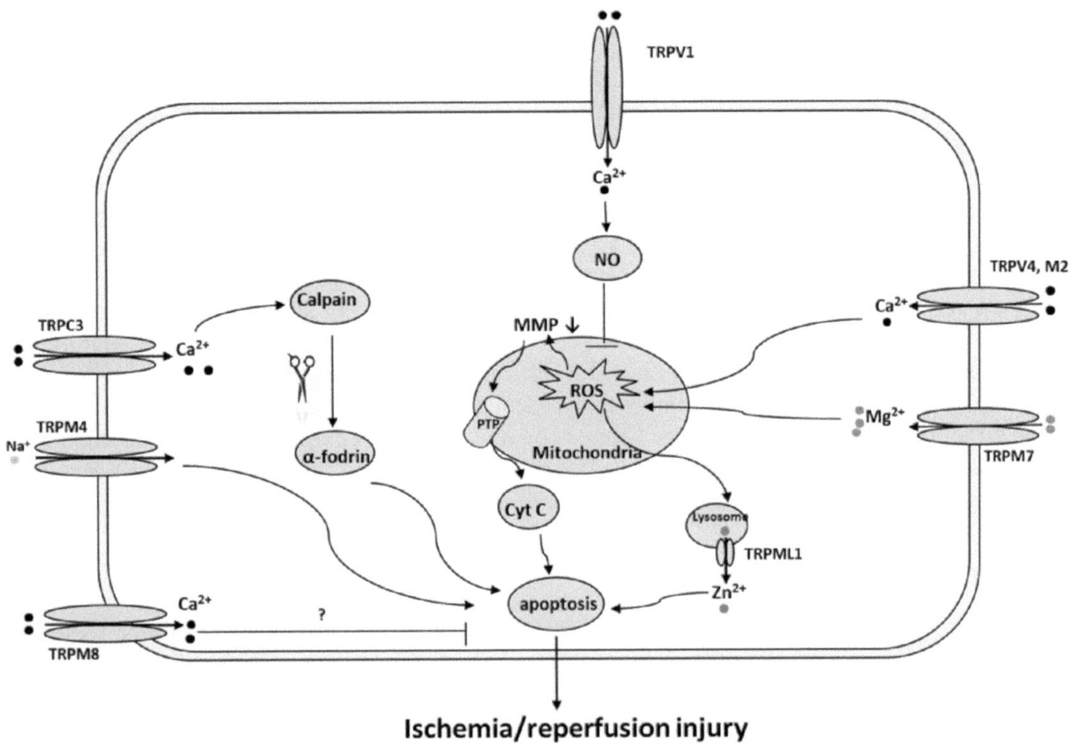

FIGURE 8.2 Role of TRP channels in myocardial ischemia/reperfusion injury. Ca^{2+} entry through TRPV4 and TRPM2 channels increases the generation of ROS in the mitochondria, leading to the decrease of mitochondrial membrane potential (MMP), the opening of mitochondrial permeability transition pore (PTP), the release of cytochrome c (Cyt C) into the cytosol, and eventually the initiation of the caspase-dependent apoptosis pathway. ROS activate lysosomal channel TRPML1 to cause the release of Zn^{2+} into the cytosol, which inhibits autophagic flux, leading to cardiomyocyte death. The TRPC3 channel promotes the activation of calpain, which cleaves α-fodrin in the heart and finally causes cell death. TRPM4-mediated Na^+ entry also causes cell death. In contrast, TRPM8-mediated Ca^{2+} entry is protective against the apoptosis of cardiomyocytes. TRPV1 activation stimulates the NO signaling pathway to reduce ROS production in mitochondira and thus exerts a cardioprotective role against myocardial ischemia/reperfusion injury.

8.5 CONCLUSIONS AND PERSPECTIVES

TRP channels are important regulators of intracellular Ca^{2+} homeostasis that support the development and function of the heart. Dysregulation of TRP channels may cause cardiac diseases, such as cardiac hypertrophy, fibrosis, arrhythmia, and myocardial I/R injury. With regard to cardiac hypertrophy, excessive activity of TRPC1, C3, C4, C6, C7, V2, and V4 has pro-hypertrophic roles, whereas TRPC5 has a cardioprotective role against hypertrophy. TRPM4 may be pro-hypertrophic under pressure stress but anti-hypertrophic under abnormal neurohormonal stimuli. Cardiac arrhythmias may arise via abnormal impulse generation, conduction disturbance, and/or abnormal refractory period. Therefore, it is not surprising that alteration of cardiac TRP channels, either decrease or increase in activity/expression, may cause cardiac arrhythmias. Indeed, abnormal activity of TRPC3, C6, M4, M6, M7, and V4 is demonstrated to contribute to cardiac arrhythmias. In I/R injury, excessive activities of TRPC3, C6, C7, M4, V4, and ML1 are detrimental, whereas TRPM8 and V1 have protective roles. Notably, in some cases, there are controversies as to whether a specific TRP isoform has a cardiac protective role against hypertrophy and/or I/R injury or plays a detrimental role in these processes. Such discrepancies could result from experimental artifacts related to cell culture, low specificity of some TRP channel agonists/antagonists, compensatory changes in the expression of other TRP when one TRP isoform is knocked out, different genetic backgrounds of different mice, etc. Functional redundancy of TRP channels can also complicate the interpretation of their phenotypes. One of the future directions is to develop more specific pharmacological activators/inhibitors for different TRP isoforms. The recent progress in understanding the structure of TRP channels shall aid the development of specific modulators. Another future direction is to use cardiac cell-specific TRP knockout mice, which can reduce the complications seen in global TRP knockout mice. Furthermore, recent development in human pluripotent stem cell-derived cardiomyocytes provides new opportunities to extensively study human cardiomyocytes because scientists can use pluripotent stem cell-derived cardiomyocytes to generate an unlimited supply of human cardiomyocytes for disease modeling and drug screening. Moreover, the properties of these cells are based on humans rather than rodents. This is important because there are considerable differences in cardiac physiology between human and rodent cardiomyocytes. Development of other recent technologies such as mini-heart/organoids generated from stem cells, patient iPSCs, and single-cell sequencing should also promote new development in cardiac research.

ACKNOWLDEGMENTS

This work was supported by grants from the Hong Kong Research Grant Committee (14100619, RIF/R4005-18F) and the Hong Kong Innovation and Technology Fund (ITS 212/21).

REFERENCES

Abramowitz, J., Birnbaumer, L. Physiology and pathophysiology of canonical transient receptor potential channels. *FASEB J*, 2009, 23: 297–328.

Alvarez, J., Coulombe, A., Cazorla, O., Ugur, M., Rauzier, J. M., Magyar, J., Mathieu, E.L., Boulay, G., Souto, R., Bideaux, P., Salazar, G., Rassendren, F., Lacampagne, A., Fauconnier, J., Vassort, G. ATP/UTP activate cation-permeable channels with TRPC3/7 properties in rat cardiomyocytes. *Am J Physiol Heart Circ Physiol*, 2008, 295: H21–H28.

Amarouch, M. Y., El Hilaly, J. Inherited cardiac arrhythmia syndromes: Focus on molecular mechanisms underlying TRPM4 channelopathies. *Cardiovasc Ther*, 2020, 2020: 6615038.

Arking, D. E., Pulit, S. L., Crotti, L., van der Harst, P., Munroe, P. B., Koopmann, T. T., Sotoodehnia, N., Rossin, E.J., Morley, M., Wang, X., Johnson, A.D., Lundby, A., Gudbjartsson, D.F., Noseworthy, P.A., Eijgelsheim, M., Bradford, Y., Tarasov, K.V., Dorr, M., Muller-Nurasyid, M., Lahtinen, A.M., Nolte, I.M., Smith, A.V., Bis, J.C., Isaacs, A., Newhouse, S.J., Evans, D.S., Post, W.S., Waggott, D., Lyytikainen, L.P., Hicks, A.A., Eisele, L., Ellinghaus, D., Hayward, C., Navarro, P., Ulivi, S., Tanaka, T., Tester, D.J., Chatel, S., Gustafsson, S., Kumari, M., Morris, R.W., Naluai, A.T., Padmanabhan, S., Kluttig, A., Strohmer, B., Panayiotou, A.G., Torres, M., Knoflach, M., Hubacek, J.A., Slowikowski, K., Raychaudhuri, S., Kumar, R.D., Harris, T.B., Launer, L.J., Shuldiner, A.R., Alonso, A., Bader, J.S., Ehret, G., Huang, H., Kao, W.H., Strait, J.B., Macfarlane, P.W., Brown, M., Caulfield, M.J., Samani, N.J., Kronenberg, F., Willeit, J., Consortium, C.A., Consortium, C., Smith, J.G., Greiser, K.H., Meyer Zu Schwabedissen, H., Werdan, K., Carella, M., Zelante, L., Heckbert, S.R., Psaty, B.M., Rotter, J.I., Kolcic, I., Polasek, O., Wright, A.F., Griffin, M., Daly, M.J., Dcct/Edic, Arnar, D.O., Holm, H., Thorsteinsdottir, U., e, M.C., Denny, J.C., Roden, D.M., Zuvich, R.L., Emilsson, V., Plump, A.S., Larson, M.G., O'Donnell, C.J., Yin, X., Bobbo, M., D'Adamo, A.P., Iorio, A., Sinagra, G., Carracedo, A., Cummings, S.R., Nalls, M.A., Jula, A., Kontula, K.K., Marjamaa, A., Oikarinen, L., Perola, M., Porthan, K., Erbel, R., Hoffmann, P., Jockel, K.H., Kalsch, H., Nothen, M.M., Consortium, H., den Hoed, M., Loos, R.J., Thelle, D.S., Gieger, C., Meitinger, T., Perz, S., Peters, A., Prucha, H., Sinner, M.F., Waldenberger, M., de Boer, R.A., Franke, L., van der Vleuten, P.A., Beckmann, B.M., Martens, E., Bardai, A., Hofman, N., Wilde, A.A., Behr, E.R., Dalageorgou, C., Giudicessi, J.R., Medeiros-Domingo, A., Barc, J., Kyndt, F., Probst, V., Ghidoni, A., Insolia, R., Hamilton, R.M., Scherer, S.W., Brandimarto, J., Margulies, K., Moravec, C.E., del Greco, M.F., Fuchsberger, C., O'Connell, J.R., Lee, W.K., Watt, G.C., Campbell, H., Wild, S.H., El Mokhtari, N.E., Frey, N., Asselbergs, F.W., Mateo

Leach, I., Navis, G., van den Berg, M.P., van Veldhuisen, D.J., Kellis, M., Krijthe, B.P., Franco, O.H., Hofman, A., Kors, J.A., Uitterlinden, A.G., Witteman, J.C., Kedenko, L., Lamina, C., Oostra, B.A., Abecasis, G.R., Lakatta, E.G., Mulas, A., Orru, M., Schlessinger, D., Uda, M., Markus, M.R., Volker, U., Snieder, H., Spector, T.D., Arnlov, J., Lind, L., Sundstrom, J., Syvanen, A.C., Kivimaki, M., Kahonen, M., Mononen, N., Raitakari, O.T., Viikari, J.S., Adamkova, V., Kiechl, S., Brion, M., Nicolaides, A.N., Paulweber, B., Haerting, J., Dominiczak, A.F., Nyberg, F., Whincup, P.H., Hingorani, A.D., Schott, J.J., Bezzina, C.R., Ingelsson, E., Ferrucci, L., Gasparini, P., Wilson, J.F., Rudan, I., Franke, A., Muhleisen, T.W., Pramstaller, P.P., Lehtimaki, T.J., Paterson, A.D., Parsa, A., Liu, Y., van Duijn, C.M., Siscovick, D.S., Gudnason, V., Jamshidi, Y., Salomaa, V., Felix, S.B., Sanna, S., Ritchie, M.D., Stricker, B.H., Stefansson, K., Boyer, L.A., Cappola, T.P., Olsen, J.V., Lage, K., Schwartz, P.J., Kaab, S., Chakravarti, A., Ackerman, M.J., Pfeufer, A., de Bakker, P.I., Newton-Cheh, C. Genetic association study of QT interval highlights role for calcium signaling pathways in myocardial repolarization. *Nat Genet*, 2014, 46: 826–836.

Bernardo, B. C., Weeks, K. L., Pretorius, L., McMullen, J. R. Molecular distinction between physiological and pathological cardiac hypertrophy: Experimental findings and therapeutic strategies. *Pharmacol Ther*, 2010, 128: 191–227.

Bode, F., Sachs, F., Franz, M. R. Tarantula peptide inhibits atrial fibrillation. *Nature*, 2001, 409: 35–36.

Boukenna, M., Rougier, J. S., Aghagolzadeh, P., Pradervand, S., Guichard, S., Hammerli, A. F., Pedrazzini, T., Abriel, H. Multiomics uncover the proinflammatory role of Trpm4 deletion after myocardial infarction in mice. *Am J Physiol Heart Circ Physiol*, 2023, 324: H504–H518.

Bush, E. W., Hood, D. B., Papst, P. J., Chapo, J. A., Minobe, W., Bristow, M. R., Olson, E.N., McKinsey, T.A. Canonical transient receptor potential channels promote cardiomyocyte hypertrophy through activation of calcineurin signaling. *J Biol Chem*, 2006, 281: 33487–33496.

Camacho Londono, J. E., Tian, Q., Hammer, K., Schroder, L., Camacho Londono, J., Reil, J. C., He, T., Oberhofer, M., Mannebach, S., Mathar, I., Philipp, S.E., Tabellion, W., Schweda, F., Dietrich, A., Kaestner, L., Laufs, U., Birnbaumer, L., Flockerzi, V., Freichel, M., Lipp, P. A background Ca^{2+} entry pathway mediated by TRPC1/TRPC4 is critical for development of pathological cardiac remodelling. *Eur Heart J*, 2015, 36: 2257–2266.

Camacho, P., Fan, H., Liu, Z., He, J. Q. Small mammalian animal models of heart disease. *Am J Cardiovasc Dis*, 2016, 6: 70–80.

Carden, D. L., Granger, D. N. Pathophysiology of ischaemia-reperfusion injury. *J Pathol*, 2000, 190: 255–266.

Castrejon-Tellez, V., Del Valle-Mondragon, L., Perez-Torres, I., Guarner-Lans, V., Pastelin-Hernandez, G., Ruiz-Ramirez, A., Diaz-Juarez, J.A., Varela-Lopez, E., Oidor-Chan, V.H., Vargas-Gonzalez, A., Martinez-Memije, R., Flores-Chavez, P., Leon-Ruiz, B., Arriaga-Carrillo, S., Torres-Narvaez, J.C. TRPV1 contributes to modulate the nitric oxide pathway and oxidative stress in the isolated and perfused rat heart during ischemia and reperfusion. *Molecules*, 2022, 27: 1031.

Chen, M., Xin, J., Liu, B., Luo, L., Li, J., Yin, W., Li, M. Mitogen-activated protein kinase and intracellular polyamine signaling is involved in TRPV1 activation-induced cardiac hypertrophy. *J Am Heart Assoc*, 2016. 5: e003718.

Cheng, Q. Y., Yang, M. C., Wu, J., Jia, X. L., Xiao, C., Lian, T., Zhang, S.Z. Reduced cardiac ischemia/reperfusion injury by hypothermic reperfusion via activation of transient receptor potential M8 channel. *Life Sci*, 2019, 232: 116658.

Demion, M., Thireau, J., Gueffier, M., Finan, A., Khoueiry, Z., Cassan, C., Serafini, N., Aimond, F., Granier, M., Pasquie, J.L., Launay, P., Richard, S. Trpm4 gene invalidation leads to cardiac hypertrophy and electrophysiological alterations. *PLoS One*, 2014, 9: e115256.

Doleschal, B., Primessnig, U., Wolkart, G., Wolf, S., Schernthaner, M., Lichtenegger, M., Glasnov, T.N., Kappe, C.O., Mayer, B., Antoons, G., Heinzel, F., Poteser, M., Groschner, K. TRPC3 contributes to regulation of cardiac contractility and arrhythmogenesis by dynamic interaction with NCX1. *Cardiovasc Res*, 2015, 106: 163–173.

Dolmetsch, R. E., Lewis, R. S., Goodnow, C. C., Healy, J. I. Differential activation of transcription factors induced by Ca^{2+} response amplitude and duration. *Nature*, 1997, 386: 855–858.

Earley, S., Brayden, J. E. Transient receptor potential channels in the vasculature. *Physiol Rev*, 2015, 95: 645–690.

Eder, P., Molkentin, J. D. TRPC channels as effectors of cardiac hypertrophy. *Circ Res*, 2011, 108: 265–272.

Falcon, D., Galeano-Otero, I., Martin-Bornez, M., Fernandez-Velasco, M., Gallardo-Castillo, I., Rosado, J. A., Ordonez, A., Smani, T., TRPC channels: Dysregulation and Ca^{2+} mishandling in ischemic heart disease. *Cells*, 2020, 9: 173.

Gao, F., Liang, Y., Wang, X., Lu, Z., Li, L., Zhu, S., Liu, D., Yan, Z., Zhu, Z. TRPV1 activation attenuates high-salt diet-induced cardiac hypertrophy and fibrosis through PPAR-delta upregulation. *PPAR Res*, 2014, 2014: 491963.

Gueffier, M., Zintz, J., Lambert, K., Finan, A., Aimond, F., Chakouri, N., Hedon, C., Granier, M., Launay, P., Thireau, J., Richard, S., Demion, M. The TRPM4 channel is functionally important for the beneficial cardiac remodeling induced by endurance training. *J Muscle Res Cell Motil*, 2017, 38: 3–16.

Guinamard, R., Chatelier, A., Demion, M., Potreau, D., Patri, S., Rahmati, M., Bois, P. Functional characterization of a Ca^{2+}-activated non-selective cation channel in human atrial cardiomyocytes. *J Physiol*, 2004, 558: 75–83.

Guo, Y., Yu, Z. Y., Wu, J., Gong, H., Kesteven, S., Iismaa, S. E., Chan, A.Y., Holman, S., Pinto, S., Pironet, A., Cox, C.D., Graham, R.M., Vennekens, R., Feneley, M.P., Martinac, B. The Ca^{2+}-activated cation channel TRPM4 is a positive regulator of pressure overload-induced cardiac hypertrophy. *Elife*, 2021, 10: e66582.

Han, L., Li, J. Canonical transient receptor potential 3 channels in atrial fibrillation. *Eur J Pharmacol*, 2018, 837: 1–7.

Harada, M., Luo, X., Qi, X. Y., Tadevosyan, A., Maguy, A., Ordog, B., Ledoux, J., Kato, T., Naud, P., Voigt, N., Shi, Y., Kamiya, K., Murohara, T., Kodama, I., Tardif, J.C., Schotten, U., Van Wagoner, D.R., Dobrev, D., Nattel, S. Transient receptor potential canonical-3 channel-dependent fibroblast regulation in atrial fibrillation. *Circulation*, 2012, 126: 2051–2064.

He, X., Li, S., Liu, B., Susperreguy, S., Formoso, K., Yao, J., Kang, J., Shi, A., Birnbaumer, L., Liao, Y. Major contribution of the 3/6/7 class of TRPC channels to myocardial ischemia/reperfusion and cellular hypoxia/reoxygenation injuries. *Proc Natl Acad Sci U S A*, 2017, 114: E4582–E4591.

Hedon, C., Lambert, K., Chakouri, N., Thireau, J., Aimond, F., Cassan, C., Bideaux, P., Richard, S., Faucherre, A., Le Guennec, J.Y., Demion, M. New role of TRPM4 channel in the cardiac excitation-contraction coupling in response to physiological and pathological hypertrophy in mouse. *Prog Biophys Mol Biol*, 2021, 159: 105–117.

Hiroi, T., Wajima, T., Negoro, T., Ishii, M., Nakano, Y., Kiuchi, Y., Mori, Y., Shimizu, S. Neutrophil TRPM2 channels are implicated in the exacerbation of myocardial ischaemia/reperfusion injury. *Cardiovasc Res*, 2013, 97: 271–281.

Hof, T., Chaigne, S., Recalde, A., Salle, L., Brette, F., Guinamard, R. Transient receptor potential channels in cardiac health and disease. *Nat Rev Cardiol*, 2019, 16: 344–360.

Hof, T., Simard, C., Rouet, R., Salle, L., Guinamard, R. Implication of the TRPM4 nonselective cation channel in mammalian sinus rhythm. *Heart Rhythm*, 2013, 10: 1683–1689.

Hoffman, N. E., Miller, B. A., Wang, J., Elrod, J. W., Rajan, S., Gao, E., Song, J., Zhang, X.Q., Hirschler-Laszkiewicz, I., Shanmughapriya, S., Koch, W.J., Feldman, A.M., Madesh, M., Cheung, J.Y. Ca^{2+} entry via Trpm2 is essential for cardiac myocyte bioenergetics maintenance. *Am J Physiol Heart Circ Physiol*, 2015, 308: H637–H650.

Houser, S. R., Molkentin, J. D. Does contractile Ca^{2+} control calcineurin-NFAT signaling and pathological hypertrophy in cardiac myocytes? *Sci Signal* 2008, 1: pe31.

Hu, Y., Kaschitza, D. R., Essers, M., Arullampalam, P., Fujita, T., Abriel, H., Inoue, R. Pathological activation of CaMKII induces arrhythmogenicity through TRPM4 overactivation. *Pflugers Arch*, 2021, 473: 507–519.

Iwata, Y., Ohtake, H., Suzuki, O., Matsuda, J., Komamura, K., Wakabayashi, S. Blockade of sarcolemmal TRPV2 accumulation inhibits progression of dilated cardiomyopathy. *Cardiovasc Res*, 2013, 99: 760–768.

Iwata, Y., Wakabayashi, S., Ito, S., Kitakaze, M. Production of TRPV2-targeting functional antibody ameliorating dilated cardiomyopathy and muscular dystrophy in animal models. *Lab Invest*, 2020, 100: 324–337.

Jacobs, G., Oosterlinck, W., Dresselaers, T., Geenens, R., Kerselaers, S., Himmelreich, U., Herijgers, P., Vennekens, R. Enhanced beta-adrenergic cardiac reserve in Trpm4(-)/(-) mice with ischaemic heart failure. *Cardiovasc Res*, 2015, 105: 330–339.

Jeong, E. M., Liu, M., Sturdy, M., Gao, G., Varghese, S. T., Sovari, A. A., Dudley Jr, S.C. Metabolic stress, reactive oxygen species, and arrhythmia. *J Mol Cell Cardiol*, 2012, 52: 454–463.

Jiang, X. X., Liu, G. Y., Lei, H., Li, Z. L., Feng, Q. P., Huang, W. Activation of transient receptor potential vanilloid 1 protects the heart against apoptosis in ischemia/reperfusion injury through upregulating the PI3K/Akt signaling pathway. *Int J Mol Med*, 2018, 41: 1724–1730.

Jin, J., Wu, L. J., Jun, J., Cheng, X., Xu, H., Andrews, N. C., Clapham, D.E. The channel kinase, TRPM7, is required for early embryonic development. *Proc Natl Acad Sci U S A*, 2012, 109: E225–E233.

Ju, Y. K., Lee, B. H., Trajanovska, S., Hao, G., Allen, D. G., Lei, M., Cannell, M.B. The involvement of TRPC3 channels in sinoatrial arrhythmias. *Front Physiol*, 2015, 6: 86.

Kato, Y., Nishiyama, K., Man Lee, J., Ibuki, Y., Imai, Y., Noda, T., Kamiya, N., Kusakabe, T., Kanda, Y., Nishida, M. TRPC3-Nox2 protein complex formation increases the risk of SARS-CoV-2 spike protein-induced cardiomyocyte dysfunction through ACE2 upregulation. *Int J Mol Sci*, 2022, 24: 102.

Kecskes, M., Jacobs, G., Kerselaers, S., Syam, N., Menigoz, A., Vangheluwe, P., Freichel, M., Flockerzi, V., Voets, T., Vennekens, R. The Ca^{2+}-activated cation channel TRPM4 is a negative regulator of angiotensin II-induced cardiac hypertrophy. *Basic Res Cardiol*, 2015, 110: 43.

Keefe, J. A., Moore, O. M., Ho, K. S., Wehrens, X. H. T. Role of Ca^{2+} in healthy and pathologic cardiac function: From normal excitation-contraction coupling to mutations that cause inherited arrhythmia. *Arch Toxicol*, 2023, 97: 73–92.

Kinoshita, H., Kuwahara, K., Nishida, M., Jian, Z., Rong, X., Kiyonaka, S., Kuwabara, Y., Kurose, H., Inoue, R., Mori, Y., Li, Y., Nakagawa, Y., Usami, S., Fujiwara, M., Yamada, Y., Minami, T., Ueshima, K., Nakao, K. Inhibition of TRPC6 channel activity contributes to the antihypertrophic effects of natriuretic peptides-guanylyl cyclase-A signaling in the heart. *Circ Res*, 2010, 106: 1849–1860.

Kitajima, N., Numaga-Tomita, T., Watanabe, M., Kuroda, T., Nishimura, A., Miyano, K., Yasuda, S., Kuwahara, K., Sato, Y., Ide, T., Birnbaumer, L., Sumimoto, H., Mori, Y., Nishida, M. TRPC3 positively regulates reactive oxygen species driving maladaptive cardiac remodeling. *Sci Rep*, 2016, 6: 37001.

Koch, S. E., Mann, A., Jones, S., Robbins, N., Alkhattabi, A., Worley, M. C., Gao, X., Lasko-Roiniotis, V.M., Karani, R., Fulford, L., Jiang, M., Nieman, M., Lorenz, J.N., Rubinstein, J. Transient receptor potential vanilloid 2 function regulates cardiac hypertrophy via stretch-induced activation. *J Hypertens*, 2017, 35: 602–611.

Koitabashi, N., Aiba, T., Hesketh, G. G., Rowell, J., Zhang, M., Takimoto, E., Tomaselli, G.F., Kass, D.A. Cyclic GMP/PKG-dependent inhibition of TRPC6 channel activity and expression negatively regulates cardiomyocyte NFAT activation novel mechanism of cardiac stress modulation by PDE5 inhibition. *J Mol Cell Cardiol*, 2010, 48: 713–724.

Kuwahara, K., Wang, Y., McAnally, J., Richardson, J. A., Bassel-Duby, R., Hill, J. A., Olson, E.N. TRPC6 fulfills a calcineurin signaling circuit during pathologic cardiac remodeling. *J Clin Invest*, 2006, 116: 3114–3126.

Kwan, H. Y., Huang, Y., Yao, X. Regulation of canonical transient receptor potential isoform 3 (TRPC3) channel by protein kinase g. *Proc Natl Acad Sci U S A*, 2004, 101: 2625–2630.

Kwan, H. Y., Huang, Y., Yao, X. Protein kinase c can inhibit TRPC3 channels indirectly via stimulating protein kinase G. *J Cell Physiol*, 2006, 207: 315–321.

Lang, H., Li, Q., Yu, H., Li, P., Lu, Z., Xiong, S., Yang, T., Zhao, Y., Huang, X., Gao, P., Zhang, H., Shang, Q., Liu, D., Zhu, Z. Activation of TRPV1 attenuates high salt-induced cardiac hypertrophy through improvement of mitochondrial function. *Br J Pharmacol*, 2015, 172: 5548–5558.

Liao, J., Wu, Q., Qian, C., Zhao, N., Zhao, Z., Lu, K., Zhang, S., Dong, Q., Chen, L., Li, Q., Du, Y. TRPV4 blockade suppresses atrial fibrillation in sterile pericarditis rats. *JCI Insight*, 2020, 5: e137528.

Liu, C., Montell, C. Forcing open TRP channels: Mechanical gating as a unifying activation mechanism. *Biochem Biophys Res Commun*, 2015, 460: 22–25.

Liu, Y., Li, L., Wang, Z., Zhang, J., Zhou, Z. Myocardial ischemia-reperfusion injury; Molecular mechanisms and prevention. *Microvasc Res*, 2023, 149: 104565.

Lopez, J. J., Jardin, I., Sanchez-Collado, J., Salido, G. M., Smani, T., Rosado, J. A. TRPC channels in the SOCE scenario. *Cells*, 2020, 9: 126.

Lu, J., Lee, Y. K., Ran, X., Lai, W. H., Li, R. A., Keung, W., Tse, K., Tse, H.F., Yao, X. An abnormal TRPV4-related cytosolic Ca^{2+} rise in response to uniaxial stretch in induced pluripotent stem cells-derived cardiomyocytes from dilated cardiomyopathy patients. *Biochim Biophys Acta Mol Basis Dis*, 2017, 1863: 2964–2972.

Ma, T., Lin, S., Wang, B., Wang, Q., Xia, W., Zhang, H., Cui, Y., He, C., Wu, H., Sun, F., Zhao, Z., Gao, P., Zhu, Z., Liu, D. TRPC3 deficiency attenuates high salt-induced cardiac hypertrophy by alleviating cardiac mitochondrial dysfunction. *Biochem Biophys Res Commun*, 2019, 519: 674–681.

Mathar, I., Vennekens, R., Meissner, M., Kees, F., Van der Mieren, G., Camacho Londono, J. E., Uhl, S., Voets, T., Hummel, B., van den Bergh, A., Herijgers, P., Nilius, B., Flockerzi, V., Schweda, F., Freichel, M. Increased catecholamine secretion contributes to hypertension in TRPM4-deficient mice. *J Clin Invest*, 2010, 120: 3267–3279.

Miller, B. A., Hoffman, N. E., Merali, S., Zhang, X. Q., Wang, J., Rajan, S., et al. TRPM2 channels protect against cardiac ischemia-reperfusion injury: Role of mitochondria. *J Biol Chem*, 2014, 289: 7615–7629.

Miller, B. A., Wang, J., Hirschler-Laszkiewicz, I., Gao, E., Song, J., Zhang, X. Q., Shanmughapriya, S., Gao, E., Barrero, C.A., Mallilankaraman, K., Song, J., Gu, T., Hirschler-Laszkiewicz, I., Koch, W.J., Feldman, A.M., Madesh, M., Cheung, J.Y. The second member of transient receptor potential-melastatin channel family protects hearts from ischemia-reperfusion injury. *Am J Physiol Heart Circ Physiol*, 2013, 304: H1010–H1022.

Nakamura, M., Sadoshima, J. Mechanisms of physiological and pathological cardiac hypertrophy. *Nat Rev Cardiol*, 2018, 15: 387–407.

Nakayama, H., Wilkin, B. J., Bodi, I., Molkentin, J. D. Calcineurin-dependent cardiomyopathy is activated by TRPC in the adult mouse heart. *FASEB J*, 2006, 20: 1660–1670.

Ohba, T., Watanabe, H., Murakami, M., Takahashi, Y., Iino, K., Kuromitsu, S., Mori, Y., Ono, K., Iijima, T., Ito, H. Upregulation of TRPC1 in the development of cardiac hypertrophy. *J Mol Cell Cardiol*, 2007, 42: 498–507.

Owsianik, G., Talavera, K., Voets, T., Nilius, B. Permeation and selectivity of TRP channels. *Annu Rev Physiol*, 2006, 68: 685–717.

Peana, D., Polo-Parada, L., Domeier, T. L. Arrhythmogenesis in the aged heart following ischaemia-reperfusion: Role of transient receptor potential vanilloid 4. *Cardiovasc Res*, 2022, 118: 1126–1137.

Pertusa, M., Rivera, B., Gonzalez, A., Ugarte, G., Madrid, R. Critical role of the pore domain in the cold response of TRPM8 channels identified by ortholog functional comparison. *J Biol Chem*, 2018, 293: 12454–12471.

Piao, H., Takahashi, K., Yamaguchi, Y., Wang, C., Liu, K., Naruse, K. Transient receptor potential melastatin-4 is involved in hypoxia-reoxygenation injury in the cardiomyocytes. *PLoS One*, 2015, 10: e0121703.

Pironet, A., Vandewiele, F., Vennekens, R. Exploring the role of TRPM4 in calcium-dependent triggered activity and cardiac arrhythmias. *J Physiol*, 2024, 602: 1605–1621.

Qi, Z., Wong, C. K., Suen, C. H., Wang, J., Long, C., Sauer, H., et al. TRPC3 regulates the automaticity of embryonic stem cell-derived cardiomyocytes. *Int J Cardiol*, 2016, 203: 169–181.

Randhawa, P. K., Jaggi, A. S. Investigating the involvement of glycogen synthase kinase-3beta and gap junction signaling in TRPV(1) and remote hind preconditioning-induced cardioprotection. *Eur J Pharmacol*, 2017, 814: 9–17.

Roberge, S., Roussel, J., Andersson, D. C., Meli, A. C., Vidal, B., Blandel, F., Lanner, J.T., Le Guennec, J.Y., Katz, A., Westerblad, H., Lacampagne, A., Fauconnier, J. TNF-alpha-mediated caspase-8 activation induces ROS production and TRPM2 activation in adult ventricular myocytes. *Cardiovasc Res*, 2014, 103: 90–99.

Sabourin, J., Antigny, F., Robin, E., Frieden, M., Raddatz, E. Activation of transient receptor potential canonical 3 (TRPC3)-mediated Ca^{2+} entry by A1 adenosine receptor in cardiomyocytes disturbs atrioventricular conduction. *J Biol Chem*, 2012, 287: 26688–26701.

Seo, K., Rainer, P. P., Lee, D. I., Hao, S., Bedja, D., Birnbaumer, L., Cingolani, O.H., Kass, D.A. Hyperactive adverse mechanical stress responses in dystrophic heart are coupled to transient receptor potential canonical 6 and blocked by cGMP-protein kinase G modulation. *Circ Res*, 2014a, 114: 823–832.

Seo, K., Rainer, P. P., Shalkey Hahn, V., Lee, D. I., Jo, S. H., Andersen, A., Liu, T., Xu, X., Willette, R.N., Lepore, J.J., Marino, J.P., Jr., Birnbaumer, L., Schnackenberg, C.G., Kass, D.A. Combined TRPC3 and TRPC6 blockade by selective small-molecule or genetic deletion inhibits pathological cardiac hypertrophy. *Proc Natl Acad Sci U S A*, 2014b, 111: 1551–1556.

Seth, M., Zhang, Z. S., Mao, L., Graham, V., Burch, J., Stiber, J., Tsiokas, L., Winn, M., Abramowitz, J., Rockman, H.A., Birnbaumer, L., Rosenberg, P. TRPC1 channels are critical for hypertrophic signaling in the heart. *Circ Res*, 2009, 105: 1023–1030.

Shan, D., Marchase, R. B., Chatham, J. C. Overexpression of TRPC3 increases apoptosis but not necrosis in response to ischemia-reperfusion in adult mouse cardiomyocytes. *Am J Physiol Cell Physiol*, 2008, 294: C833–C841.

Shattock, M. J., Ottolia, M., Bers, D. M., Blaustein, M. P., Boguslavskyi, A., Bossuyt, J., Bridge, J.H., Chen-Izu, Y., Clancy, C.E., Edwards, A., Goldhaber, J., Kaplan, J., Lingrel, J.B., Pavlovic, D., Philipson, K., Sipido, K.R., Xie, Z.J. $Na+/Ca^{2+}$ exchange and Na+/K+-ATPase in the heart. *J Physiol*, 2015, 593: 1361–1382.

Shimauchi, T., Numaga-Tomita, T., Ito, T., Nishimura, A., Matsukane, R., Oda, S., Hoka, S., Ide, T., Koitabashi, N., Uchida, K., Sumimoto, H., Mori, Y., Nishida, M. TRPC3-Nox2 complex mediates doxorubicin-induced myocardial atrophy. *JCI Insight*, 2017, 2: e93358.

Simard, C., Hof, T., Keddache, Z., Launay, P., Guinamard, R. The TRPM4 non-selective cation channel contributes to the mammalian atrial action potential. *J Mol Cell Cardiol*, 2013, 59: 11–19.

Simard, C., Salle, L., Rouet, R., Guinamard, R. Transient receptor potential melastatin 4 inhibitor 9-phenanthrol abolishes arrhythmias induced by hypoxia and re-oxygenation in mouse ventricle. *Br J Pharmacol*, 2012, 165: 2354–2364.

Srinivasan, N. T., Schilling, R. J. Sudden cardiac death and arrhythmias. *Arrhythm Electrophysiol Rev*, 2018, 7: 111–117.

Sui, Z., Wang, M. M., Xing, Y., Qi, J., Wang, W. Targeting MCOLN1/TRPML1 channels to protect against ischemia-reperfusion injury by restoring the inhibited autophagic flux in cardiomyocytes. *Autophagy*, 2022, 18: 3053–3055.

Sunggip, C., Shimoda, K., Oda, S., Tanaka, T., Nishiyama, K., Mangmool, S., Nishimura, A., Numaga-Tomita, T., Nishida, M. TRPC5-eNOS axis negatively regulates ATP-induced cardiomyocyte hypertrophy. *Front Pharmacol*, 2018, 9: 523.

Tang, L., Yao, F., Wang, H., Wang, X., Shen, J., Dai, B., Wu, H., Zhou, D., Guo, F., Wang, J., Li, T., Wang, H., Gong, T., Su, J., Wang, L., Liang, P. Inhibition of TRPC1 prevents cardiac hypertrophy via NF-kappaB signaling pathway in human pluripotent stem cell-derived cardiomyocytes. *J Mol Cell Cardiol*, 2019, 126: 143–154.

Vandewiele, F., Pironet, A., Jacobs, G., Kecskes, M., Wegener, J., Kerselaers, S., Hendrikx, L., Verelst, J., Philippaert, K., Oosterlinck, W., Segal, A., Van Den Broeck, E., Pinto, S., Priori, S.G., Lehnart, S.E., Nilius, B., Voets, T. TRPM4 inhibition by meclofenamate suppresses Ca^{2+}-dependent triggered arrhythmias. *Eur Heart J*, 2022, 43: 4195–4207.

Wang, C., Chen, J., Wang, M., Naruse, K., Takahashi, K. Role of the TRPM4 channel in mitochondrial function, calcium release, and ROS generation in oxidative stress. *Biochem Biophys Res Commun*, 2021, 566: 190–196.

Wang, J., Takahashi, K., Piao, H., Qu, P., Naruse, K. 9-phenanthrol, a TRPM4 inhibitor, protects isolated rat hearts from ischemia-reperfusion injury. *PLoS One*, 2013, 8: e70587.

Wang, M., Ji, P., Wang, R., Zhao, L., Xia, Z. TRPV1 agonist capsaicin attenuates lung ischemia-reperfusion injury in rabbits. *J Surg Res*, 2012, 173: 153–160.

Wang, Y., Li, X., Xu, X., Qu, X., Yang, Y. Transient receptor potential vanilloid type 1 protects against pressure overload-induced cardiac hypertrophy by promoting mitochondria-associated endoplasmic reticulum membranes. *J Cardiovasc Pharmacol*, 2022, 80: 430–441.

Watanabe, H., Murakami, M., Ohba, T., Takahashi, Y., Ito, H. TRP channel and cardiovascular disease. *Pharmacol Ther*, 2008, 118: 337–351.

Wen, H., Gwathmey, J. K., Xie, L. H. Role of transient receptor potential canonical channels in heart physiology and pathophysiology. *Front Cardiovasc Med*, 2020, 7: 24.

Wen, H., Zhao, Z., Fefelova, N., Xie, L. H. Potential arrhythmogenic role of TRPC channels and store-operated calcium entry mechanism in mouse ventricular myocytes. *Front Physiol*, 2018, 9: 1785.

Wu, Q., Lu, K., Zhao, Z., Wang, B., Liu, H., Zhang, S., Liao, J., Zeng, Y., Dong, Q., Zhao, N., Han, B., Du, Y. Blockade of transient receptor potential vanilloid 4 enhances antioxidation after myocardial Ischemia/Reperfusion. *Oxid Med Cell Longev*, 2019, 2019: 7283683.

Wu, Q. F., Qian, C., Zhao, N., Dong, Q., Li, J., Wang, B. B., et al. Activation of transient receptor potential vanilloid 4 involves in hypoxia/reoxygenation injury in cardiomyocytes. *Cell Death Dis*, 2017, 8: e2828.

Wu, X., Eder, P., Chang, B., Molkentin, J. D. TRPC channels are necessary mediators of pathologic cardiac hypertrophy. *Proc Natl Acad Sci U S A*, 2010, 107: 7000–7005.

Wu, Y., Lu, K., Lu, Y., Liao, J., Zhang, S., Yang, S., Zhao, N., Dong, Q., Chen, L., Wu, Q., Du, Y. Transient receptor potential vanilloid 4 (TRPV4) in neutrophils enhances myocardial ischemia/reperfusion injury. *J Leukoc Biol*, 2023, 114: 266–279.

Xing, Y., Sui, Z., Liu, Y., Wang, M. M., Wei, X., Lu, Q., Wang, X., Liu, N., Lu, C., Chen, R., Wu, M., Wang, Y., Zhao, Y.H., Guo, F., Cao, J.L., Qi, J., Wang, W. Blunting TRPML1 channels protects myocardial ischemia/reperfusion injury by restoring impaired cardiomyocyte autophagy. *Basic Res Cardiol*, 2022, 117: 20.

Yang, K. T., Chang, W. L., Yang, P. C., Chien, C. L., Lai, M. S., Su, M. J., Wu, M.L. Activation of the transient receptor potential M2 channel and poly(ADP-ribose) polymerase is involved in oxidative stress-induced cardiomyocyte death. *Cell Death Differ*, 2006, 13: 1815–1826.

Yanni, J., Tellez, J. O., Maczewski, M., Mackiewicz, U., Beresewicz, A., Billeter, R., Dobrzynski, H., Boyett, M.R. Changes in ion channel gene expression underlying heart failure-induced sinoatrial node dysfunction. *Circ Heart Fail*, 2011, 4: 496–508.

Yue, Z., Xie, J., Yu, A. S., Stock, J., Du, J., Yue, L. Role of TRP channels in the cardiovascular system. *Am J Physiol Heart Circ Physiol*, 2015, 308: H157–H182.

Zhang, S., Lu, K., Yang, S., Wu, Y., Liao, J., Lu, Y., Wu, Q., Zhao, N., Dong, Q., Chen, L., Du, Y. Activation of transient receptor potential vanilloid 4 exacerbates myocardial ischemia-reperfusion injury via JNK-CaMKII phosphorylation pathway in isolated mice hearts. *Cell Calcium*, 2021, 100: 102483.

Zhang, Y. H., Sun, H. Y., Chen, K. H., Du, X. L., Liu, B., Cheng, L. C., Li, X., Jin, M.W., Li, G.R. Evidence for functional expression of TRPM7 channels in human atrial myocytes. *Basic Res Cardiol*, 2012, 107: 282.

Zhang, Y. J., Ma, N., Su, F., Liu, H., Mei, J. Increased TRPM6 expression in atrial fibrillation patients contribute to atrial fibrosis. *Exp Mol Pathol*, 2015, 98: 486–490.

Zhong, B., Ma, S., Wang, D. H. Protective effects of TRPV1 activation against cardiac Ischemia/Reperfusion injury is blunted by diet-induced obesity. *Cardiovasc Hematol Disord Drug Targets*, 2020, 20: 122–130.

Zhong, B., Rubinstein, J., Ma, S., Wang, D. H. Genetic ablation of TRPV1 exacerbates pressure overload-induced cardiac hypertrophy. *Biomed Pharmacother*, 2018, 99: 261–270.

Zhou, Y., Wang, X. C., Wei, J. H., Xue, H. M., Sun, W. T., He, G. W., Yang, Q. Soluble epoxide hydrolase and TRPC3 channels jointly contribute to homocysteine-induced cardiac hypertrophy: Interrelation and regulation by C/EBPbeta. *Biochim Biophys Acta Mol Basis Dis*, 2023, 1869: 166643.

Zou, Y., Zhang, M., Wu, Q., Zhao, N., Chen, M., Yang, C., Du, Y., Han, B. Activation of transient receptor potential vanilloid 4 is involved in pressure overload-induced cardiac hypertrophy. *Elife*, 2022, 11: e74519.

9 Vascular Function Mediators

TRP Channels in Endothelial and Smooth Muscle Cells

Li Geng and Xin Ma

9.1 INTRODUCTION

Calcium (Ca^{2+}) is a fundamental element in the intricate orchestration of physiological processes within the vasculature, having profound importance in maintaining cardiovascular health. The entry of Ca^{2+} (or Ca^{2+} influx) plays crucial roles in regulating the function and survival of both endothelial cells (ECs) and smooth muscle cells (SMCs).[1–3] Ca^{2+} enters the cell mainly through two types of ion channels, highly Ca^{2+}-selective channels (e.g., voltage-gated Ca^{2+} channels, receptor-operated channels, and store-operated channels) and nonselective cation channels (NSCCs). The influx of extracellular Ca^{2+} in ECs and SMCs represents a process that exhibits multimodal activation or modulation in response to factors such as receptor stimulation, mechanical stress, temperature changes, and lipid secondary messengers originating from diverse sources. This intricate Ca^{2+} entry mechanism plays a role in both acute vasomotor regulation and long-term vascular remodeling. Nevertheless, our ability to advance our understanding of this entry pathway has been hindered by the challenge of identifying the genes responsible for various NSCCs in ECs and SMCs.

The transient receptor potential (TRP) channels are thought to constitute one of the most important Ca^{2+} entry pathways in ECs and SMCs, playing vital roles in regulating vascular functions under physiological conditions.[4] The recognition that TRP channels have relevance in the diastolic, contractile, and proliferative functions of ECs and SMCs not only enhances our comprehension of vascular hemostasis but also provides a better understanding of the pathophysiology of cardiovascular diseases.[5] The expression of various isoforms of the TRP family has been widely studied in previous publications. Table 9.1 lists the TRP isoforms expressed in ECs and SMCs across different vascular beds in humans, mice, and rats. In both ECs and SMCs, TRP channels present remarkable diversity in functions and structures among different organs and species. The mechanism underlying this diversity remains largely unexplored. Notably, TRP channels are broadly expressed across almost all sizes of arteries including aortic, coronary, and cerebral arteries, and small-resistance arteries. A number of chemical and physical stimuli from extracellular and intracellular sides have been reported to activate and/or modulate TRP channels, through which they regulate vascular functions such as vasodilation, vasoconstriction, angiogenesis, and vascular permeability.[4,6–8] In light of the pivotal roles played by TRP channels in regulating vascular functions, a novel class of therapeutic compounds tailored to target these channels has emerged for the treatment of cardiovascular diseases.

9.2 TRP CHANNELS IN ANGIOGENESIS

Angiogenesis is responsible for the development of blood vessels in the later stage of embryogenesis and in adulthood,[102] Relying on various vascular endothelial growth factors (VEGFs), which are thought to initiate the sprouting of the vessel through extracellular matrix degradation, directional EC migration, and proliferation, TRP channels have been identified as important contributors to the process of angiogenesis under physiological and pathological conditions.[6] Strong evidence exists

DOI: 10.1201/9781003402732-9

TABLE 9.1
Expressions of TRP Isoforms across Different Vascular Beds in Human, Mouse, and Rat

Endothelial Cell	TRPC	TRPV	TRPM	TRPA
Human Coronary Artery	C1,[9] C3,[9,10] C4,[9, 11] C5,[9] C6,[9] C7[9]	V4[12,13]		
Human Aortic Artery	C4[14]	V4[15]	M2[16]	
Human Pulmonary Artery	C1,[17,18] C3,[19] C4,[20] C6[18]	V1,[21] V2,[21] V4[21]	M2[22,23]	
Human Mesenteric Artery		V1,[24] V4[25,26]		
Human Cerebral Artery	C1,[9] C3,[9] C4,[9] C5,[9] C6[9]	V1,[21,27] V3,[28] V4[21,29]		A1[30]
Human Umbilical Vein	C1,[31–33] C3,[34,35] C4,[35] C5,[35] C6[34]	V4[36]	M2,[37] M4,[38] M7[39,40]	
Mouse Coronary Artery		V1[41]		
Mouse Aortic Artery	C1,[19,36] C4,[42] C5,[43] C6[44,45]	V1,[24,46,47] V4[48,49]	M2,[50] M4[51]	
Mouse Pulmonary Artery	C1,[52] C4,[52,53] C6[54]	V4[55]		
Mouse Mesenteric Artery	C1,[56] C5[57]	V1,[24,47] V4[48,56,58]		
Mouse Cerebral Artery	C3[59]	V4[60]		A1[30]
Rat Coronary Artery	C1[61]			
Rat Aortic Artery	C6[62]	V4[63]		
Rat Pulmonary Artery	C1,[64,65] C4,[64,65] C5,[65] C6[62,66]	V4[55, 67]	M2[68]	
Rat Mesenteric Artery	C1,[56] C3[69]	V1,[70–72] V4[56, 58, 73]		
Rat Cerebral Artery		V3[28], V4[74,75]	M4[38]	A1[25,30,76,77]
Smooth Muscle Cell				
Human Coronary Artery	C1,[78] C3,[78] C4,[78] C5,[78] C6[78]			
Human Pulmonary Artery		V1,[79,80] V2,[80] V3,[80] V4[80]		
Mouse Mesenteric Artery	C1,[81,82] C5[81]	V2,[83] V4[84]		
Mouse Cerebral Artery	C3,[85] C6[86]			
Rat Aortic Artery	C4,[87] C5,[87] C6[88]	V1,[89] V2,[90] V4[90]	M2,[91] M4,[92] M8[92]	
Rat Pulmonary Artery	C1,[93] C3,[93] C4,[93] C5,[93] C6[94]	V1,[95] V3,[96] V4[95]	M6,[97] M7[97]	
Rat Mesenteric Artery		V2,[90] V4[90]	M4,[90] M7[92]	
Rat Cerebral Artery	C1,[98] C3,[99,100] C6[99]	V2,[90] V4[101]	M4[90]	

to support the mechanism that angiogenic growth factors activate TRP channels, leading to an increase in intracellular Ca^{2+} concentration ($[Ca^{2+}]_i$) within ECs. This rise in $[Ca^{2+}]_i$, in turn, influences the signal transduction pathways that orchestrate and promote angiogenesis.[103,104] In addition to working as homotetrameric channels with subunits consisting of four identical isoforms, some TRP isoforms are thought to be associated with other isoforms to form heteromeric channels to support angiogenesis.[29,58,105–107]

9.2.1 TRPC1

A proangiogenic role of TRPC1 was initially described *in vivo* in zebrafish.[108] Genetic knockout of TRPC1 in endothelial tip cells led to disruption of the angiogenic sprouting of intersegmental vessels. Furthermore, this study showed that TRPC1 mediates the VEGF-induced signaling of the mitogen-activated protein kinase-extracellular signal-regulated kinase (MAPK/ERK) cascade, suggesting a possible mechanism of how TRPC1 regulates angiogenesis. Another study demonstrated that endothelial progenitor cells from TRPC1 knockout mice show impairment of migration and tube formation, adding more evidence that TRPC1 plays an important role in angiogenesis.[109] However, other studies argue that TRPC1 is not required for angiogenesis. For instance, suppression by an anti-TRPC1 blocking antibody failed to alter VEGF-induced Ca^{2+} transients in Human

Umbilical Vein ECs (HUVECs) and tubulogenesis.[110] In addition, development of the vasculature was not affected by the genetic silencing of TRPC1 in mice.[111] Therefore, further studies are required to verify the role of TRPC1 in angiogenesis.

9.2.2 TRPC3 AND TRPC6

TRPC3 has been found to mediate VEGF-triggered ERK1/2 phosphorylation and $[Ca^{2+}]_i$ transients in HUVECs. TRPC3 siRNA significantly curbed endothelial tube formation.[34] This research suggests that TRPC3 activation in HUVECs is driven by diacylglycerols (DAG) downstream of VEGF receptors, leading to Na^+ influx and subsequent $[Ca^{2+}]_i$ rise due to activation of the Na^+/Ca^{2+} exchanger in reverse mode, ultimately contributing to angiogenesis. Similarly, TRPC6 plays a pivotal role in angiogenesis and Ca^{2+} entry in response to VEGF and 1-oleoyl-2-acetyl-sn-glycerol, a membrane-permeant DAG analog, in human microvascular ECs and HUVECs. In Matrigel assays, a dominant-negative TRPC6 mutant with three pore region mutations reduced EC proliferation, migration, and sprouting.[112] Similar outcomes were reported in HUVECs, where a dominant-negative form of TRPC6 hindered VEGF-induced cation currents, cell growth, proliferation, and VEGF-driven capillary formation *in* vitro.[113] The involvement of TRPC6 in EC proliferation and tube formation was also evident when 11,12-epoxyeicosatrienoic acid (EET) was used to stimulate ECs.[114] Interestingly, an earlier report indicated the involvement of a heteromeric TRPC3/TRPC6 channel in VEGF-induced Ca^{2+} influx in human microvascular ECs.[115] The initiation of TRPC3 and TRPC6 gating was attributed to the phospholipase C (PLC) gamma 1 metabolite, DAG, in a protein kinase C (PKC)-independent manner, as described previously.[116]

9.2.3 TRPC5 AND TRPC4

TRPC5 has been found to influence angiogenesis in various vascular systems. Our group found that knocking down TRPC5 through specific siRNA disrupted hypoxia-induced capillary sprouting and tube formation in primary mouse intestinal mesenteric vascular ECs.[57] In a mouse model lacking TRPC5, hypoxia-induced retinal neovascularization was compromised.[57] Similarly, in the context of TRPC4, oxygen depletion increased its expression in the mouse retina, while intravitreal infusion of a specific TRPC4 siRNA hindered the subsequent retinal neovascularization process.[117] In previous research on human pulmonary artery ECs, it has been found that hypoxia upregulated TRPC4 through AP-1 transcription factors and that TRPC4 facilitated store-operated Ca^{2+} entry (SOCE).[20] Moreover, the absence of TRPC4 also prevented VEGF-induced migration and tube formation by interfering with the ERK and Akt signaling pathways in human retinal microvascular ECs.[117]

9.2.4 TRPV1

TRPV1 has been found to contribute to angiogenesis.[6,118] Intraperitoneal administration of evodiamine, a TRPV1 ligand, increased vascularization in Matrigel plugs *in* vivo.[119] Conversely, deletion of TRPV1 significantly reduced angiogenesis.[119] TRPV1's role in angiogenesis involves mediating extracellular Ca^{2+} entry, which promotes PI3K/Akt/calmodulin-dependent protein kinase II (CaMKII) signaling. This signaling cascade leads to TRPV1 phosphorylation, physical interaction between endothelial nitric oxide synthase (eNOS) and TRPV1, eNOS phosphorylation, and ultimately, the release of nitric oxide (NO).[119] The formation of the TRPV1-eNOS complex also leads to NO production and the formation of angiogenesis-inducing tubes *in* vitro.[120] In addition, TRPV1 appears to be crucial for 14,15-EET-induced Ca^{2+} influx, NO production, and angiogenesis, as assessed through tube formation and in vivo Matrigel assays.[121] Furthermore, in mouse retinal microvascular ECs, TRPV1 may form a heteromeric complex with TRPV4. This complex promotes VEGF-induced extracellular Ca^{2+} entry, leading to angiogenesis, EC proliferation, and the formation of capillary-like tubes.[122] Hofmann et al. demonstrated that TRPV1 mediates the uptake of the endogenous cannabinoid anandamide in a Ca^{2+}-independent manner, which, in turn, promotes cell proliferation and tube formation.[123] These findings strongly suggest that stimulating TRPV1 is a promising pathway to induce angiogenesis, especially in the presence of cardiovascular risk factors.

9.2.5 TRPV4

TRPV4 plays a well-established role in mediating the proliferation and migration of ECs.[124] Activation of TRPV4 by 4α-phorbol 12,13-didecanoate (4α-PDD) has been shown to stimulate proliferation in human brain ECs.[125] Moreover, TRPV4 functions as a mechanosensor, regulating changes in the cytoskeleton that are associated with cell adhesion, proliferation, and motility of ECs.[126–129] In cases of increased fluid shear stress due to femoral artery ligature, TRPV4 protein expression is upregulated, leading to the development of new collateral arteries.[130] TRPV4-induced Ca^{2+} influx has also been implicated in pulmonary arterial hypertension and cancer. Oxidative stress has been shown to disrupt endothelial permeability by activating TRPV4 through the Src kinase Fyn in human and mouse lung microvascular ECs.[131] More recent research from the same group has revealed that higher levels of reactive oxygen species (ROS) led to the continuous opening of TRPV4 channels, increasing resting intracellular Ca^{2+} levels and promoting abnormal EC proliferation and migration.[132] In breast tumor-derived ECs, TRPV4 is notably upregulated, resulting in an increased rate of EC migration and motility.[133] Notably, the absence of TRPV4 in knockout mice was associated with heightened basal Rho/Rho kinase activity, a significant increase in EC proliferation, migration, and abnormal tube formation *in* vitro.[129]

9.2.6 TRPMs

TRPM2, TRPM4, TRPM6, and TRPM7 have been implicated in angiogenesis. TRPM2-induced Ca^{2+} influx, activated by VEGF in a ROS-dependent manner, plays a role in the migration of human pulmonary artery ECs.[22] Interestingly, the same study demonstrated a significant impairment in vessel formation in TRPM2 knockout mice compared to wild-type (WT) mice.[22] Hypoxia and ischemia lead to an increase in TRPM4 expression in vascular endothelium, as well as in HUVECs under oxygen-glucose deprivation. Blocking TRPM4 with pharmacological agents enhanced tube formation on Matrigel and improved capillary integrity *in vivo*.[38] Both TRPM6 and TRPM7 facilitate the entry of magnesium (Mg^{2+}) and play a vital role in the proliferation of ECs as well as the process of angiogenesis.[134,135] Previous research has verified that TRPM7 is activated when intracellular Mg^{2+} levels decrease in HUVECs. This activation leads to an increase in HUVEC proliferation through ERK phosphorylation and the release of NO.[39] Conversely, the absence of TRPM7 enhances HUVEC adhesion to the substrate, promotes migration in wound-healing assays, and facilitates tubulogenic activity in Matrigel.[40] Nevertheless, further experiments are still necessary to confirm the proangiogenic role of TRPM7.

9.3 TRP CHANNELS IN ENDOTHELIUM-DEPENDENT VASODILATION

The endothelium plays a pivotal role in detecting both physical and chemical stimuli, leading to a significant relaxation effect on SMCs that helps maintain the optimal vascular tone. Three mechanisms are involved in mediating the signals from the endothelium to relax the SMCs and to cause vasodilation.[4,136] Traditionally, NO and prostacyclin (PGI2) are known to be produced by increased Ca^{2+} levels in the endothelium and diffuse to SMCs. In addition, compelling evidence supports the notion that changes in endothelial Ca^{2+} signaling, whether occurring globally or locally, trigger the activation of Ca^{2+}-sensitive potassium (K^+) channels. This activation leads to membrane hyperpolarization within ECs, which is subsequently transmitted to vascular SMCs via myoendothelial gap junctions. This mechanism introduces a novel concept known as endothelium-dependent hyperpolarization (EDH), which is recognized as the principal mechanism responsible for vasodilation in small resistance arteries. To date, at least TRPC3, TRPC4, TRPV1, TRPV3, TRPV4, and TRPA1 have been reported to be involved in endothelium-dependent vasodilation.

9.3.1 TRPC3

TRPC3 is expressed in ECs and plays a pivotal role in endothelium-dependent vasodilation. An early study demonstrated that genetic silencing of TRPC3 significantly reduced vasodilation induced by both flow and bradykinin.[69] Subsequently, a report showed that inhibiting TRPC3 with Pyr3 resulted in a marked reduction in acetylcholine (ACh)-induced vasodilation in human internal mammary arteries.[137] In line with these findings, another study reported a significant decrease in EC hyperpolarization and vasorelaxation when ACh was administered in the presence of the selective TRPC3 blocker Pyr3.[138] Moreover, this research indicated that combining Pyr3 with an intermediate conductance Ca^{2+}-activated potassium (IK) channel blocker or a small-conductance Ca^{2+}-activated potassium (SK) channel blocker eliminated the vasodilatory responses to ACh in mesenteric arteries, implying a close association between TRPC3 channels and EDH. Another study demonstrated that TRPC3 channels contribute to EDH-mediated, SK, and IK channel-dependent dilation of cerebral arteries in response to ATP.[59] In summary, TRPC3-mediated Ca^{2+} influx significantly contributes to EDH-mediated vasodilation in small resistance arteries.

9.3.2 TRPC4

TRPC4 channels are believed to become active *via* G protein-coupled receptor (GPCR) pathways, facilitating Ca^{2+} influx in the endothelium through SOCE.[4,139] Subsequent studies have similarly shown that the deletion of TRPC4 in ECs attenuates ACh-induced Ca^{2+} entry and SOCE, ultimately impairing endothelium-dependent vasodilation in mouse aortic arteries. This impairment is rooted in the fact that SOCE, mediated by TRPC4, stimulates eNOS activity and NO production, leading to vasodilation.[14,140] However, the contribution of TRPC4 to SOCE in ECs still needs to be confirmed due to the failure of other reports to present consistent conclusions.[139,141]

9.3.3 OTHER TRPCs

Other TRPC channels, specifically TRPC1 and TRPC5, have been implicated in the mediation of endothelium-dependent vasodilation. TRPC1 has the capability to co-localize with other TRP channels, contributing to vasorelaxation. For instance, the formation of heteromultimeric channels between TRPC1 and TRPC3 is known to be involved in muscarinic agonist-induced vasodilation.[142] Furthermore, the co-localization of TRPC1 and TRPV4, activated by the Ca^{2+}-sensing receptor, induces vasorelaxation mediated by NO.[143] The precise role of TRPC5 in endothelium-dependent vasodilation awaits confirmation in future studies. An early report suggested that S-nitrosylation, involving the covalent attachment of NO to cysteine groups, enhances the activity of TRPC5 channels and promotes Ca^{2+} influx in cultured ECs.[143] However, a subsequent study did not find any changes in the vasodilatory response to ACh in TRPC5 knockout arteries.[43]

9.3.4 TRPV1

TRPV1 is recognized as playing a role in endothelium-dependent vasodilation. In an early study, TRPV1 activation was shown to increase NO levels in the endothelium of rat mesenteric arteries.[70] Subsequently, Yang et al. discovered that capsaicin-induced Ca^{2+} influx led to the phosphorylation of eNOS and an increase in NO production, resulting in the endothelium-dependent relaxation of isolated mesenteric arteries.[24] Importantly, this effect was not found in TRPV1-deficient mice. Moreover, this study demonstrated that capsaicin significantly reduced blood pressure in spontaneously hypertensive rats, suggesting that TRPV1 channels are involved in the regulation of vascular tone and arterial pressure. In a later investigation, it was found that capsaicin-induced Ca^{2+} entry through TRPV1 channels activates the Akt-CaMKII-eNOS pathway. Consequently, the phosphorylated eNOS physically interacts with TRPV1 channels to produce NO.[119] Collectively, these studies lead to the conclusion that Ca^{2+} influx through TRPV1 channels enhances the production of endothelium-derived NO, resulting in vasodilation.

9.3.5 TRPV4

TRPV4 channels have been extensively investigated in the context of endothelium-dependent vasodilation (Fig. 9.1). These channels can be activated by a wide range of stimuli, including endogenous substances such as EETs (epoxyeicosatrienoic acids) and ACh, pharmacological compounds like 4α-PDD, carbachol, and GSK1016790A (GSK101), as well as physical factors including shear stress and warm temperature.[4]

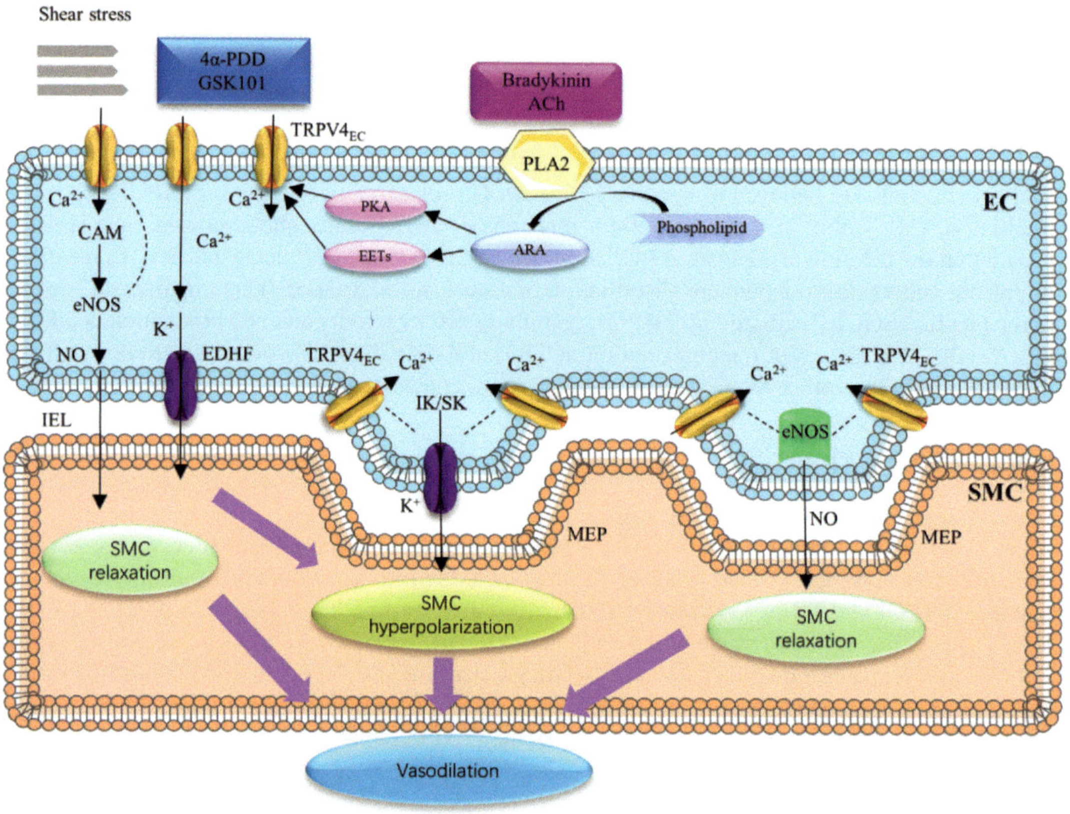

FIGURE 9.1 Role of TRPV4 channels in endothelium-dependent vasodilation. TRPV4-mediated Ca^{2+} influx triggered by diverse stimuli elicits vasodilatory responses through NO and/or EDHF pathways across various vascular beds. TRPV4 forms a channel complex in MEPs to dilate small resistance arteries. In small mesenteric arteries, TRPV4 sparklets activate IK/SK channels to cause EC and SMC hyperpolarization, ultimately leading to vasodilation. In small pulmonary arteries, TRPV4 sparklets enhance eNOS activity, promoting the release of NO and consequent vasodilation. ACh: Acetylcholine; ARA: Arachidonic acid; CAM: Calmodulin-dependent protein; EC: Endothelial cell; EDHF: Endothelium-dependent hyperpolarization factor; EET: Epoxyeicosatrienoic acid; eNOS: Endothelial nitric oxide synthase; IEL: Internal elastic lamina; MEP: Myoendothelial projections; NO: Nitric oxide; PKA: Protein kinase A; PLA2: Phospholipase A2; SMC: Smooth muscle cell.

TRPV4 channels are extensively studied for their involvement in shear stress-induced dilation. Application of shear stress induces dilation of rat gracilis arteries, a process that was inhibited by the nonselective TRPV blocker ruthenium red.[126] In mesenteric arteries, TRPV4 channels activated by shear stress led to increased Ca^{2+} entry into ECs, resulting in vasodilation. This vasodilatory response is significantly suppressed in TRPV4-deficient mice.[144] Thus, the mechanoactivation of TRPV4 channels was believed to induce vasodilation through the production of NO. However, in the presence of NOS and cyclooxygenase (COX) inhibitors, the perfusion of 4α-PDD and 11,12-EET still evoked vasodilation in mesenteric arteries, indicating alternative pathways for TRPV4-mediated dilation. Interestingly, some studies have suggested that TRPV4-mediated Ca^{2+} signaling, activated by shear stress, leads to endothelium-dependent hyperpolarization factor (EDHF)-dependent vasodilation in small resistance arteries.[126,145] These findings imply that different physiological or pharmacological forms of TRPV4 channel activation may result in distinct downstream signaling for vasodilation in various vascular beds.

It is essential to note that it remains uncertain whether TRPV4 channels are directly activated by shear stress or respond to signaling pathways triggered by flow-sensing apparatus. Loot et al. demonstrated that the inhibition of cytochrome P450 (CYP) epoxygenase attenuated the flow-induced dilation of mouse carotid arteries, suggesting that EETs produced in response to shear stress are responsible for TRPV4-mediated dilation.[29] Similarly, another study found that inhibition of phospholipase A2 impaired flow-induced dilation in mouse carotid arteries.[146] Furthermore, a recent study revealed that Piezo1 activation by shear stress initiates TRPV4-mediated Ca^{2+} influx in HUVECs.[147] As a result, the precise mechanisms underlying TRPV4's role in flow-induced vasodilation still require further investigation and confirmation.

Endothelial TRPV4 activation by EETs plays a crucial role in vasodilation. EETs, including 5,6-EET, 8,9-EET, 11,12-EET, and 14,15-EET, are generated from arachidonic acid (AA) by CYP epoxygenase enzymes, and they have anti-hypertensive effects in the cardiovascular system.[148,149] A wealth of evidence has shown that EETs stimulate TRPV4 in mouse aortic ECs, leading to vasodilatory responses. In particular, one study demonstrated that application of exogenous 11,12-EET resulted in dilation of isolated mesenteric arteries in WT mice, but this effect was absent in TRPV4-deficient mice.[25] Interestingly, the vasodilation associated with 11,12-EET was blocked in the presence of inhibitors of BK, IK, or SK channels but not by the inhibition of NOS and COX, suggesting that the underlying mechanisms involve EDH. A subsequent study revealed that the removal of extracellular Ca^{2+} or pretreatment with CYP inhibitors significantly inhibited TRPV4-denpendent Ca^{2+} influx induced by AA, suggesting that CYP-dependent metabolite generation from AA increases extracellular Ca^{2+} entry through endothelial TRPV4 channels. This process subsequently requires IK and SK-mediated hyperpolarization to induce vasodilation in human coronary arteries.[150]

TRPV4 activity can be potentiated by muscarinic and purinergic receptor agonists, leading to endothelium-dependent relaxation in various systemic arteries. A study has demonstrated that dilatory responses induced by ACh were abolished in mesenteric arteries from $TRPV4^{-/-}$ animals.[48] This group has also established a strong correlation between TRPV4-mediated Ca^{2+} influx by ACh and NO production. Studies by both our group and others have indicated that GSK101-induced endothelial Ca^{2+} influx and vasodilation of third-order mesenteric arteries were hindered by IK and SK channel inhibitors. This suggests that TRPV4-mediated Ca^{2+} influx results in the dilation of small resistance arteries through an EDHF or EDH mechanism.[26,145,151] Furthermore, GSK101 has been found to selectively activate eNOS, not IK/SK channels, to induce dilation in fourth-order pulmonary arteries, highlighting the close association between TRPV4 and eNOS.[151] These findings suggest that TRPV4-associated vasodilation may involve different pathways in distinct small resistance arteries. However, ACh-induced relaxation of pulmonary arteries was not affected by blockade of TRPV4 channels, indicating that TRPV4 channels do not play a role in muscarinic agonist-induced responses in pulmonary arteries.[152] Furthermore, some other studies have noted that elevated levels of endothelial TRPV4 activation result in vasoconstriction through endothelial COX-dependent prostanoid production, implying a contractile role of TRPV4 in the endothelium.[84,153]

Significantly, the Sonkusare and Nelson research team made a novel discovery by observing TRPV4 sparklets.[26] These sparklets represent a new concept of endothelial Ca^{2+} influx through TRPV4 channels, which were spatially identified at myoendothelial projections (MEPs) in intact mesenteric arteries. Their findings revealed that TRPV4 channels exhibit quantal events of Ca^{2+} influx, which can be stimulated by compounds like 4α-PDD, 11,12-EET, or GSK101. They found that having three to four TRPV4 sparklet sites per EC is sufficient for achieving maximal vasodilation in third-order mesenteric arteries. Furthermore, they discovered that the clustered TRPV4 channels, in the form of a four-channel metastructure and locally distributed at MEPs, exhibit cooperative gating. This cooperative gating is influenced by the presence of protein kinase C-anchoring protein 150 and by intracellular Ca^{2+} levels.[154] When a single TRPV4 channel opens, it leads to minimal Ca^{2+} influx and activates only a small number of IK and/or SK channels. However, the increased cooperativity of these clustered TRPV4 channels results in quantal events and higher Ca^{2+} amplitudes, which lead to enhanced hyperpolarization in ECs and subsequently in SMCs, ultimately causing vasodilation. Moreover, subsequent studies from the same research group have identified a crucial role for endothelial TRPV4 channels in various pathophysiological circumstances within the cardiovascular system.[155,156]

9.3.6 TRPA1

TRPA1 channels are present in ECs and exhibit a unique distribution exclusively in cerebral arteries but not in mesenteric, coronary, dermal, or renal arteries.[30] Initially, allyl isothiocyanate (AITC), a compound derived from mustard oil, induced endothelium-dependent dilation and hyperpolarization of SMC membranes in pressurized cerebral arteries. This response was attenuated by the selective TRPA1 inhibitor.[76] Significantly, the study went on to reveal that the AITC-induced effects remained unaffected when NOS and COX enzymes were blocked, but they were markedly reduced by the inhibition of SK, IK, or inward-rectifier K^+ channels. In addition, this investigation highlighted that TRPA1 channels are expressed in MEPs and are co-localized with IK channels. Beyond AITC, another study from the same research group delved into the ROS-mediated activation of TRPA1 channels in cerebral arteries.[30] This study uncovered that TRPA1 co-localized with NOX2 (NADPH oxidase isoform 2) and recorded TRPA1 sparklets within the ECs of cerebral arteries. Importantly, they found that two endothelial TRPA1 channels opened simultaneously, resulting in twice the amplitude produced by TRPV4 sparklets. When exogenous NADPH was administered, it led to vasodilation, a response that was significantly attenuated in the presence of a TRPA1 inhibitor or an IK channel inhibitor. Therefore, it was demonstrated that NADPH-NOX signaling played a pivotal role in stimulating Ca^{2+} entry through endothelial TRPA1 channels, subsequently leading to IK channel-dependent vasodilation. Notably, these responses were contingent on the generation of extracellular H_2O_2, hydroxyl radicals, and lipid peroxidation. This suggested that NOX-derived H_2O_2 contributed to the production of lipid peroxidation metabolites, which, in turn, stimulated EC TRPA1 Ca^{2+} sparklets, ultimately resulting in vasodilation in cerebral arteries.

9.4 TRP CHANNELS IN VASCULAR SMOOTH MUSCLE CONTRACTILITY

Vascular resistance is determined by the contractile condition of vascular SMCs, which encircle the inner lumen of all blood vessels. Vascular smooth muscle contraction is highly reliant on the balance of myosin light chain (MLC) kinase and MLC phosphatase activity, which, in turn, dictates the phosphorylation status of the regulatory light chain.[157] The activity of MLC phosphatase is intricately regulated by intracellular signaling pathways, playing a critical role in adjusting the sensitivity of the contractile apparatus to intracellular Ca^{2+}.[4] While the L-type voltage-gated Ca^{2+} channels (VGCCs) are typically regarded as the principal pathway for Ca^{2+} influx, which is responsible for membrane depolarization and the subsequent contraction of vascular SMCs, it is important to note that the heightened cation permeability linked to TRP channel activation also leads to membrane depolarization and Ca^{2+} influx. These processes are the primary mechanisms by which TRP channels induce vasoconstriction.

9.4.1 TRPC1

TRPC1 channels have been the subject of extensive research in the context of both SOCE and receptor-operated Ca^{2+} entry (ROCE) in vascular SMCs. While their role in SOCE has been in debate, evidence exists to suggest that TRPC1 channels might contribute to ROCE. Studies have investigated the receptor-operated regulation of TRPC1 and heteromultimeric channels containing TRPC1 as subunits, demonstrating downstream activation of PLC- and PKC-dependent pathways.[81,158] It has been proposed that an association between TRPC1, inositol trisphosphate receptors (IP3Rs), and stromal interaction molecule 1 may account for Ca^{2+} influx and vascular smooth muscle contraction independently of SOCE.[159] However, supporting experimental evidence for this model remains limited. The biophysical properties of TRPC1 channels suggest a small fractional Ca^{2+} component under physiological conditions,[4] arguing against direct Ca^{2+} influx. Instead, TRPC1-mediated cation influx could potentially lead to membrane depolarization and Ca^{2+} influx *via* VGCCs in excitable cells. Nevertheless, it is notable that vasoconstriction responses to various stimuli remain unaltered in TRPC1$^{-/-}$ mice, indicating complexities in TRPC1's role in vascular contractility.[82,111] The situation is further complicated by findings of interactions of TRPC1 with other TRP isoforms and non-TRP ion channels in vascular SMCs, suggesting a multifaceted role for TRPC1 in regulating vascular contractility.

9.4.2 TRPC3

TRPC3 channels play an important role in mediating contractile responses in various types of arteries, particularly in response to receptor-ligand interactions. Their activation has been reported in response to GPCR stimulation by ligands such as pyrimidine, endothelin-1, and angiotensin II.[99,160] TRPC3 channels contribute to receptor-mediated vasoconstriction in cerebral, carotid, and aortic arteries.[99,160,161] The primary mechanism of TRPC3 activation downstream of receptor stimulation was initially thought to involve DAG.[162] Still, it was later discovered that TRPC3 channels functionally interact with IP3Rs in vascular SMCs.[98,100,163] This interaction is mediated by IP3 and results in the activation of TRPC3-mediated cation currents, leading to membrane depolarization, Ca^{2+} entry through VGCCs, and myocyte contraction. Close physical association between IP3Rs and TRPC3 channels is facilitated by the integral membrane protein caveolin-1.[98] Additionally, heteromultimeric TRPC3/TRPC7 channels activated by ET-1 in coronary artery myocytes also involve phosphoinositide metabolism products.[161] Specifically, phosphatidylinositol-3-phosphate, generated by specific phosphatidylinositol 3-kinase isoforms, seems to directly activate this channel.[164] Furthermore, regulation of TRPC3 channels in native vascular SMCs involves phosphatidylinositol 4,5-bisphosphate (PIP2), the primary substrate for phosphoinositide metabolism, as demonstrated in expression systems.[165,166] However, whether PIP2 regulation holds true in native vascular SMCs remains to be determined. Generally speaking, TRPC3 channels participate in GPCR-mediated vasoconstriction in various arteries through multiple mechanisms, including direct activation by DAG and interaction with IP3Rs, and their regulation involves the interplay of various intracellular signaling pathways.

9.4.3 TRPC6

TRPC6 channels are involved in cellular responses to GPCR activation, with their diverse roles extending to native vascular smooth muscle. Initial evidence for the involvement of TRPC6 channels in native vascular smooth muscle was provided by Inoue et al.[167] Subsequent studies have further illuminated their intricate functions, particularly in response to various GPCR ligands. Notably, TRPC6 channels share biophysical and pharmacological characteristics with previously characterized nonselective cation channels in vascular SMCs, exhibiting activation by vasopressin, angiotensin II, and other GPCR ligands.[168–170] Their activation involves intricate downstream signaling *via* PLC and the generation of DAG.[171] Moreover, these channels exhibit direct sensitivity to DAG, independently of PKC activation.[172] The intricate interplay between receptor-mediated signaling and other cellular pathways results in the amplification of TRPC6 channel activity. For instance, transmembrane Ca^{2+} mobilization *via* TRPC6 activation is potentiated by combined receptor and mechanical stimulation, involving contributions from both PLC and phospholipase A2.[173]

In addition, TRPC6 channels have been extensively described in arterial smooth muscle, where they play a pivotal role in contractile function downstream of PLC activation. Notably, TRPC6 currents are initiated by the activation of GPCRs, and prominent inducers include vasopressin and angiotensin II.[88] An important finding is that PIP2, the precursor of DAG, exerts a unique inhibitory influence on TRPC6 channel function in arterial SMCs. This inhibition by PIP2 is specific to arterial cells and does not manifest in other expression systems.[166] The dynamic interplay between PIP2 and DAG serves as a coregulatory mechanism, where PLC activation serves a dual role by alleviating the TRPC6 channel inhibition imposed by PIP2 while concurrently facilitating channel activation through DAG. Further investigations have underscored the synergistic relationship between the activation of TRPC6 channels by both DAG and IP3, although IP3 alone does not have any impact on channel opening.[174]

TRPC6 channels have been functionally implicated in the regulation of pressure-induced (myogenic) vasoconstriction in vascular SMCs. The activation of TRPC6 channels can be initiated by mechanical forces such as cell swelling, membrane suction, or pressure-induced depolarization.[171,175] However, the precise mechanisms underlying the mechanical activation of TRPC6 channels are still debated. Some studies suggest direct mechanical activation, while others propose indirect activation through mechanosensitive signaling pathways, possibly involving GPCRs and the formation of DAG.[86,176] Data from research by Welsh et al. support the idea that TRPC6 channels play a crucial role in the myogenic response in cerebral arteries.[175] Their experiments showed that reducing TRPC6 expression led to decreased swelling-activated cation currents, intravascular pressure-induced SMC depolarization, and vasoconstriction. However, it is important to note that the absence of TRPC6 in mice did not lead to a reduction in agonist-evoked contraction or myogenic tone.[85,86] This discrepancy could be attributed to a potential compensatory mechanism involving the upregulation of constitutively active TRPC3 channels in $TRPC6^{-/-}$ mice, providing an alternative depolarization and vasoconstrictor pathway.

9.4.4 TRPV4

An early study uncovered the surprising effect of TRPV4-mediated Ca^{2+} influx, which induces vasodilation rather than vasoconstriction in cerebral arterial myocytes.[101] The vasodilation mechanism is facilitated by the action of 11,12-EET. TRPV4 triggers ryanodine receptors positioned on the sarcoplasmic reticulum, eliciting Ca^{2+} release through Ca^{2+} sparks. Subsequently, BK channels are activated, leading to myocyte hyperpolarization and decreased Ca^{2+} entry through VGCCs, ultimately resulting in vasodilation. A similar role for TRPV4 channels in regulating the contractile activity of mesenteric arteries has been reported, contributing approximately 50% of 11,12-EET-induced vasodilation from the endothelium and the remaining 50% from vascular SMCs.[25]

While TRPV4 channels are widely recognized for their role in promoting vasorelaxation, there are instances in which their activation induces vasoconstriction. Notably, in pulmonary artery SMCs, these channels are implicated in adrenergic vascular contractility and respond to hypoxia.[177,178] TRPV4 contributes to hypoxia-induced pulmonary hypertension by enhancing myogenic tone and is also involved in 5-hydroxytryptamine-mediated pulmonary vasoconstriction in chronic hypoxic pulmonary hypertension.[179] Furthermore, under specific pathological conditions,

TRPV4 activation can lead to vasoconstriction. In mouse aortas, the potentiation of TRPV4 action by GPCRs induces vasoconstriction through COX-dependent prostanoid production and the activation of thromboxane receptors.[153] In addition, our research group has demonstrated the critical role of TRPV4 channels in SMCs in promoting vascular contraction in the context of obesity.[84] To reconcile these contradictory findings regarding the role of TRPV4, an important study conducted by Sonkusare's group identified two distinct, spatially separated, and functionally opposing Ca^{2+}-signaling microdomains within SMCs. These microdomains consist of α1 adrenergic receptor (α1AR)–$TRPV4_{SMC}$, which elevates blood pressure, and $TRPV4_{SMC}$–BK channel microdomains, which lower blood pressure.[180] The imbalanced increase in α1AR–$TRPV4_{SMC}$ signaling and decrease in $TRPV4_{SMC}$–BK channel signaling contribute to excessive vasoconstriction and elevated blood pressure in hypertension.

9.4.5 TRPM4

TRPM4 channels play a pivotal role in the regulation of cerebral artery function, exerting an important influence on vascular tone and blood pressure.[181,182] When TRPM4 expression is reduced in cerebral arteries, it leads to substantial decreases in pressure-induced smooth muscle membrane depolarization and vasoconstriction.[183] This initial discovery gained further support through subsequent research demonstrating that the highly specific pharmacological inhibitor of TRPM4 channels, 9-phenanthrol, effectively suppresses pressure-induced depolarization and myogenic tone in isolated pial arteries.[183,184] However, these findings have faced challenges from a study suggesting that 9-phenanthrol might activate IK channels in the endothelium, leading to mesenteric artery vasodilation.[185]

Additional studies have confirmed the role of TRPM4 channels in regulating cerebral myogenic tone, particularly within the cerebral intraparenchymal arteriolar network. The association of TRPM4 channels with pressure-induced vasoconstriction suggests a potential for mechanical activation, although there is some controversy, with one study proposing mechanosensitivity and another failing to detect inherent mechanosensitivity of TRPM4.[186,187] Recent research has suggested the involvement of purinergic receptors or angiotensin II type 1 receptor (AT1R) in arterial pressure sensing. A compelling argument has been presented, indicating that TRPM4 channels are activated downstream of a mechanosensitive, losartan-sensitive receptor, likely the AT1R, within the SMCs of cerebral arteries,[188] It has also been proposed that the mechanoactivation of P2Y4 and P2Y6 purinergic receptors leads to the downstream activation of TRPM4 channels in cerebral intraparenchymal arteriolar myocytes.[186]

9.5 TRP CHANNELS IN ENDOTHELIAL PERMEABILITY

TRP channels are implicated in vascular permeability, a process determined by the balance of adhesive and cytoskeletal contractile forces in ECs.[7] The vascular endothelium acts as a semi-permeable barrier, regulating the passage of substances between the blood and the surrounding tissues through mechanisms like passive diffusion, active transport, and bulk flow through EC clefts. The permeability of the endothelium is dynamically regulated in response to GPCR agonists like thrombin, bradykinin, and histamine through Ca^{2+}-dependent reorganization of the cytoskeleton of interendothelial cell junctions. Of all TRP channels expressed in ECs, some TRPC channels and TRPV4 channels have been reported to influence endothelial permeability.

9.5.1 TRPC1

TRPC1 channels play an important role in modulating endothelial permeability in the pulmonary vasculature, primarily by regulating SOCE. Knockdown of TRPC1 expression in pulmonary artery ECs leads to a reduction in SOCE, underscoring its involvement in this mechanism.[189] Furthermore, inhibition of RhoA activity in cultured ECs not only decreased SOCE but also revealed a physical interaction between RhoA, IP3Rs, and TRPC1 channels.[17] In human ECs, the phosphorylation of TRPC1 by PKCα is pivotal for initiating SOCE in response to thrombin.[190] TRPC1 expression can be increased by TNF-α in human pulmonary artery ECs, which promotes Ca^{2+} influx and enhances endothelial permeability.[191] When exposed to thrombin, RhoA facilitates the association of TRPC1 with IP3Rs, resulting in its translocation to the plasma membrane, as well as increased Ca^{2+} entry and enhanced endothelial permeability.[17] Moreover, the pathway through which TRPC1-mediated Ca^{2+} entry affects adherens junctions may involve sphingosine kinase 1 (SPHK1). TRPC1 inhibits SPHK1, potentially influencing the response of inflammatory agonists to vascular leakage.[192,193]

9.5.2 TRPC4

In addition to TRPC1, TRPC4 has been proposed as an important contributor to SOCE and endothelial permeability. Knockout models for TRPC4 have confirmed the channel's role *in vivo.* Ca^{2+} influx triggered by thrombin was significantly reduced in isolated-perfused TRPC4$^{-/-}$ mouse lungs and in cultured ECs from the same model, leading to impaired thrombin-induced actin stress fiber formation and EC retraction.[52] Cioffi et al. demonstrated that the interaction between spectrin-protein 4.1 complexes and TRPC4 channels is essential for SOCE in pulmonary artery EC cultures.[194] This interaction promotes the recruitment of TRPC4 channels to the plasma membrane. The authors speculated that Ca^{2+} influx through TRPC4 may play a role in the maturation of junctional complexes.[195]

9.5.3 TRPV4

TRPV4 plays a significant role in regulating pulmonary vascular permeability.[196] Research led by Townsley and colleagues demonstrated this role by exposing isolated lung preparations from WT mice to TRPV4 agonists, including 4α-PDD, 5,6-EET, and 14,15-EET, resulting in an increase in microvascular permeability.[55] This increase was specifically attributed to TRPV channels, with a focus on TRPV4, as it was attenuated by the nonselective TRPV blocker, ruthenium red. Interestingly, this effect remained unaltered in response to 4α-PDD in lungs from TRPV4$^{-/-}$ mice. Notably, SOCE-induced permeability responses showed no significant differences between TRPV4$^{-/-}$ and WT mice. The overall increases in permeability induced by 4α-PDD and EET were comparable to those triggered by SOCE, but they affected distinct vascular segments. TRPV4 agonists primarily targeted alveolar septal microvessels, whereas SOCE had a more pronounced effect on extra-alveolar vessels. Furthermore, the study highlighted that TRPV4 agonists led to disruptions in the endothelial barrier and blebbing of ECs, while SOCE resulted in the formation of gaps at cellular junctions. This research underscores the differential effects of TRPV4-mediated Ca^{2+} influx and SOCE on the pulmonary endothelium. Additionally, Yin et al. provided evidence that TRPV4 channels play a role in increasing EC [Ca^{2+}] in response to elevations in left atrial pressure.[197] The authors proposed a mechanism where pressure-induced activation of TRPV4 channels facilitated Ca^{2+} influx, subsequently enhancing endothelial permeability and promoting NO production. NO, operating through cGMP, functioned in a negative-feedback mechanism to regulate TRPV4-mediated Ca^{2+} influx. In summary, TRPV4 is a key player in regulating pulmonary vascular permeability, and its activation can be influenced by mechanical pressure changes, ultimately affecting endothelial integrity.

9.6 SUMMARY

TRP channels play a crucial role in maintaining normal vascular function and are involved in various physiological responses. Endothelial TRP channels contribute to essential processes like endothelium-dependent vasodilation, the regulation of vascular permeability, and angiogenesis.

Concurrently, in vascular smooth muscle, these channels are involved in controlling contractility and proliferation. The versatility of TRP channels is evident in their ability to respond to a wide array of physical and chemical stimuli, leading to Na^+ and Ca^{2+} influx. Significant evidence underscores the involvement of TRP channels in various vascular diseases such as hypertension, oxidative stress responses, and pulmonary edema. Dysfunctional TRP channels are implicated in abnormal contractility, vascular hypertrophy, hyperplasia, and other forms of vascular wall remodeling in these disease mechanisms. While our understanding of vascular TRP channel biology has grown significantly over the last two decades, many questions remain regarding their structural, functional, and mechanistic aspects. Particularly intriguing is the organization and interactions of closely related TRP channels within vascular cells, which still require extensive exploration. The development of selective pharmacological compounds targeting TRP channels may hold great promise for the creation of novel therapies aimed at preventing and treating vascular-related diseases.

ACKNOWLEDGMENTS

This work was supported by the National Natural Science Foundation of China (82100416).

REFERENCES

1. Tran, Q. K., Ohashi, K., Watanabe, H. Calcium signalling in endothelial cells. *Cardiovasc Res*, Oct 2000, 48(1): 13–22. doi:10.1016/s0008-6363(00)00172-3
2. Hill-Eubanks, D. C., Werner, M. E., Heppner, T. J., Nelson, M. T. Calcium signaling in smooth muscle. *Cold Spring Harb Perspect Biol*, Sep 1 2011, 3(9): a004549. doi:10.1101/cshperspect.a004549
3. Dora, K. A., Doyle, M. P., Duling, B. R. Elevation of intracellular calcium in smooth muscle causes endothelial cell generation of NO in arterioles. *Proc Natl Acad Sci U S A*, June 10 1997, 94(12): 6529–6534. doi:10.1073/pnas.94.12.6529
4. Earley, S., Brayden, J. E. Transient receptor potential channels in the vasculature. *Physiol Rev*, 2015, 95(2): 645–690. doi:10.1152/physrev.00026.2014
5. Watanabe, H., Murakami, M., Ohba, T., Takahashi, Y., Ito, H. TRP channel and cardiovascular disease. *Pharmacol Ther*, 2008, 118(3): 337–351. doi:10.1016/j.pharmthera.2008.03.008
6. Smani, T., Gomez, L. J., Regodon, S., Woodard, G. E., Siegfried, G., Khatib, A. M., Rosado, J. A. TRP channels in angiogenesis and other endothelial functions. *Front Physiol*, 2018, 9: 1731. doi:10.3389/fphys.2018.01731
7. Genova, T., Gaglioti, D., Munaron, L. Regulation of vessel permeability by TRP channels. *Front Physiol*, 2020, 11: 421. doi:10.3389/fphys.2020.00421
8. Hong, K. S., Lee, M. G. Endothelial Ca^{2+} signaling-dependent vasodilation through transient receptor potential channels. *Korean J Physiol Pharmacol*, Jul 1 2020, 24(4): 287–298. doi:10.4196/kjpp.2020.24.4.287
9. Yip, H., Chan, W. Y., Leung, P. C., Kwan, H. Y., Liu, C., Huang, Y., Michel, V., Yew, D. T., Yao, X. Expression of TRPC homologs in endothelial cells and smooth muscle layers of human arteries. *Histochem Cell Biol*, Dec 2004, 122(6): 553–561. doi:10.1007/s00418-004-0720-y
10. Gao, G., Bai, X.-Y., Xuan, C., Liu, X. C., Jing, W. B., Novakovic, A., Yang, Q., He, G. W. Role of TRPC3 channel in human internal mammary artery. *Arch Med Res*, Aug 2012, 43(6): 431–437. doi: 10.1016/j.arcmed.2012.08.010
11. Qin, W., Xie, W., Xia, N., He, Q., Sun, T. Silencing of transient receptor potential channel 4 alleviates oxLDL-induced angiogenesis in human coronary artery endothelial cells by inhibition of VEGF and NF-κB. *Med Sci Monit*, Mar 21 2016, 22: 930–936. doi:10.12659/msm.897634
12. Cao, S., Anishkin, A., Zinkevich, N. S., Nishijima, Y., Korishettar, A., Wang, Z., Fang, J., Wilcox, D. A., Zhang, D. X. Transient receptor potential vanilloid 4 (TRPV4) activation by arachidonic acid requires protein kinase a-mediated phosphorylation. *J Biol Chem*, Apr 6 2018, 293(14): 5307–5322. doi:10.1074/jbc.M117.811075
13. Bubolz, A. H., Mendoza, S. A., Zheng, X., Zinkevich, N. S., Li, R., Gutterman, D. D., Zhang, D. X. Activation of endothelial TRPV4 channels mediates flow-induced dilation in human coronary arterioles: Role of Ca^{2+} entry and mitochondrial ROS signaling. *Am J Physiol Heart Circ Physiol*, Feb 1 2012, 302(3): H634–H642. doi:10.1152/ajpheart.00717.2011

14. Freichel, M., Suh, S. H., Pfeifer, A., Schweig, U., Trost, C., Weissgerber, P., Biel, M., Philipp, S., Freise, D., Droogmans, G., Hofmann, F., Flockerzi, V., Nilius, B. Lack of an endothelial store-operated Ca^{2+} current impairs agonist-dependent vasorelaxation in TRP4-/- mice. *Nat Cell Biol*, Feb 2001, 3(2): 121–127. doi:10.1038/35055019
15. Watanabe, H., Davis, J. B., Smart, D., Jerman, J. C., Smith, G. D., Hayes, P., Vriens, J., Cairns, W., Wissenbach, U., Prenen, J., Flockerzi, V., Droogmans, G., Benham, C. D., Nilius, B. Activation of TRPV4 channels (hVRL-2/mTRP12) by phorbol derivatives. *J Biol Chem*, Apr 19 2002, 277(16): 13569–13577. doi:10.1074/jbc.M200062200
16. Neves, K. B., Alves-Lopes, R., Montezano, A. C., Touyz, R. M. Role of PARP and TRPM2 in VEGF inhibitor-induced vascular dysfunction. *J Am Heart Assoc*, Feb 21 2023, 12(4): e027769. doi:10.1161/jaha.122.027769
17. Mehta, D., Ahmmed, G. U., Paria, B. C., Holinstat, M., Voyno-Yasenetskaya, T., Tiruppathi, C., Minshall, R. D., Malik, A. B. RhoA interaction with inositol 1,4,5-trisphosphate receptor and transient receptor potential channel-1 regulates Ca^{2+} entry. Role in signaling increased endothelial permeability. *J Biol Chem*, Aug 29 2003, 278(35): 33492–33500. doi:10.1074/jbc.M302401200
18. Singh, I., Knezevic, N., Ahmmed, G. U., Kini, V., Malik, A. B., Mehta, D. Galphaq-TRPC6-mediated Ca^{2+} entry induces RhoA activation and resultant endothelial cell shape change in response to thrombin. *J Biol Chem*, Mar 16 2007, 282(11): 7833–7843. doi:10.1074/jbc.M608288200
19. Moccia, F., Berra-Romani, R., Tanzi, F. Update on vascular endothelial Ca^{2+} signalling: A tale of ion channels, pumps and transporters. *World J Biol Chem*, Jul 26 2012, 3(7): 127–158. doi:10.4331/wjbc.v3.i7.127
20. Fantozzi, I., Zhang, S., Platoshyn, O., Remillard, C. V., Cowling, R. T., Yuan, J. X. Hypoxia increases AP-1 binding activity by enhancing capacitative Ca^{2+} entry in human pulmonary artery endothelial cells. *Am J Physiol Lung Cell Mol Physiol*, Dec 2003, 285(6): L1233–L1245. doi:10.1152/ajplung.00445.2002
21. Willette, R. N., Bao, W., Nerurkar, S., Yue, T. L., Doe. C. P., Stankus, G., Turner, G. H., Ju, H., Thomas, H., Fishman, C. E., Sulpizio, A., Behm, D. J., Hoffman, S., Lin, Z., Lozinskaya, I., Casillas, L. N., Lin, M., Trout, R. E., Votta, B. J., Thorneloe, K., Lashinger, E. S., Figueroa, D. J., Marquis, R., Xu, X. Systemic activation of the transient receptor potential vanilloid subtype 4 channel causes endothelial failure and circulatory collapse: Part 2. *J Pharmacol Exp Ther*, Aug 2008, 326(2): 443–452. doi:10.1124/jpet.107.134551
22. Mittal, M., Urao, N., Hecquet, C. M., Zhang, M., Sudhahar, V., Gao, X. P., Komarova, Y., Ushio-Fukai, M., Malik, A. B. Novel role of reactive oxygen species-activated Trp melastatin channel-2 in mediating angiogenesis and postischemic neovascularization. *Arterioscler Thromb Vasc Biol*, Apr 2015, 35(4): 877–887. doi:10.1161/ATVBAHA.114.304802
23. Hecquet, C. M., Ahmmed, G. U., Vogel, S. M., Malik, A. B. Role of TRPM2 channel in mediating H2O2-induced Ca^{2+} entry and endothelial hyperpermeability. *Circ Res*, Feb 15 2008, 102(3): 347–355. doi:10.1161/circresaha.107.160176
24. Yang, D., Luo, Z., Ma, S., Wong, W. T., Ma, L., Zhong, J., He, H., Zhao, Z., Cao, T., Yan, Z., Liu, D. Activation of TRPV1 by dietary capsaicin improves endothelium-dependent vasorelaxation and prevents hypertension. *Cell Metab*, Aug 4 2010, 12(2): 130–141. doi:10.1016/j.cmet.2010.05.015
25. Earley, S., Pauyo, T., Drapp, R., Tavares, M. J., Liedtke, W., Brayden, J. E. TRPV4-dependent dilation of peripheral resistance arteries influences arterial pressure. *Am J Physiol Heart Circ Physiol*, Sep 2009, 297(3): H1096–H1102. doi:10.1152/ajpheart.00241.2009
26. Sonkusare, S. K., Bonev, A. D., Ledoux, J., Liedtke, W., Kotlikoff, M. I., Heppner, T. J., Hill-Eubanks, D. C., Nelson, M. T. Elementary Ca^{2+} signals through endothelial TRPV4 channels regulate vascular function. *Science*, May 4 2012, 336(6081): 597–601. doi:10.1126/science.1216283
27. Golech, S. A., McCarron, R. M., Chen, Y., Bembry, J., Lenz, F., Mechoulam, R., Shohami, E., Spatz, M. Human brain endothelium: Coexpression and function of vanilloid and endocannabinoid receptors. *Mol Brain Res*, 2004/12/06/2004, 132(1): 87–92. doi: 10.1016/j.molbrainres.2004.08.025
28. Earley, S., Gonzales, A. L., Garcia, Z. I. A dietary agonist of transient receptor potential cation channel V3 elicits endothelium-dependent vasodilation. *Mol Pharmacol*, Apr 2010, 77(4): 612–620. doi:10.1124/mol.109.060715
29. Loot, A. E., Popp, R., Fisslthaler, B., Vriens, J., Nilius, B., Fleming, I. Role of cytochrome P450-dependent transient receptor potential V4 activation in flow-induced vasodilatation. *Cardiovasc Res*, Dec 1 2008, 80(3): 445–452. doi:10.1093/cvr/cvn207
30. Sullivan, M. N., Gonzales, A. L., Pires, P. W., Bruhl, A., Leo, M. D., Li, W., Oulidi, A., Boop, F. A., Feng, Y., Jaggar, J. H., Welsh, D. G., Earley, S. Localized TRPA1 channel Ca^{2+} signals stimulated by reactive oxygen species promote cerebral artery dilation. *Sci Signal*, Jan 6 2015, 8(358): ra2. doi:10.1126/scisignal.2005659

31. Qu, Y. Y., Wang, L. M., Zhong, H., Liu, Y. M., Tang, N., Zhu, L. P., He, F., Hu, Q. H. TRPC1 stimulates calcium-sensing receptor-induced store-operated Ca^{2+} entry and nitric oxide production in endothelial cells. *Mol Med Rep*, Oct 2017, 16(4): 4613–4619. doi:10.3892/mmr.2017.7164
32. Paria, B. C., Bair, A. M., Xue, J., Yu, Y., Malik, A. B., Tiruppathi, C. Ca^{2+} influx induced by protease-activated receptor-1 activates a feed-forward mechanism of TRPC1 expression via nuclear factor-κB activation in endothelial Cells. *J Biol Chem*, Jul 28 2006, 281(30): 20715–20727. doi: 10.1074/jbc.M600722200
33. Jho, D., Mehta, D., Ahmmed, G., Gao, X. P., Tiruppathi, C., Broman, M., Malik, A. B. Angiopoietin-1 opposes VEGF-induced increase in endothelial permeability by inhibiting TRPC1-dependent Ca2 influx. *Circ Res*, Jun 24 2005, 96(12): 1282–1290. doi: 10.1161/01.RES.0000171894.03801.03
34. Andrikopoulos, P., Eccles, S. A., Yaqoob, M. M. Coupling between the TRPC3 ion channel and the NCX1 transporter contributed to VEGF-induced ERK1/2 activation and angiogenesis in human primary endothelial cells. *Cell Signal*, Sep 2017, 37: 12–30. doi:10.1016/j.cellsig.2017.05.013
35. Antigny, F., Girardin, N., Frieden, M. Transient receptor potential canonical channels are required for in vitro endothelial tube formation. *J Biol Chem*, Feb 17 2012, 287(8): 5917–5927. doi:10.1074/jbc.M111.295733
36. Ma, X., Cheng, K. T., Wong, C. O., O'Neil, R. G., Birnbaumer, L., Ambudkar, I. S., Yao, X. Heteromeric TRPV4-C1 channels contribute to store-operated Ca(2+) entry in vascular endothelial cells. *Cell Calcium*, Dec 2011, 50(6): 502–509. doi:10.1016/j.ceca.2011.08.006
37. Abuarab, N., Munsey, T. S., Jiang, L. H., Li, J., Sivaprasadarao, A. High glucose-induced ROS activates TRPM2 to trigger lysosomal membrane permeabilization and Zn(2+)-mediated mitochondrial fission. *Sci Signal*, Aug 1 2017, 10(490): eaal4161. doi:10.1126/scisignal.aal4161
38. Loh, K. P., Ng, G., Yu, C. Y., Fhu, C. K., Yu, D., Vennekens, R., Nilius, B., Soong, T. W., Liao, P. TRPM4 inhibition promotes angiogenesis after ischemic stroke. *Pflugers Arch*, Mar 2014, 466(3): 563–576. doi:10.1007/s00424-013-1347-4
39. Inoue, K., Xiong, Z. G. Silencing TRPM7 promotes growth/proliferation and nitric oxide production of vascular endothelial cells via the ERK pathway. *Cardiovasc Res*, 2009, 83(3): 547–557. doi:10.1093/cvr/cvp153
40. Zeng, Z., Inoue, K., Sun, H., Leng, T., Feng, X., Zhu, L., Xiong, Z. G. TRPM7 regulates vascular endothelial cell adhesion and tube formation. *Am J Physiol Cell Physiol*, Feb 15 2015, 308(4): C308–C318. doi:10.1152/ajpcell.00275.2013
41. Guarini, G., Ohanyan, V. A., Kmetz, J. G., DelloStritto, D. J., Thoppil, R. J., Thodeti, C. K., Meszaros, J. G., Damron, D. S., Bratz, I. N. Disruption of TRPV1-mediated coupling of coronary blood flow to cardiac metabolism in diabetic mice: Role of nitric oxide and BK channels. *Am J Physiol Heart Circ Physiol*, Jul 15 2012, 303(2): H216–H223. doi:10.1152/ajpheart.00011.2012
42. Freichel, M., Vennekens, R., Olausson, J., Hoffmann, M., Müller, C., Stolz, S,, Scheunemann, J., Weissgerber, P., Flockerzi, V. Functional role of TRPC proteins in vivo: Lessons from TRPC-deficient mouse models. *Biochem Biophys Res Commun*, Oct 1 2004, 322(4): 1352–1358. doi:10.1016/j.bbrc.2004.08.041
43. Li, Z., Guo, G., Wang, H., Si, X., Zhou, G., Xiong, Y., Li, S., Dai, R., Yang, C. TRPC5 channel modulates endothelial cells senescence. *Eur J Pharmacol*, May 5 2017, 802: 27–35. doi:10.1016/j.ejphar.2017.02.037
44. Rosenbaum, M. A., Chaudhuri, P., Graham, L. M. Hypercholesterolemia inhibits re-endothelialization of arterial injuries by TRPC channel activation. *J Vasc Surg*, Oct 2015, 62(4): 1040–1047.e2. doi:10.1016/j.jvs.2014.04.033
45. Chaudhuri, P., Colles, S. M., Bhat, M., Van Wagoner, D. R., Birnbaumer, L., Graham, L. M. Elucidation of a TRPC6-TRPC5 channel cascade that restricts endothelial cell movement. *Mol Biol Cell*, Aug 2008, 19(8): 3203–3211. doi:10.1091/mbc.e07-08-0765
46. Ohanyan, V. A., Guarini, G., Thodeti, C. K., Talasila, P. K., Raman, P., Haney, R. M., Meszaros, J. G., Damron, D. S., Bratz, I. N. Endothelin-mediated in vivo pressor responses following TRPV1 activation. *Am J Physiol Heart Circ Physiol*, Sep 2011, 301(3): H1135–H1142. doi:10.1152/ajpheart.00082.2011
47. Sun, J., Pu, Y., Wang, P., Chen, S., Zhao, Y., Liu, C., Shang, Q., Zhu, Z., Liu, D. TRPV1-mediated UCP2 upregulation ameliorates hyperglycemia-induced endothelial dysfunction. *Cardiovasc Diabetol*, Apr 22 2013, 12: 69. doi:10.1186/1475-2840-12-69
48. Zhang, D. X., Mendoza, S. A., Bubolz, A. H., Mizuno, A., Ge, Z. D., Li, R., Warltier, D. C., Suzuki, M., Gutterman, D. D. Transient receptor potential vanilloid type 4-deficient mice exhibit impaired endothelium-dependent relaxation induced by acetylcholine in vitro and in vivo. *Hypertension*, Mar 2009, 53(3): 532–538. doi:10.1161/HYPERTENSIONAHA.108.127100

49. Gao, P., Li, L., Wei, X., Wang, M., Hong, Y., Wu, H., Shen, Y., Ma, T., Wei, X., Zhang, Q., Fang, X., Wang, L., Yan, Z., Du, G. H., Zheng, H., Yang, G., Liu, D., Zhu, Z. Activation of transient receptor potential channel vanilloid 4 by DPP-4 (Dipeptidyl peptidase-4) inhibitor vildagliptin protects against diabetic endothelial dysfunction. *Hypertension*, Jan 2020, 75(1): 150–162. doi: 10.1161/HYPERTENSIONAHA.119.13778
50. Sun, L., Liu, Y. L., Ye, F., Xie, J. W., Zeng, J. W., Qin, L., Xue, J., Wang, Y. T., Guo, K. M., Ma, M. M., Tang, Y. B., Li, X. Y., Gao, M. Free fatty acid-induced h(2)O(2) activates TRPM2 to aggravate endothelial insulin resistance via Ca^{2+}-dependent PERK/ATF4/TRB3 cascade in obese mice. *Free Radic Biol Med*, Nov 1 2019, 143: 288–299. doi:10.1016/j.freeradbiomed.2019.08.018
51. Mathar, I., Vennekens, R., Meissner, M., Kees, F., Van der Mieren, G., Camacho Londoño, J. E., Uhl, S., Voets, T., Hummel, B., van den Bergh, A., Herijgers, P., Nilius, B., Flockerzi, V., Schweda, F., Freichel, M. Increased catecholamine secretion contributes to hypertension in TRPM4-deficient mice. *J Clin Invest*, Sep 2010, 120(9): 3267–3279. doi: 10.1172/JCI41348
52. Tiruppathi, C., Freichel, M., Vogel, S. M., Paria, B. C., Mehta, D., Flockerzi, V., Malik, A. B. Impairment of store-operated Ca^{2+} entry in TRPC4(-/-) mice interferes with increase in lung microvascular permeability. *Circ Res*, Jul 12 2002, 91(1): 70–76. doi: 10.1161/01.res.0000023391.40106.a8
53. Sundivakkam, P. C., Freichel, M., Singh, V., Yuan, J. P., Vogel, S. M., Flockerzi, V., Malik, A. B., Tiruppathi, C. The Ca^{2+} Sensor stromal interaction molecule 1 (STIM1) is necessary and sufficient for the store-operated Ca^{2+} Entry function of transient receptor potential canonical (TRPC) 1 and 4 channels in endothelial cells. *Mol Pharmacol*, 2012, 81(4): 510. doi:10.1124/mol.111.074658
54. Weissmann, N., Sydykov, A., Kalwa, H., Storch, U., Fuchs, B., Mederos y Schnitzler, M., Brandes, R. P., Grimminger, F., Meissner, M., Freichel, M., Offermanns, S., Veit, F., Pak, O., Krause, K. H., Schermuly, R. T., Brewer, A. C., Schmidt, H. H., Seeger, W., Shah, A. M., Gudermann, T., Ghofrani, H. A., Dietrich, A. Activation of TRPC6 channels is essential for lung ischaemia-reperfusion induced oedema in mice. *Nat Commun*, Jan 31 2012, 3: 649. doi:10.1038/ncomms1660
55. Alvarez, D. F., King, J. A., Weber, D., Addison, E., Liedtke, W., Townsley, M. I. Transient receptor potential vanilloid 4-mediated disruption of the alveolar septal barrier: A novel mechanism of acute lung injury. *Circ Res*, Oct 27 2006, 99(9): 988–995. doi:10.1161/01.RES.0000247065.11756.19
56. Du, J., Ma, X., Shen, B., Huang, Y., Birnbaumer, L., Yao, X. TRPV4, TRPC1, and TRPP2 assemble to form a flow-sensitive heteromeric channel. *FASEB J*, Nov 2014, 28(11): 4677–4685. doi:10.1096/fj.14-251652
57. Zhu, Y., Gao, M., Zhou, T., Xie, M., Mao, A., Feng, L., Yao, X., Wong, W. T., Ma, X. The TRPC5 channel regulates angiogenesis and promotes recovery from ischemic injury in mice. *J Biol Chem*, Jan 4 2019, 294(1): 28–37. doi:10.1074/jbc.RA118.005392
58. Ma, X., Du, J., Zhang, P., Deng, J., Liu, J., Lam, F. F., Li, R. A., Huang, Y., Jin, J., Yao, X. Functional role of TRPV4-KCa2.3 signaling in vascular endothelial cells in normal and streptozotocin-induced diabetic rats. *Hypertension*, Jul 2013, 62(1): 134–139. doi:10.1161/HYPERTENSIONAHA.113.01500
59. Kochukov, M. Y., Balasubramanian, A., Abramowitz, J., Birnbaumer, L., Marrelli, S. P. Activation of endothelial transient receptor potential C3 channel is required for small conductance calcium-activated potassium channel activation and sustained endothelial hyperpolarization and vasodilation of cerebral artery. *J Am Heart Assoc*, Aug 20 2014, 3(4): e000913 doi:10.1161/JAHA.114.000913
60. Zhang, L., Papadopoulos, P., Hamel, E. Endothelial TRPV4 channels mediate dilation of cerebral arteries: Impairment and recovery in cerebrovascular pathologies related to Alzheimer's disease. *Br J Pharmacol*, Oct 2013, 170(3): 661–670. doi:10.1111/bph.12315
61. Huang, H., Wang, W., Liu, P., Jiang, Y.0, Zhao, Y., Wei, H., Niu, W. TRPC1 expression and distribution in rat hearts. *Eur J Histochem*, Dec 29 2009, 53(4): e26. doi:10.4081/ejh.2009.e26
62. Gomart, S., Gaudreau-Ménard, C., Jespers, P., Dilek, O. G., Hupkens, E., Hanthazi, A., Naeije, R., Melot, C., Labranche, N., Dewachter, L., Mc Entee, K. Leptin-induced endothelium-independent vasoconstriction in thoracic aorta and pulmonary artery of spontaneously hypertensive rats: Role of calcium channels and stores. *PLoS One*, 2017, 12(1): e0169205. doi:10.1371/journal.pone.0169205
63. Shamsaldeen, Y. A., Lione, L. A., Benham, C. D. Dysregulation of TRPV4, eNOS and caveolin-1 contribute to endothelial dysfunction in the streptozotocin rat model of diabetes. *Eur J Pharmacol*, Dec 5 2020, 888: 173441. doi:10.1016/j.ejphar.2020.173441
64. Li, Y., Chen, X., Zeng, X., Chen, S., Yang, X., Zhang, L. Galectin-3 mediates pulmonary vascular endothelial cell dynamics via TRPC1/4 under acute hypoxia. *J Biochem Mol Toxicol*, May 2020, 34(5): e22463. doi:10.1002/jbt.22463

65. Xu, N., Ayers, L., Pastukh, V., Alexeyev, M., Stevens, T., Tambe, D. T. Impact of Na+ permeation on collective migration of pulmonary arterial endothelial cells. *PLoS One*, 2021, 16(4): e0250095. doi:10.1371/journal.pone.0250095
66. Samapati, R., Yang, Y., Yin, J., Stoerger, C., Arenz, C., Dietrich, A., Gudermann, T., Adam, D., Wu, S., Freichel, M., Flockerzi, V., Uhlig, S., Kuebler, W. M. Lung endothelial Ca^{2+} and permeability response to platelet-activating factor is mediated by acid sphingomyelinase and transient receptor potential classical 6. *Am J Respir Crit Care Med*, Jan 15 2012, 185(2): 160–170. doi:10.1164/rccm.201104-0717OC
67. Addison, M. P., Singh, T. U., Parida, S., Choudhury, S., Kasa, J. K., Sukumaran, S. V., Darzi, S. A., Kandasamy, K., Singh, V., Kumar, D., Mishra, S. K. NO synthase inhibition attenuates EDHF-mediated relaxation induced by TRPV4 channel agonist GSK1016790A in the rat pulmonary artery: Role of TxA2. *Pharmacol Rep*, Jun 2016, 68(3): 620–626. doi:10.1016/j.pharep.2016.01.003
68. Lu, S., Xiang, L., Clemmer, J. S., Mittwede, P. N., Hester, R. L. Oxidative stress increases pulmonary vascular permeability in diabetic rats through activation of transient receptor potential melastatin 2 channels. *Microcirculation*, Nov 2014, 21(8): 754–760. doi:10.1111/micc.12158
69. Liu, C. L., Huang, Y., Ngai, C. Y., Leung, Y. K., Yao, X. Q. TRPC3 is involved in flow- and bradykinin-induced vasodilation in rat small mesenteric arteries. *Acta Pharmacol Sin*, Aug 2006, 27(8): 981–990. doi:10.1111/j.1745-7254.2006.00354.x
70. Poblete, I. M., Orliac, M. L., Briones, R., Adler-Graschinsky, E., Huidobro-Toro, J. P. Anandamide elicits an acute release of nitric oxide through endothelial TRPV1 receptor activation in the rat arterial mesenteric bed. *J Physiol*, Oct 15 2005, 568(Pt 2): 539–551. doi:10.1113/jphysiol.2005.094292
71. Zhang, Y., Chen, Q., Sun, Z., Han, J., Wang, L., Zheng, L. Impaired capsaicin-induced relaxation in diabetic mesenteric arteries. *J Diabetes Complications*, Aug 2015, 29(6): 747–754. doi:10.1016/j.jdiacomp.2015.05.005
72. Lobato, N. S., Filgueira, F. P., Prakash, R., Giachini, F. R., Ergul, A., Carvalho, M. H., Webb, R. C., Tostes, R. C., Fortes, Z. B. Reduced endothelium-dependent relaxation to anandamide in mesenteric arteries from young obese Zucker rats. *PLoS One*, 2013, 8(5): e63449. doi:10.1371/journal.pone.0063449
73. Ye, L., Xu, M., Hu, M., Zhang, H., Tan, X., Li, Q., Shen, B., Huang, J. TRPV4 is involved in irisin-induced endothelium-dependent vasodilation. *Biochem Biophys Res Commun*, Jan 1 2018, 495(1): 41–45. doi:10.1016/j.bbrc.2017.10.160
74. Marrelli, S. P., O'Neil, R. G., Brown, R. C., Bryan, R. M. Jr. PLA2 and TRPV4 channels regulate endothelial calcium in cerebral arteries. *Am J Physiol Heart Circ Physiol*, Mar 2007, 292(3): H1390–H1397. doi:10.1152/ajpheart.01006.2006
75. Han, J., Xu, H. H., Chen, X. L., Hu, H. R., Hu, K. M., Chen, Z. W., He, G. W. Total flavone of Rhododendron improves cerebral ischemia injury by activating vascular TRPV4 to induce endothelium-derived hyperpolarizing factor-mediated responses. *Evid Based Complement Alternat Med*, 2018, 2018: 8919867. doi:10.1155/2018/8919867
76. Earley, S., Gonzales, A. L., Crnich, R. Endothelium-dependent cerebral artery dilation mediated by TRPA1 and Ca^{2+}-activated K+ channels. *Circ Res*, Apr 24 2009, 104(8): 987–994. doi:10.1161/CIRCRESAHA.108.189530
77. Qian, X., Francis, M., Solodushko, V., Earley, S., Taylor, M. S. Recruitment of dynamic endothelial Ca^{2+} signals by the TRPA1 channel activator AITC in rat cerebral arteries. *Microcirculation*, Feb 2013, 20(2): 138–148. doi:10.1111/micc.12004
78. Zeng, W., Ji, Y., Zhang, H., Chen, L., Du, L., Guo, R. Downregulation of TRPC4 and TRPC5 inhibits smooth muscle cell proliferation without affecting endothelial cell proliferation. *Genet Res (Camb)*, 2021, 2021: 2949986. doi:10.1155/2021/2949986
79. Song, S., Ayon, R. J., Yamamura, A., Yamamura, H., Dash, S., Babicheva, A., Tang, H., Sun, X., Cordery, A. G., Khalpey, Z., Black, S. M., Desai, A. A., Rischard, F., McDermott, K. M., Garcia, J. G., Makino, A., Yuan, J. X. Capsaicin-induced Ca^{2+} signaling is enhanced via upregulated TRPV1 channels in pulmonary artery smooth muscle cells from patients with idiopathic PAH. *Am J Physiol Lung Cell Mol Physiol*, Mar 1 2017, 312(3): L309–1325. doi:10.1152/ajplung.00357.2016
80. Wang, Y. X., Wang, J., Wang, C., Liu, J., Shi, L. P., Xu, M., Wang, C. Functional expression of transient receptor potential vanilloid-related channels in chronically hypoxic human pulmonary arterial smooth muscle cells. *J Membr Biol*, Jun 2008, 223(3): 151–159. doi:10.1007/s00232-008-9121-9
81. Shi, J., Ju, M., Abramowitz, J., Large, W. A., Birnbaumer, L., Albert, A. P. TRPC1 proteins confer PKC and phosphoinositol activation on native heteromeric TRPC1/C5 channels in vascular smooth muscle: Comparative study of wild-type and TRPC1-/- mice. *FASEB J*, Jan 2012, 26(1): 409–419. doi:10.1096/fj.11-185611

82. Dietrich, A., Kalwa, H., Storch, U., Mederos y Schnitzler, M., Salanova, B., Pinkenburg, O., Dubrovska, G., Essin, K., Gollasch, M., Birnbaumer, L., Gudermann, T. Pressure-induced and store-operated cation influx in vascular smooth muscle cells is independent of TRPC1. *Pflugers Arch*, Dec 2007, 455(3): 465–477. doi:10.1007/s00424-007-0314-3
83. Muraki, K., Iwata, Y., Katanosaka, Y., Ito, T., Ohya, S., Shigekawa, M., Imaizumi, Y. TRPV2 is a component of osmotically sensitive cation channels in murine aortic myocytes. *Circ Res*, Oct 31 2003, 93(9): 829–838. doi: 10.1161/01.RES.0000097263.10220.0C
84. Zhu, Y., Chu, Y., Wang, S., Tang, J., Li, H., Feng, L., Yu, F., Ma, X. Vascular smooth muscle TRPV4 (Transient receptor potential vanilloid family member 4) channels regulate vasoconstriction and blood pressure in obesity. *Hypertension*, Apr 2023, 80(4): 757–770. doi:10.1161/HYPERTENSIONAHA.122.20109
85. Dietrich, A., Mederos, Y. S. M., Gollasch, M., Gollasch, M., Gross, V., Storch, U., Dubrovska, G., Obst, M., Yildirim, E., Salanova, B., Kalwa, H., Essin, K., Pinkenburg, O., Luft, F. C., Gudermann, T., Birnbaumer, L. Increased vascular smooth muscle contractility in TRPC6-/- mice. *Mol Cell Biol*, Aug 2005, 25(16): 6980–6989. doi:10.1128/MCB.25.16.6980-6989.2005
86. Schleifenbaum, J., Kassmann, M., Szijarto, I. A., Hercule, H. C., Tano, J. Y., Weinert, S., Heidenreich, M., Pathan, A. R., Anistan, Y. M., Alenina, N., Rusch, N.J., Bader, M., Jentsch, T. J., Gollasch, M. Stretch-activation of angiotensin II type 1a receptors contributes to the myogenic response of mouse mesenteric and renal arteries. *Circ Res*, Jul 7 2014, 115(2): 263–272. doi:10.1161/CIRCRESAHA.115.302882
87. Facemire, C. S., Mohler, P. J., Arendshorst, W. J. Expression and relative abundance of short transient receptor potential channels in the rat renal microcirculation. *Am J Physiol Renal Physiol*, 2004, 286(3): F546–F551. doi:10.1152/ajprenal.00338.2003
88. Jung, S., Strotmann, R., Schultz, G., Plant, T. D. TRPC6 is a candidate channel involved in receptor-stimulated cation currents in A7r5 smooth muscle cells. *Am J Physiol Cell Physiol*, 2002, 282(2): C347–C359. doi:10.1152/ajpcell.00283.2001
89. Hopps, J. J., Dunn, W. R., Randall, M. D. Vasorelaxation to capsaicin and its effects on calcium influx in arteries. *Eur J Pharmacol*, Apr 15 2012, 681(1-3): 88–93. doi:10.1016/j.ejphar.2012.02.019
90. Inoue, R., Jensen, L. J., Shi, J., Morita, H., Nishida, M., Honda, A., Ito, Y. Transient receptor potential channels in cardiovascular function and disease. *Circ Res*, Jul 21 2006, 99(2): 119–131. doi:10.1161/01.RES.0000233356.10630.8a
91. Zhang, Z., Wang, M., Fan, X. H., Chen, J. H., Guan, Y. Y., Tang, Y. B. Upregulation of TRPM7 channels by angiotensin II triggers phenotypic switching of vascular smooth muscle cells of ascending aorta. *Circ Res*, Oct 12 2012, 111(9): 1137–1146. doi: 10.1161/CIRCRESAHA.112.273755
92. Yang, X. R., Lin, M. J., McIntosh, L. S., Sham, J. S. Functional expression of transient receptor potential melastatin- and vanilloid-related channels in pulmonary arterial and aortic smooth muscle. *Am J Physiol Lung Cell Mol Physiol*, Jun 2006, 290(6): L1267–L1276. doi:10.1152/ajplung.00515.2005
93. Ng, L. C., Gurney, A. M. Store-operated channels mediate Ca^{2+} influx and contraction in rat pulmonary artery. *Circ Res*, Nov 9 2001, 89(10): 923–929. doi:10.1161/hh2201.100315
94. Kaymak, E., Akin, A. T., Tufan, E., Başaran, K. E., Taheri, S., Özdamar, S,, Yakan, B. The effect of chloroquine on the TRPC1, TRPC6, and CaSR in the pulmonary artery smooth muscle cells in hypoxia-induced experimental pulmonary artery hypertension. *J Biochem Mol Toxicol*, Feb 2021, 35(2): e22636. doi:10.1002/jbt.22636
95. Parpaite, T., Cardouat, G., Mauroux, M., Gillibert-Duplantier, J., Robillard, P., Quignard, J. F., Marthan, R., Savineau, J. P., Ducret, T. Effect of hypoxia on TRPV1 and TRPV4 channels in rat pulmonary arterial smooth muscle cells. *Pflugers Arch*, Jan 2016, 468(1): 111–130. doi:10.1007/s00424-015-1704-6
96. Zhang, Q., Cao, Y., Luo, Q., Wang, P., Shi, P., Song, C., E, M., Ren, J., Fu, B., Sun, H. The transient receptor potential vanilloid-3 regulates hypoxia-mediated pulmonary artery smooth muscle cells proliferation via PI3K/AKT signaling pathway. *Cell Prolif*, Jun 2018, 51(3): e12436. doi:10.1111/cpr.12436
97. Touyz, R. M., He, Y., Montezano, A. C., Yao, G., Chubanov, V., Gudermann, T., Callera, G. E. Differential regulation of transient receptor potential melastatin 6 and 7 cation channels by ANG II in vascular smooth muscle cells from spontaneously hypertensive rats. *Am J Physiol Regul Integr Comp Physiol*, Jan 2006, 290(1): R73–R78. doi:10.1152/ajpregu.00515.2005
98. Adebiyi, A., Narayanan, D., Jaggar, J. H. Caveolin-1 assembles type 1 inositol 1,4,5-trisphosphate receptors and canonical transient receptor potential 3 channels into a functional signaling complex in arterial smooth muscle cells. *J Biol Chem*, Feb 11 2011, 286(6): 4341–4348. doi:10.1074/jbc.M110.179747

99. Reading, S. A., Earley, S., Waldron, B. J., Welsh, D. G., Brayden, J. E. TRPC3 mediates pyrimidine receptor-induced depolarization of cerebral arteries. *Am J Physiol Heart Circ Physiol*, May 2005, 288(5): H2055–H2061. doi:10.1152/ajpheart.00861.2004
100. Adebiyi, A., Zhao, G., Narayanan, D., Thomas-Gatewood, C. M., Bannister, J. P., Jaggar, J. H. Isoform-selective physical coupling of TRPC3 channels to IP3 receptors in smooth muscle cells regulates arterial contractility. *Circ Res*, May 28 2010, 106(10): 1603–1612. doi:10.1161/CIRCRESAHA.110.216804
101. Earley, S., Heppner, T. J., Nelson, M. T., Brayden, J. E. TRPV4 forms a novel Ca^{2+} signaling complex with ryanodine receptors and BKCa channels. *Circ Res*, 2005, 97(12): 1270–1279. doi:10.1161/01.RES.0000194321.60300.d6
102. Wong, C. O., Yao, X. TRP channels in vascular endothelial cells. *Adv Exp Med Biol*, 2011, 704: 759–780. doi:10.1007/978-94-007-0265-3_40
103. Pocock, T. M., Foster, R. R., Bates, D. O. Evidence of a role for TRPC channels in VEGF-mediated increased vascular permeability in vivo. *Am J Physiol Heart Circ Physiol*, Mar 2004, 286(3): H1015–H1026. doi:10.1152/ajpheart.00826.2003
104. Antoniotti, S., Lovisolo, D., Fiorio Pla, A., Munaron, L. Expression and functional role of bTRPC1 channels in native endothelial cells. *FEBS Lett*, Jan 16 2002, 510(3): 189–195. doi:10.1016/s0014-5793(01)03256-2
105. Antoniotti, S., Fiorio Pla, A., Pregnolato, S., Mottola, A., Lovisolo, D., Munaron, L. Control of endothelial cell proliferation by calcium influx and arachidonic acid metabolism: A pharmacological approach. *J Cell Physiol*, 2003, 197(3): 370–378. doi:10.1002/jcp.10359
106. Nauli, S. M., Kawanabe, Y., Kaminski, J. J., Pearce, W. J., Ingber, D. E., Zhou, J. Endothelial cilia are fluid shear sensors that regulate calcium signaling and nitric oxide production through polycystin-1. *Circulation*, Mar 4 2008, 117(9): 1161–1171. doi:10.1161/CIRCULATIONAHA.107.710111
107. He, D., Pan, Q., Chen, Z., Sun, C., Zhang, P., Mao, A., Zhu, Y., Li, H., Lu, C., Xie, M., Zhou, Y., Shen, D., Tang, C., Yang, Z., Jin, J., Yao, X., Nilius, B., Ma, X. Treatment of hypertension by increasing impaired endothelial TRPV4-KCa2.3 interaction. *EMBO Mol Med*, Nov 2017, 9(11): 1491–1503. doi:10.15252/emmm.201707725
108. Yu, P. C., Gu, S. Y., Bu, J. W., Du, J. L. TRPC1 is essential for in vivo angiogenesis in zebrafish. *Circ Res*, Apr 16 2010, 106(7): 1221–1232. doi:10.1161/CIRCRESAHA.109.207670
109. Du, L. L., Shen, Z., Li, Z., Ye, X., Wu, M., Hong, L., Zhao, Y. TRPC1 deficiency impairs the endothelial progenitor cell function via inhibition of Calmodulin/eNOS pathway. *J Cardiovasc Transl Res*, Aug 2018, 11(4): 339–345. doi:10.1007/s12265-018-9798-9
110. Li, J., Cubbon, R. M., Wilson, L. A., Amer, M. S., McKeown, L., Hou, B., Majeed, Y., Tumova, S., Seymour, V. A., Taylor, H., Stacey, M., O'Regan, D., Foster, R., Porter, K. E., Kearney, M. T., Beech, D. J. Orai1 and CRAC channel dependence of VEGF-activated Ca^{2+} entry and endothelial tube formation. *Circ Res*, May 13 2011, 108(10): 1190–1198. doi:10.1161/CIRCRESAHA.111.243352
111. Schmidt, K., Dubrovska, G., Nielsen, G.,Fesüs, G., Uhrenholt, T. R., Hansen, P. B., Gudermann, T., Dietrich, A., Gollasch, M., de Wit, C., Köhler, R. Amplification of EDHF-type vasodilatations in TRPC1-deficient mice. *Br J Pharmacol*, Dec 2010, 161(8): 1722–1733. doi:10.1111/j.1476-5381.2010.00985.x
112. Hamdollah Zadeh, M. A., Glass, C. A., Magnussen, A., Hancox, J. C., Bates, D. O. VEGF-mediated elevated intracellular calcium and angiogenesis in human microvascular endothelial cells in vitro are inhibited by dominant negative TRPC6. *Microcirculation*, 2008, 15(7): 605–614. doi:10.1080/10739680802220323
113. Ge, R., Tai, Y., Sun, Y., Zhou, K., Yang, S., Cheng, T., Zou, Q., Shen, F., Wang, Y. Critical role of TRPC6 channels in VEGF-mediated angiogenesis. *Cancer Lett*, Sep 28 2009, 283(1): 43–51. doi:10.1016/j.canlet.2009.03.023
114. Ding, Y., Fromel, T., Popp, R., Falck, J. R., Schunck, W. H., Fleming, I. The biological actions of 11,12-epoxyeicosatrienoic acid in endothelial cells are specific to the R/S-enantiomer and require the G(s) protein. *J Pharmacol Exp Ther*, Jul 2014, 350(1): 14–21. doi:10.1124/jpet.114.214254
115. Cheng, H. W., James, A. F., Foster, R. R., Hancox, J. C., Bates, D. O. VEGF activates receptor-operated cation channels in human microvascular endothelial cells. *Arterioscler Thromb Vasc Biol*, Aug 2006, 26(8): 1768–1776. doi:10.1161/01.ATV.0000231518.86795.0f
116. Moccia, F., Lucariello, A., Guerra, G. TRPC3-mediated Ca^{2+} signals as a promising strategy to boost therapeutic angiogenesis in failing hearts: The role of autologous endothelial colony forming cells. *J Cell Physiol*, May 2018, 233(5): 3901–3917. doi:10.1002/jcp.26152
117. Song, H. B., Jun, H. O., Kim, J. H., Fruttiger, M., Kim, J. H. Suppression of transient receptor potential canonical channel 4 inhibits vascular endothelial growth factor-induced retinal neovascularization. *Cell Calcium*, Feb 2015, 57(2): 101–108. doi:10.1016/j.ceca.2015.01.002

118. Thakore, P., Earley, S. Transient receptor potential channels and endothelial cell calcium signaling. *Compr Physiol*, Jun 12 2019, 9(3): 1249–1277. doi:10.1002/cphy.c180034
119. Ching, L. C., Kou, Y. R., Shyue, S. K., Su, K. H., Wei, J., Cheng, L. C., Yu, Y. B., Pan, C. C., Lee, T. S. Molecular mechanisms of activation of endothelial nitric oxide synthase mediated by transient receptor potential vanilloid type 1. *Cardiovasc Res*, Aug 1 2011, 91(3): 492–501. doi:10.1093/cvr/cvr104
120. Su, K. H., Lin, S. J., Wei, J., Lee, K. I., Zhao, J. F., Shyue, S. K., Lee, T. S. The essential role of transient receptor potential vanilloid 1 in simvastatin-induced activation of endothelial nitric oxide synthase and angiogenesis. *Acta Physiol (Oxf)*, Nov 2014, 212(3): 191–204. doi:10.1111/apha.12378
121. Su, K. H., Lee, K. I., Shyue, S. K., Chen, H. Y., Wei, J., Lee, T. S. Implication of transient receptor potential vanilloid type 1 in 14,15-epoxyeicosatrienoic acid-induced angiogenesis. *Int J Biol Sci*, 2014, 10(9): 990–996. doi:10.7150/ijbs.9832
122. O'Leary, C., McGahon, M. K., Ashraf, S., McNaughten, J., Friedel, T., Cincolà, P., Barabas, P., Fernandez, J. A., Stitt, A. W., McGeown, J. G., Curtis, T. M. Involvement of TRPV1 and TRPV4 channels in retinal angiogenesis. *Invest Ophthalmol Vis Sci*, Aug 1 2019, 60(10): 3297–3309. doi:10.1167/iovs.18-26344
123. Hofmann, N. A., Barth, S., Waldeck-Weiermair, M., Klec, C., Strunk, D., Malli, R., Graier, W. F. TRPV1 mediates cellular uptake of anandamide and thus promotes endothelial cell proliferation and network-formation. *Biol Open*, Nov 13 2014, 3(12): 1164–1172. doi:10.1242/bio.20149571
124. Moccia, F. Endothelial Ca^{2+} signaling and the resistance to anticancer treatments: Partners in crime. *Int J Mol Sci*, Jan 11 2018, 19(1): 217. doi:10.3390/ijms19010217
125. Hatano, N., Suzuki, H., Itoh, Y., Muraki, K. TRPV4 partially participates in proliferation of human brain capillary endothelial cells. *Life Sci*, Mar 12 2013, 92(4-5): 317–324. doi:10.1016/j.lfs.2013.01.002
126. Kohler, R., Heyken, W. T., Heinau, P., Schubert, R., Si, H., Kacik, M., Busch, C., Grgic, I., Maier, T., Hoyer, J. Evidence for a functional role of endothelial transient receptor potential V4 in shear stress-induced vasodilatation. *Arterioscler Thromb Vasc Biol*, Jul 2006, 26(7): 1495–1502. doi:10.1161/01.ATV.0000225698.36212.6a
127. Reddy, K., Khaliq, A., Henning, R. J. Recent advances in the diagnosis and treatment of acute myocardial infarction. *World J Cardiol*, May 26 2015, 7(5): 243–276. doi:10.4330/wjc.v7.i5.243
128. Adapala, R. K., Thoppil, R. J., Ghosh, K., Cappelli, H. C., Dudley, A. C., Paruchuri, S., Keshamouni, V., Klagsbrun, M., Meszaros, J. G., Chilian, W. M., Ingber, D. E., Thodeti, C. K. Activation of mechanosensitive ion channel TRPV4 normalizes tumor vasculature and improves cancer therapy. *Oncogene*, Jan 21 2016, 35(3): 314–322. doi:10.1038/onc.2015.83
129. Thoppil, R. J., Cappelli, H. C., Adapala, R. K., Kanugula, A. K., Paruchuri, S., Thodeti, C. K. TRPV4 channels regulate tumor angiogenesis via modulation of Rho/Rho kinase pathway. *Oncotarget*, May 3 2016, 7(18): 25849–25861. doi:10.18632/oncotarget.8405
130. Troidl, C., Troidl, K., Schierling, W., Cai, W. J., Nef, H., Möllmann, H., Kostin, S., Schimanski, S., Hammer, L., Elsässer, A., Schmitz-Rixen, T., Schaper, W. Trpv4 induces collateral vessel growth during regeneration of the arterial circulation. *J Cell Mol Med*, Aug 2009, 13(8B): 2613–2621. doi:10.1111/j.1582-4934.2008.00579.x
131. Suresh, K., Servinsky, L., Reyes, J., Baksh, S., Undem, C., Caterina, M., Pearse, D. B., Shimoda, L. A. Hydrogen peroxide-induced calcium influx in lung microvascular endothelial cells involves TRPV4. *Am J Physiol Lung Cell Mol Physiol*, Dec 15 2015, 309(12): L1467–L1477. doi:10.1152/ajplung.00275.2015
132. Suresh, K., Servinsky, L., Jiang, H., Bigham, Z., Yun, X., Kliment, C., Huetsch, J., Damarla, M., Shimoda, L. A. Reactive oxygen species induced Ca^{2+} influx via TRPV4 and microvascular endothelial dysfunction in the SU5416/hypoxia model of pulmonary arterial hypertension. *Am J Physiol Lung Cell Mol Physiol*, May 1 2018, 314(5): L893–L907. doi:10.1152/ajplung.00430.2017
133. Fiorio Pla, A., Ong, H. L., Cheng, K. T., Brossa, A., Bussolati, B., Lockwich, T., Paria, B., Munaron, L., Ambudkar, I. S. TRPV4 mediates tumor-derived endothelial cell migration via arachidonic acid-activated actin remodeling. *Oncogene*, Jan 12 2012, 31(2): 200–212. doi:10.1038/onc.2011.231
134. Yao, X., Garland, C. J. Recent developments in vascular endothelial cell transient receptor potential channels. *Circ Res*, Oct 28 2005, 97(9): 853–863. doi:10.1161/01.RES.0000187473.85419.3e
135. Nilius, B., Owsianik, G., Voets, T., Peters, J. A. Transient receptor potential cation channels in disease. *Physiol Rev*, Jan 2007, 87(1): 165–217. doi:10.1152/physrev.00021.2006
136. Liu, L., Guo, M., Lv, X., Wang, Z., Yang, J., Li, Y., Yu, F., Wen, X., Feng, L., Zhou, T. Role of transient receptor potential vanilloid 4 in vascular function. *Front Mol Biosci*, 2021, 8: 677661. doi:10.3389/fmolb.2021.677661

137. Gao, G., Bai, X. Y., Xuan, C., Liu, X. C., Jing, W. B., Novakovic, A., Yang, Q., He, G. W. Role of TRPC3 channel in human internal mammary artery. *Arch Med Res*, Aug 2012, 43(6): 431–437. doi:10.1016/j.arcmed.2012.08.010
138. Senadheera, S., Kim, Y., Grayson, T. H., Toemoe, S., Kochukov, M. Y., Abramowitz, J., Housley, G. D., Bertrand, R. L., Chadha, P. S., Bertrand, P. P., Murphy, T. V., Tare, M., Birnbaumer, L., Marrelli, S. P., Sandow, S. L. Transient receptor potential canonical type 3 channels facilitate endothelium-derived hyperpolarization-mediated resistance artery vasodilator activity. *Cardiovasc Res*, Sep 1 2012, 95(4): 439–447. doi:10.1093/cvr/cvs208
139. Abdullaev, I. F., Bisaillon, J. M., Potier, M., Gonzalez, J. C., Motiani, R. K., Trebak, M. Stim1 and Orai1 mediate CRAC currents and store-operated calcium entry important for endothelial cell proliferation. *Circ Res*, Nov 21 2008, 103(11): 1289–1299. doi:10.1161/01.RES.0000338496.95579.56
140. Freichel, M., Tsvilovskyy, V., Camacho-Londono, J. E. TRPC4- and TRPC4-containing channels. *Handb Exp Pharmacol*, 2014, 222: 85–128. doi:10.1007/978-3-642-54215-2_5
141. Shinde, A. V., Motiani, R. K., Zhang, X., Abdullaev, I. F., Adam, A. P., González-Cobos, J. C., Zhang, W., Matrougui, K., Vincent, P. A., Trebak, M. STIM1 controls endothelial barrier function independently of Orai1 and Ca^{2+} entry. *Sci Signal*, Mar 19 2013, 6(267): ra18. doi:10.1126/scisignal.2003425
142. Kochukov, M. Y., Balasubramanian, A., Noel, R. C., Marrelli, S. P. Role of TRPC1 and TRPC3 channels in contraction and relaxation of mouse thoracic aorta. *J Vasc Res*, 2013, 50(1): 11–20. doi:10.1159/000342461
143. Greenberg, H. Z. E., Carlton-Carew, S. R. E., Zargaran, A. K., Jahan, K. S., Birnbaumer, L., Albert, A. P. Heteromeric TRPV4/TRPC1 channels mediate calcium-sensing receptor-induced relaxations and nitric oxide production in mesenteric arteries: Comparative study using wild-type and TRPC1(-/-) mice. *Channels (Austin)*, 2019, 13(1): 410–423. doi:10.1080/19336950.2019.1673131
144. Mendoza, S. A., Fang, J., Gutterman, D. D., Wilcox, D. A., Bubolz, A. H., Li, R., Suzuki, M., Zhang, D. X. TRPV4-mediated endothelial Ca^{2+} influx and vasodilation in response to shear stress. *Am J Physiol Heart Circ Physiol*, Feb 2010, 298(2): H466–H476. doi:10.1152/ajpheart.00854.2009
145. Geng, L., Zhang, C., He, C., Zhang, K., Kan, H., Mao, A., Ma, X. Physiological levels of fluid shear stress modulate vascular function through TRPV4 sparklets. *Acta Biochim Biophys Sin (Shanghai)*, Sep 25 2022, 54(9): 1268–1277. doi:10.3724/abbs.2022118
146. Hartmannsgruber, V., Heyken, W. T., Kacik, M., Kaistha, A., Grgic, I., Harteneck, C., Liedtke, W., Hoyer, J., Köhler, R. Arterial response to shear stress critically depends on endothelial TRPV4 expression. *PLoS One*, Sep 5 2007, 2(9): e827. doi:10.1371/journal.pone.0000827
147. Swain, S. M., Liddle, R. A. Piezo1 acts upstream of TRPV4 to induce pathological changes in endothelial cells due to shear stress. *J Biol Chem.*, 2021, 296: 100171. doi:10.1074/jbc.RA120.015059
148. Imig, J. D. Epoxides and soluble epoxide hydrolase in cardiovascular physiology. *Physiol Rev*, Jan 2012, 92(1): 101–130. doi:10.1152/physrev.00021.2011
149. Yang, L., Maki-Petaja, K., Cheriyan, J., McEniery, C., Wilkinson, I. B. The role of epoxyeicosatrienoic acids in the cardiovascular system. *Br J Clin Pharmacol*, Jul 2015, 80(1): 28–44. doi:10.1111/bcp.12603
150. Zheng, X., Zinkevich, N. S., Gebremedhin, D., Gauthier, K. M., Nishijima, Y., Fang, J., Wilcox, D. A., Campbell, W. B., Gutterman, D. D., Zhang, D. X. Arachidonic acid-induced dilation in human coronary arterioles: Convergence of signaling mechanisms on endothelial TRPV4-mediated Ca^{2+} entry. *J Am Heart Assoc*, Apr 25 2013, 2(3): e000080. doi:10.1161/JAHA.113.000080
151. Ottolini, M., Daneva, Z., Chen, Y. L., Cope, E. L., Kasetti, R. B., Zode, G. S., Sonkusare, S. K. Mechanisms underlying selective coupling of endothelial Ca^{2+} signals with eNOS vs. IK/SK channels in systemic and pulmonary arteries. *J Physiol*, Sep 2020, 598(17): 3577–3596. doi:10.1113/JP279570
152. Pankey, E. A., Zsombok, A., Lasker, G. F., Kadowitz, P. J. Analysis of responses to the TRPV4 agonist GSK1016790A in the pulmonary vascular bed of the intact-chest rat. *Am J Physiol Heart Circ Physiol*, Jan 1 2014, 306(1): H33–H40. doi:10.1152/ajpheart.00303.2013
153. Saifeddine, M., El-Daly, M., Mihara, K., Bunnett, N. W., McIntyre, P., Altier, C., Hollenberg, M. D., Ramachandran, R. GPCR-mediated EGF receptor transactivation regulates TRPV4 action in the vasculature. *Br J Pharmacol*, May 2015, 172(10): 2493–2506. doi:10.1111/bph.13072
154. Sonkusare, S. K., Dalsgaard, T., Bonev, A. D., Hill-Eubanks, D. C., Kotlikoff, M. I., Scott, J. D., Santana, L. F., Nelson, M. T. AKAP150-dependent cooperative TRPV4 channel gating is central to endothelium-dependent vasodilation and is disrupted in hypertension. *Sci Signal*, Jul 8 2014, 7(333): ra66. doi:10.1126/scisignal.2005052

155. Daneva, Z., Marziano, C., Ottolini, M., Chen, Y. L., Baker, T. M., Kuppusamy, M., Zhang, A., Ta, H. Q., Reagan, C. E., Mihalek, A. D., Kasetti, R. B., Shen, Y., Isakson, B. E., Minshall, R. D., Zode, G. S., Goncharova, E. A., Laubach, V. E., Sonkusare, S. K. Caveolar peroxynitrite formation impairs endothelial TRPV4 channels and elevates pulmonary arterial pressure in pulmonary hypertension. *Proc Natl Acad Sci U S A*, Apr 27 2021, 118(17): e2023130118. doi:10.1073/pnas.2023130118
156. Ottolini, M., Hong, K., Cope, E. L., Daneva, Z., DeLalio, L. J., Sokolowski, J. D., Marziano, C., Nguyen, N. Y., Altschmied, J., Haendeler, J., Johnstone, S. R., Kalani, M. Y., Park, M. S., Patel, R. P., Liedtke, W., Isakson, B. E., Sonkusare, S. K. Local peroxynitrite impairs endothelial transient receptor potential vanilloid 4 channels and elevates blood pressure in obesity. *Circulation*, Apr 21 2020, 141(16): 1318–1333. doi:10.1161/CIRCULATIONAHA.119.043385
157. Liu, Z., Khalil, R. A. Evolving mechanisms of vascular smooth muscle contraction highlight key targets in vascular disease. *Biochem Pharmacol*, Jul 2018, 153: 91–122. doi:10.1016/j.bcp.2018.02.012
158. Shi, J., Birnbaumer, L., Large, W. A., Albert, A. P. Myristoylated alanine-rich C kinase substrate coordinates native TRPC1 channel activation by phosphatidylinositol 4,5-bisphosphate and protein kinase C in vascular smooth muscle. *FASEB J*, Jan 2014, 28(1): 244–255. doi:10.1096/fj.13-238022
159. Bolotina, V. M. Orai, STIM1 and iPLA2beta: A view from a different perspective. *J Physiol*, Jul 1 2008, 586(13): 3035–3042. doi:10.1113/jphysiol.2008.154997
160. Liu, D., Yang, D., He, H., Chen, X., Cao, T., Feng, X., Ma, L., Luo, Z., Wang, L., Yan, Z., Zhu, Z., Tepel, M. Increased transient receptor potential canonical type 3 channels in vasculature from hypertensive rats. *Hypertension*, Jan 2009, 53(1): 70–76. doi:10.1161/HYPERTENSIONAHA.108.116947
161. Peppiatt-Wildman, C. M., Albert, A. P., Saleh, S. N., Large, W. A. Endothelin-1 activates a Ca^{2+}-permeable cation channel with TRPC3 and TRPC7 properties in rabbit coronary artery myocytes. *J Physiol*, May 1 2007, 580(Pt.3): 755–764. doi:10.1113/jphysiol.2006.126656
162. Albert, A. P., Pucovsky, V., Prestwich, S. A., Large, W. A. TRPC3 properties of a native constitutively active Ca^{2+}-permeable cation channel in rabbit ear artery myocytes. *J Physiol*, Mar 1 2006, 571(2): 361–369. doi:10.1113/jphysiol.2005.102780
163. Adebiyi, A., Thomas-Gatewood, C. M., Leo, M. D., Kidd, M. W., Neeb, Z. P., Jaggar, J. H. An elevation in physical coupling of type 1 inositol 1,4,5-trisphosphate (IP3) receptors to transient receptor potential 3 (TRPC3) channels constricts mesenteric arteries in genetic hypertension. *Hypertension.*, Nov 2012, 60(5): 1213–1219. doi:10.1161/HYPERTENSIONAHA.112.198820
164. Shi, J., Ju, M., Large, W. A., Albert, A. P. Pharmacological profile of phosphatidylinositol 3-kinases and related phosphatidylinositols mediating endothelin(A) receptor-operated native TRPC channels in rabbit coronary artery myocytes. *Br J Pharmacol*, Aug 2012, 166(7): 2161–2175. doi:10.1111/j.1476-5381.2012.01937.x
165. Imai, Y., Itsuki, K., Okamura, Y., Inoue, R., Mori, M. X. A self-limiting regulation of vasoconstrictor-activated TRPC3/C6/C7 channels coupled to PI(4,5)P(2)-diacylglycerol signalling. *J Physiol.*, Mar 1 2012, 590(5): 1101–1119. doi:10.1113/jphysiol.2011.221358
166. Lemonnier, L., Trebak, M., Putney, J. W. Jr Complex regulation of the TRPC3, 6 and 7 channel subfamily by diacylglycerol and phosphatidylinositol-4,5-bisphosphate. *Cell Calcium*, May 2008, 43(5): 506–514. doi:10.1016/j.ceca.2007.09.001
167. Inoue, R., Okada, T., Onoue, H., Hara, Y., Shimizu, S., Naitoh, S., Ito, Y., Mori, Y. The transient receptor potential protein homologue TRP6 is the essential component of vascular alpha(1)-adrenoceptor-activated Ca^{2+}-permeable cation channel. *Circ Res*, Feb 16 2001, 88(3): 325–332. doi:10.1161/01.res.88.3.325
168. Soboloff, J., Spassova, M., Xu, W., He, L. P., Cuesta, N., Gill, D. L. Role of endogenous TRPC6 channels in Ca^{2+} signal generation in A7r5 smooth muscle cells. *J Biol Chem.*, Dec 2 2005, 280(48): 39786–39794. doi:10.1074/jbc.M506064200
169. Saleh, S. N., Albert, A. P., Peppiatt, C. M., Large, W. A. Angiotensin II activates two cation conductances with distinct TRPC1 and TRPC6 channel properties in rabbit mesenteric artery myocytes. *J Physiol*, Dec 1 2006, 577(2): 479–495. doi:10.1113/jphysiol.2006.119305
170. Mani, B. K., Brueggemann, L. I., Cribbs, L. L., Byron, K. L. Opposite regulation of KCNQ5 and TRPC6 channels contributes to vasopressin-stimulated calcium spiking responses in A7r5 vascular smooth muscle cells. *Cell Calcium*, Apr 2009, 45(4): 400–411. doi:10.1016/j.ceca.2009.01.004
171. Spassova, M. A., Hewavitharana, T., Xu, W., Soboloff, J., Gill, D. L. A common mechanism underlies stretch activation and receptor activation of TRPC6 channels. *Proc Natl Acad Sci U S A*, Oct 31 2006, 103(44): 16586–16591. doi:10.1073/pnas.0606894103
172. Hofmann, T., Obukhov, A. G., Schaefer, M., Harteneck, C., Gudermann, T., Schultz, G. Direct activation of human TRPC6 and TRPC3 channels by diacylglycerol. *Nature*, Jan 21 1999, 397(6716): 259–263. doi:10.1038/16711

173. Inoue, R., Jensen, L. J., Jian, Z., Shi, J., Hai, L., Lurie, A. I., Henriksen, F. H., Salomonsson, M., Morita, H., Kawarabayashi, Y., Mori, M., Mori, Y., Ito, Y. Synergistic activation of vascular TRPC6 channel by receptor and mechanical stimulation via phospholipase C/diacylglycerol and phospholipase A2/omega-hydroxylase/20-HETE pathways. *Circ Res*, Jun 19 2009, 104(12): 1399–1409. doi:10.1161/CIRCRESAHA.108.193227
174. Ju, M., Shi, J., Saleh, S. N., Albert, A. P., Large, W. A. Ins(1,4,5)P3 interacts with PIP2 to regulate activation of TRPC6/C7 channels by diacylglycerol in native vascular myocytes. *J Physiol*, May 1 2010, 588(Pt 9): 1419–1433. doi:10.1113/jphysiol.2009.185256
175. Welsh, D. G., Morielli, A. D., Nelson, M. T., Brayden, J. E. Transient receptor potential channels regulate myogenic tone of resistance arteries. *Circ Res*, Feb 22 2002, 90(3): 248–250. doi:10.1161/hh0302.105662
176. Mederos y Schnitzler, M., Storch, U., Meibers, S., Nurwakagari, P., Breit, A., Essin, K., Gollasch, M., Gudermann, T. Gq-coupled receptors as mechanosensors mediating myogenic vasoconstriction. *EMBO J*, Dec 3 2008, 27(23): 3092–3103. doi:10.1038/emboj.2008.233
177. Yang, X. R., Lin, A. H., Hughes, J. M., Flavahan, N. A., Cao, Y. N., Liedtke, W., Sham, J. S. Upregulation of osmo-mechanosensitive TRPV4 channel facilitates chronic hypoxia-induced myogenic tone and pulmonary hypertension. *Am J Physiol Lung Cell Mol Physiol*, Mar 15 2012, 302(6): L555–L568. doi:10.1152/ajplung.00005.2011
178. Goldenberg, N. M., Wang, L., Ranke, H., Liedtke, W., Tabuchi, A., Kuebler, W. M. TRPV4 is required for hypoxic pulmonary vasoconstriction. *Anesthesiology*, Jun 2015, 122(6): 1338–1348. doi:10.1097/ALN.0000000000000647
179. Xia, Y., Fu, Z., Hu, J., Huang, C., Paudel, O., Cai, S., Liedtke, W., Sham, J. S. TRPV4 channel contributes to serotonin-induced pulmonary vasoconstriction and the enhanced vascular reactivity in chronic hypoxic pulmonary hypertension. *Am J Physiol Cell Physiol*, Oct 1 2013, 305(7): C704–C715. doi:10.1152/ajpcell.00099.2013
180. Chen, Y. L., Daneva, Z., Kuppusamy, M., Ottolini, M., Baker, T. M., Klimentova, E., Shah, S. A., Sokolowski, J. D., Park, M. S., Sonkusare, S. K. Novel smooth muscle Ca^{2+}-signaling nanodomains in blood pressure regulation. *Circulation*, Aug 16 2022, 146(7): 548–564. doi:10.1161/CIRCULATIONAHA.121.058607
181. Gonzales, A. L., Garcia, Z. I., Amberg, G. C., Earley, S. Pharmacological inhibition of TRPM4 hyperpolarizes vascular smooth muscle. *Am J Physiol Cell Physiol*, Nov 2010, 299(5): C1195–C1202. doi:10.1152/ajpcell.00269.2010
182. Earley, S. TRPM4 channels in smooth muscle function. *Pflugers Arch*, Sep 2013, 465(9): 1223–1231. doi:10.1007/s00424-013-1250-z
183. Earley, S., Waldron, B. J., Brayden, J. E. Critical role for transient receptor potential channel TRPM4 in myogenic constriction of cerebral arteries. *Circ Res*, Oct 29 2004, 95(9): 922–929. doi:10.1161/01.RES.0000147311.54833.03
184. Grand, T., Demion, M., Norez, C., Mettey, Y., Launay, P., Becq, F., Bois, P., Guinamard, R. 9-phenanthrol inhibits human TRPM4 but not TRPM5 cationic channels. *Br J Pharmacol*, Apr 2008, 153(8): 1697–1705. doi:10.1038/bjp.2008.38
185. Garland, C. J., Smirnov, S. V., Bagher, P., Lim, C. S., Huang, C. Y., Mitchell, R., Stanley, C., Pinkney, A., Dora, K. A. TRPM4 inhibitor 9-phenanthrol activates endothelial cell intermediate conductance calcium-activated potassium channels in rat isolated mesenteric artery. *Br J Pharmacol*, Feb 2015, 172(4): 1114–1123. doi:10.1111/bph.12985
186. Li, Y., Baylie, R. L., Tavares, M. J., Brayden, J. E. TRPM4 channels couple purinergic receptor mechanoactivation and myogenic tone development in cerebral parenchymal arterioles. *J Cereb Blood Flow Metab.*, Oct 2014, 34(10): 1706–1714. doi:10.1038/jcbfm.2014.139
187. Brayden, J. E., Li, Y., Tavares, M. J. Purinergic receptors regulate myogenic tone in cerebral parenchymal arterioles. *J Cereb Blood Flow Metab*, Feb 2013, 33(2): 293–299. doi:10.1038/jcbfm.2012.169
188. Gonzales, A. L., Yang, Y., Sullivan, M. N., Sanders, L., Dabertrand, F., Hill-Eubanks, D. C., Nelson, M. T., Earley, S. A PLCgamma1-dependent, force-sensitive signaling network in the myogenic constriction of cerebral arteries. *Sci Signal*, May 27 2014, 7(327): ra49. doi:10.1126/scisignal.2004732
189. Brough, G. H., Wu, S., Cioffi, D., Moore, T. M., Li, M., Dean, N., Stevens, T. Contribution of endogenously expressed Trp1 to a Ca^{2+}-selective, store-operated Ca^{2+} entry pathway. *FASEB J*, Aug 2001, 15(10): 1704–1710.
190. Ahmmed, G. U., Mehta, D., Vogel, S., Holinstat, M., Paria, B. C., Tiruppathi, C., Malik, A. B. Protein kinase calpha phosphorylates the TRPC1 channel and regulates store-operated Ca^{2+} entry in endothelial cells. *J Biol Chem*, May 14 2004, 279(20): 20941–20949. doi:10.1074/jbc.M313975200

191. Paria, B. C., Vogel, S. M., Ahmmed, G. U., Alamgir, S., Shroff, J., Malik, A. B., Tiruppathi, C. Tumor necrosis factor-alpha-induced TRPC1 expression amplifies store-operated Ca^{2+} influx and endothelial permeability. *Am J Physiol Lung Cell Mol Physiol*, Dec 2004, 287(6): L1303–L1313. doi:10.1152/ajplung.00240.2004
192. Tauseef, M., Farazuddin, M., Sukriti, S., Rajput, C., Meyer, J. O., Ramasamy, S. K., Mehta, D. Transient receptor potential channel 1 maintains adherens junction plasticity by suppressing sphingosine kinase 1 expression to induce endothelial hyperpermeability. *FASEB J*, Jan 2016, 30(1): 102–110. doi:10.1096/fj.15-275891
193. Simmons, S., Erfinanda, L., Bartz, C., Kuebler, W. M. Novel mechanisms regulating endothelial barrier function in the pulmonary microcirculation. *J Physiol*, Feb 2019, 597(4): 997–1021. doi:10.1113/JP276245
194. Cioffi, D. L., Wu, S., Alexeyev, M., Goodman, S. R., Zhu, M. X., Stevens, T. Activation of the endothelial store-operated ISOC Ca^{2+} channel requires interaction of protein 4.1 with TRPC4. *Circ Res*, Nov 25 2005, 97(11): 1164–1172. doi:10.1161/01.RES.0000193597.65217.00
195. Graziani, A., Poteser, M., Heupel, W. M., Schleifer, H., Krenn, M., Drenckhahn, D., Romanin, C., Baumgartner, W., Groschner, K. Cell-cell contact formation governs Ca^{2+} signaling by TRPC4 in the vascular endothelium: Evidence for a regulatory TRPC4-beta-catenin interaction. *J Biol Chem*, Feb 5 2010, 285(6): 4213–4223. doi:10.1074/jbc.M109.060301
196. Lu, Q., Zemskov, E. A., Sun, X., Wang, H., Yegambaram, M., Wu, X., Garcia-Flores, A., Song, S., Tang, H., Kangath, A., Cabanillas, G. Z., Yuan, J. X., Wang, T., Fineman, J. R., Black, S. M. Activation of the mechanosensitive Ca^{2+} channel TRPV4 induces endothelial barrier permeability via the disruption of mitochondrial bioenergetics. *Redox Biol*, Jan 2021, 38: 101785. doi:10.1016/j.redox.2020.101785
197. Yin, J., Hoffmann, J., Kaestle, S. M., Neye, N., Wang, L., Baeurle, J., Liedtke, W., Wu, S., Kuppe, H., Pries, A. R., Kuebler, W. M. Negative-feedback loop attenuates hydrostatic lung edema via a cGMP-dependent regulation of transient receptor potential vanilloid 4. *Circ Res*, Apr 25 2008, 102(8): 966–974. doi:10.1161/CIRCRESAHA.107.168724

10 Study of TRP Channels in Stroke

Pengyu Zong and Lixia Yue

10.1 INTRODUCTION

10.1.1 Stroke Pathophysiology

Stroke is the primary leading cause of disability worldwide and the second leading cause of death.[1] A concerning trend exacerbating this issue is the rising incidence of stroke among younger individuals.[2,3] In the United States, while the stroke rate has declined among adults aged 75 and older, it has shown an alarming increase among adults aged 49 and younger.[2] Stroke can be classified into two primary types: ischemic stroke, often resulting from thrombus formation or blockage in a cerebral artery, and hemorrhagic stroke, which is characterized by the rupture of a cerebral artery.[1] Ischemic stroke is the major type of stroke, accounting for 80–90% of all cases.[2] Both types of stroke result in ischemia in brain tissue downstream from the occlusion or rupture site. Neural tissues are particularly vulnerable to hypoxia, with a complete cutoff of oxygen supply typically leading to neuronal death within minutes.[3] Thus, neuronal injury and death are considered among the earliest pathological features of stroke.[1] Current stroke studies predominantly focus on ischemic stroke, as it is the more common form. However, the focus on ischemic stroke in this chapter does not diminish the potential involvement of TRP channels in hemorrhagic stroke, which warrants future attention and consideration.

Extensive research over the past decades has focused on unraveling the molecular mechanisms underlying ischemic neuronal death (Fig. 10.1). Notably, glutamate excitotoxicity has emerged as a central player in these mechanisms.[4] Glutamate, the brain's most abundant excitatory neurotransmitter, can exert direct cytotoxic effects on neurons at high concentrations, a phenomenon known as excitotoxicity.[4] Glutamate receptors fall into two categories: cation-permeable ionotropic receptors (iGluRs) and G protein-coupled metabotropic receptors (mGluRs). iGluRs are further divided into four classes based on their response to specific agonists: α-amino-3-hydroxy-5-methyl-4-isoxazolepropionic acid receptors (AMPARs), N-methyl-D-aspartate receptors (NMDARs), kainate receptors (KARs), and GluD receptors.[5] While NMDAR-mediated glutamate excitotoxicity has been extensively investigated, it is important to note that other glutamate receptors also contribute to the neurotoxic effects of glutamate at varying degrees.[6–8] All iGluRs are composed of four subunits forming heterotetramers with a central ion-passing pore.[9] NMDARs exhibit diverse subunit compositions, including GluN1, GluN2A-D, and GluN3A-B.[5] NMDAR overactivation initiates neuronal death through Ca^{2+} overload and activating signaling pathways associated with cell death.[10] Interestingly, it appears that the cytotoxic effects of NMDARs are determined by their composition and location. During ischemic stroke, extrasynaptic NMDARs (esNMDARs) and NMDARs containing GluN2B primarily induce neuronal death, while synaptic NMDARs (sNMDARs) and NMDARs containing GluN2A offer neuronal protection against ischemia.[11,12] Regrettably, despite the extensive investigation in recent decades, all attempts aimed directly at targeting NMDARs to alleviate stroke have failed in clinical trials.[13] This emphasizes the urgent need to better understand how to differentially modulate the protective and detrimental aspects of NMDARs during stroke.[10,12,14]

In addition to the direct neurotoxic effects of ischemia, numerous factors contribute significantly to neuronal death during stroke (Fig. 10.1). Firstly, the extensive neuronal death triggers robust

DOI: 10.1201/9781003402732-10

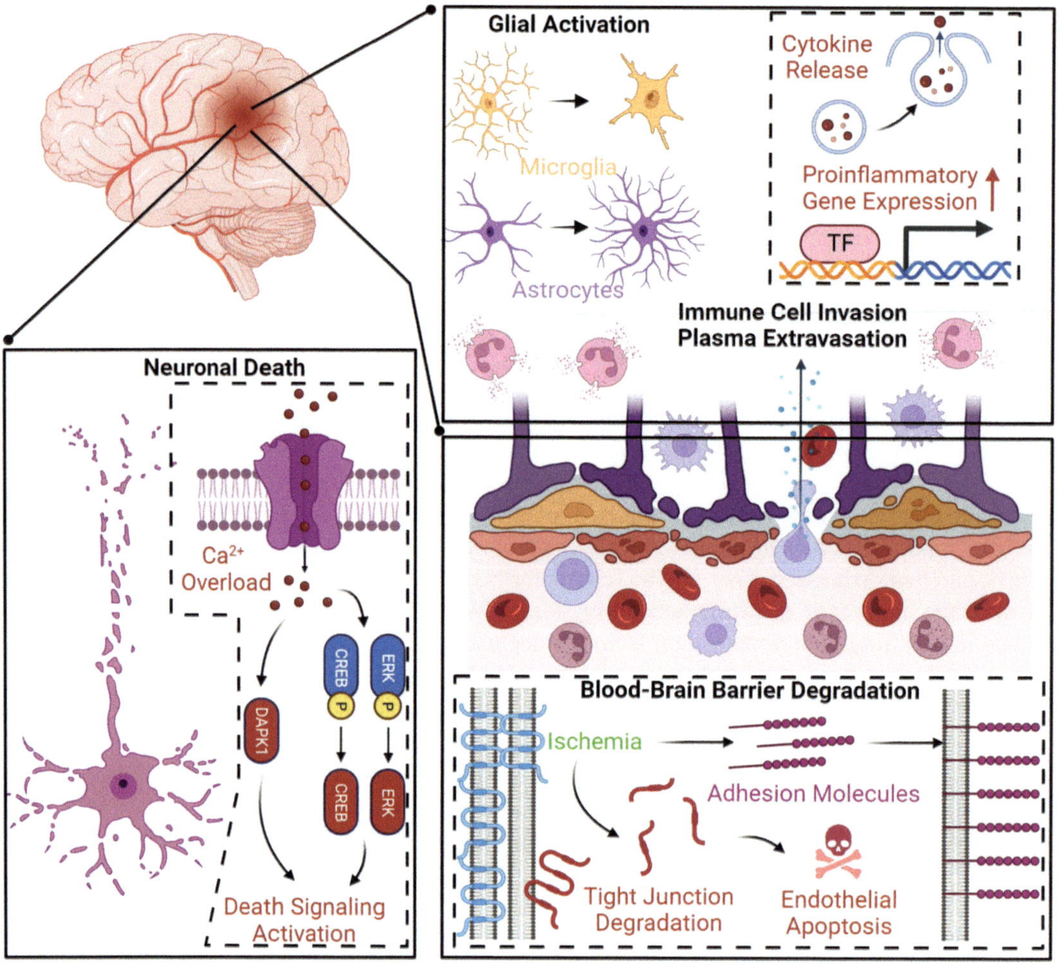

FIGURE 10.1 Stroke pathology overview. Brain ischemia caused by an ischemic or hemorrhagic stroke triggers rapid neuronal death, during which NMDAR-mediated glutamate excitotoxicity plays a pivotal role. Activation of NMDAR results in Ca^{2+} overload, initiating downstream death-related events such as DAPK1 activation and CREB/ERK dephosphorylation, ultimately leading to neuronal death. Concurrently, neuronal death induces profound tissue inflammation, prompting tight junction degradation between endothelial cells and increasing the expression of leukocyte adhesion molecules. This leads to endothelial hyperpermeability and apoptosis, compounded by astrocyte swelling and pericyte injury, collectively resulting in BBB leakage. Subsequently, BBB leakage and tissue inflammation result in glial activation and massive immune cell infiltration into the brain, exacerbating local inflammation and causing secondary neuronal death. Dysfunction of these non-neuronal cells significantly contributes to ischemic neuronal death.

inflammatory responses in the brain, activating glial cells, particularly astrocytes and microglia.[15] At the early stages of stroke, astrocyte activation exacerbates cerebral edema, while microglia activation intensifies local inflammation, which in turn causes secondary neuronal death and brain injury.[15] Secondly, ischemia disrupts the blood-brain barrier (BBB), and local inflammation accelerates its degradation.[16] This disruption allows neurotoxic blood components to leak into the brain, amplifying tissue inflammation and cerebral edema.[16] Thirdly, tissue injury prompts the recruitment of immune cells into the brain, which can be facilitated by BBB disruption.[17] The infiltration and activation of immune cells in the brain cause secondary neuronal death and further exacerbate tissue injury.[17] Therefore, neuronal death in stroke is a complex process influenced by multiple interrelated factors.[1] An ideal stroke therapy should target cellular death or dysfunction in all the cell types involved in stroke pathology.[10,18]

Current therapies for stroke, including open-skull vessel repair surgeries for hemorrhagic stroke and recanalization therapies for ischemic stroke, present inherent risks to patients.[3] Direct surgical interventions involving skull penetration and indirect mechanical thrombectomy carry high procedure risks, while pharmacological thrombolysis often proves ineffective and elevates the risk of systemic hemorrhage.[3,19] More importantly, there remains a lack of therapies specifically aiming to mitigate post-stroke infarct expansion.[1,18] This underscores the urgent need for better targets to develop pharmacological approaches that can attenuate brain injury at both the ischemic and post-ischemic stages.[18] Recently, emerging evidence highlighted TRP channels as potential candidates for targeting these two critical stages of stroke pathology.

10.1.2 Transient Receptor Potential (TRP) Channels

The mammalian TRP channel superfamily comprises twenty-eight members, which can be grouped into six subfamilies: canonical (TRPCs), vanilloid (TRPVs), melastatin (TRPMs), mucolipin (TRPMLs), polycystin (TRPPs), and ankyrin (TRPA).[20] Furthermore, the expression of TRP channel subunits undergoes alternative splicing, adding to the diversity of channel properties.[21] Structurally, TRP channels form tetrameric channels, similar to voltage-gated K^+ channels.[22] TRPC channels tend to form heterotetramers, which may significantly increase the channel diversity in this family.[22] Each TRP subunit consists of six transmembrane domains (S1–S6), an N-terminal domain (NTD), and a C-terminal domain (CTD).[23] The S5–S6 and pore-loop domains from four subunits form the central channel pore, which determines the ion selectivity of different TRP channels.[23] The majority of TRP channels are Ca^{2+}-permeable, which confers TRP channels with important physiological and pathological functions.[24,25] While the structure of S1–S4 forms a voltage-sensing domain-like domain (VSD-L), which resembles the voltage-sensing domain in voltage-gated ion channels, TRP channels generally lack voltage sensing ability.[26] Instead, TRP channel activity is commonly modulated by ligand binding, changes in phospholipid levels, oxidative stress, mechanical stretch, and temperature fluctuations.[22] Recent advancements in atomic structure analysis of TRP channels have significantly enhanced our understanding of how TRP channels are gated by various ligands,[22] yet the precise molecular mechanisms behind their temperature gating remain elusive.[27]

TRP channels are highly conserved among almost all the species, from invertebrates to humans, suggesting their indispensable functions throughout evolution.[28] The discovery of TRP channels began with their essential role in vision in Drosophila.[20] Over time, TRP channels have emerged as versatile biosensors capable of detecting a wide range of internal and external stimuli, including tissue injury, temperature fluctuations, and exposure to diverse chemicals.[29] Subsequently, the significance of TRP channels in normal development and other physiological functions was established. Mutations in many human TRP channels have been linked to inherited diseases, and knockout of certain TRP channels in mice causes early embryonic death.[30] Over the past two decades since the first cloning of human TRPC1 in 1995,[31–33] extensive studies have revealed the diverse involvement of TRP channels in numerous pathophysiological conditions, such as synaptic transmission,[34] immune cell activation,[35] inflammation,[36] pathological pain,[37] cardiac arrhythmia,[38] diabetes mellitus,[39] carcinogenesis,[40] and many vascular diseases, including ischemic stroke.[36] Remarkably, half of all TRP channels (14 out of 28) have been implicated in ischemic stroke, including TRPC1, TRPC3, TRPC4, TRPC5, TRPC6, TRPC7, TRPM2, TRPM4, TRPM7, TRPV1, TRPV2, TRPV3, TRPV4, and TRPA1.[36,41,42] Among them, the contribution of TRPM2, TRPM4, TRPC3/6/7, and TRPV1 in ischemic brain injury is best studied. The roles of TRP channels in ischemic brain injury can be beneficial, detrimental, or even dual in nature, contributing to the complexity of stroke pathology. In this chapter, our primary focus is to summarize the recent advances concerning the involvement of TRP channels in ischemic stroke.

10.2 TRP CHANNEL IN ISCHEMIC STROKE

The distinct characteristics of TRP channels, such as their ability to sense oxidative stress and facilitate Ca^{2+} influx, and their expression in diverse cell types involved in ischemic stroke, make them prime targets for therapeutic interventions against this disease. Recent investigations have increasingly highlighted the pivotal roles played by various TRP channels in ischemic stroke. Notably, TRP channels contribute to neuronal toxicity not only through conventional mechanisms involving Ca^{2+} overload and activation of cell death signaling[43] but also via recently uncovered mechanisms, specifically functional coupling with NMDARs, leading to excitotoxicity.[44,45,46]

10.2.1 TRPM Channels in Ischemic Stroke

The nomenclature for the TRPM family originates from their discovery during a screening designed to identify genes associated with melanoma metastasis, which led to the identification of melastatin (TRPM1). TRPMs are characterized by possessing four distinct melastatin homology regions (MHR1-4) in their N-terminal domain (NTD), resulting in a particularly longer NTD compared to other TRP channels; therefore, TRPMs were initially named "long TRPCs". Based on their sequence homologies and structural similarities, the eight mammalian TRPMs can be categorized into four groups, including TRPM1/3, TRPM2/8, TRPM4/5, and TRPM6/7, yet the TRPM family members exhibit considerable diversity, even within the same subgroup. For instance, TRPM2, TRPM3, TRPM4, and TRPM5 are sensitive to heat, whereas TRPM8 is sensitive to cold.[47,48] Among TRPMs, TRPM4 and TRPM5 are the sole TRP channels selectively permeable to monovalent cations.[49,50] Additionally, TRPM6 and TRPM7 demonstrate the exceptionally high permeability to Zn^{2+}. Notably, TRPM2, TRPM6, and TRPM7 are the only known ion channels featured with an enzymatic domain in their C-terminal domain (CTD), thereby being designated as "chanzymes".

TRPMs are ubiquitously expressed and are regulated by a diverse range of intracellular and extracellular stimuli. TRPM channels play pivotal roles in maintaining mineral homeostasis, influencing cardiac conductivity, modulating immune responses, and contributing to the sensation of taste and temperature. Thus far, studies have shown the involvement of TRPM2, TRPM4, and TRPM7 in ischemic brain injury.

10.2.1.1 TRPM2 in Ischemic Stroke

TRPM2 possesses a distinct feature in its C-terminus, hosting a nudix-type motif 9 (NUDT9)-homology (NUDT9-H) domain that is homologous to adenosine diphosphate ribose (ADPR) hydrolases (ADPRases).[51] Although mammalian TRPM2 lacks enzymatic function,[52] two invertebrate TRPM2 channels, the cnidarian *Nematostella vectensis* TRPM2 (nvTRPM2) and the choanoflagellate *Salpingoeca rosetta* TRPM2 (srTRPM2), exhibit active ADPRases activity to cleave ADPR.[53] The NUDT9-H domain in mammalian TRPM2 binds with ADPR to activate the channel[52,54] in the presence of Ca^{2+}, which binds to VSD-L.[55] Moreover, recent studies have demonstrated that the NUDT9-H domain is not the sole binding site for ADPR; another ADPR binding site localized at the MHR1/2 domain in the N-terminus of TRPM2 has been identified in human and drosophila TRPM2,[55,56] and suggested as a primary binding site for ADPR across TRPM2 from different species as the MHR1/2 binding site is conserved in all TRPM2 orthologues.[57]

TRPM2 is ubiquitously expressed in various cell types.[58–62] Given the substantial increase in intracellular ADPR and Ca^{2+} levels during oxidative stress, TRPM2 is often recognized as an oxidative stress sensor. Notably, TRPM2 displays plenty of splice variants, such as TRPM2-ΔN, TRPM2-ΔC, TRPM2-ΔNΔC, and TRPM2-S, albeit their specific pathophysiological functions remain elusive.[62–66] Indeed, multiple studies have demonstrated the detrimental role of TRPM2 in ischemic brain injury.[36] Its overactivation has been associated with adverse effects in different cell types: in neurons, it directly contributes to neuronal death;[45] in glial cells, it enhances local inflammation;[67]

in endothelial cells,[68] it accelerates blood-brain barrier (BBB) leakage; and in immune cells, it promotes their infiltration into the brain.[36,68]

The pivotal role of TRPM2-mediated Ca^{2+} influx in cell death induced by H_2O_2 has been well documented since 2002,[69] with reports on neuronal death emerging since 2005.[70–73] Experiments using TRPM2 knockdown via siRNA or inhibition of its activation by suppressing ADPR generation using SB-750139 have effectively mitigated neuronal death triggered by H_2O_2 and amyloid β-peptide treatment.[70] These pioneer findings, demonstrated by multiple groups, suggest that the detrimental effects of TRPM2 activation, likely through Ca^{2+} overload and mediated by ADPR overproduction, play a critical role in neuronal death under oxidative stress conditions. Subsequently, both in vitro oxygen-glucose deprivation (OGD) models and *in vivo* middle cerebral artery occlusion (MCAO) models provided further evidence of TRPM2's detrimental effects in ischemic stroke.[72,74] These studies showed that TRPM2 knockdown mitigated ischemic neuronal death and reduced brain injury in mice. Interestingly, the protective effect of inhibiting or knocking out TRPM2 seems to be observed specifically in male mice and in transient MCAO models rather than in female mice or in permanent MCAO models.[72,75,76] In line with the protective effect seen in the transient MCAO model, TRPM2-mediated Zn^{2+} influx has been reported to contribute to neuronal death during the reperfusion stage rather than the ischemic stage.[77] However, it remains challenging to fully distinguish the individual contributions of TRPM2-mediated Zn^{2+} and Ca^{2+} in neuronal death due to the similar physical and chemical properties of Zn^{2+} and Ca^{2+}. Methodological limitations also exist: Ca^{2+} chelators including BAPTA and EGTA will buffer Zn^{2+}, and Zn^{2+} fluorescence dyes such as the "highly" Zn^{2+} selective FluoZin3 also bind to Ca^{2+} when Ca^{2+} concentration is very high. Moreover, the explanation of Ca^{2+} overload and Zn^{2+} toxicity might be somewhat general when trying to pinpoint TRPM2-induced neuronal death. During the death of neurons, regardless of the underlying cause, there is typically an increase in intracellular Ca^{2+} and Zn^{2+} concentrations. This simultaneous elevation in ion concentrations is a common occurrence in various forms of neuronal death, making it unreasonable to attribute this mechanism solely to TRPM2-induced neuronal death. Therefore, it is crucial to identify more specific pathways or mechanisms influenced by TRPM2 activation that lead to neuronal death under particular conditions, beyond the general notion of ionic imbalance.

Indeed, additional molecular mechanisms contributing to TRPM2-mediated neuronal death have been demonstrated in recent studies. In a neonatal mouse model of hypoxic-ischemic injury, the protective effect observed upon TRPM2 knockout appears to be closely linked to GSK3β signaling.[78] Moreover, a recent finding highlighted that bilirubin binds directly to TRPM2, enhancing its activation.[79] This interaction was found to play a role in bilirubin-induced neuronal hyperexcitation following ischemic stroke.[79] Inhibiting the binding of bilirubin to TRPM2 by mutating the binding site on TRPM2 reduced ischemic brain injury in mice.[79] The interactions between TRPM2 and NMDAR, the culprit for excitotoxicity, are also intriguing (Fig. 10.2). A previous study showed that TRPM2 knockout resulted in an increase in GluN2A expression and a decrease in GluN2B expression, thereby inhibiting the pro-death CREB/ERK/GSK3β signaling pathway and ultimately attenuating ischemic stroke in mice.[76] However, contradictory findings were reported in another study that TRPM2 knockout did not alter NMDAR expression.[80] In our study, we also examined this relationship and discovered that under physiological conditions, TRPM2 knockout did not influence the surface expression levels of GluN2A and GluN2B. However, during ischemic stroke, TRPM2 knockout significantly curtailed the abnormal surge in their surface expression.[45] The effect of TRPM2 on NMDAR surface expression appears to be mediated by PKCγ, a well-established positive regulator of NMDAR activity.[45] This suggests that TRPM2 might selectively modulate the surface expression of GluN2A and GluN2B subunits under pathological conditions. Moreover, we found that TRPM2 is physically and functionally associated with NMDAR and PKCγ.[45] We identified that the EE_3 motif within the NTD of TRPM2 directly binds to the KKR motif located in the CTD of GluN2A and GluN2B.[45] This interaction significantly enhanced NMDAR activity and consequently amplified NMDAR-mediated excitotoxicity during ischemic stroke.[45] In line with these findings, disrupting the TRPM2-NMDAR binding by a specific interfering peptide exerted a

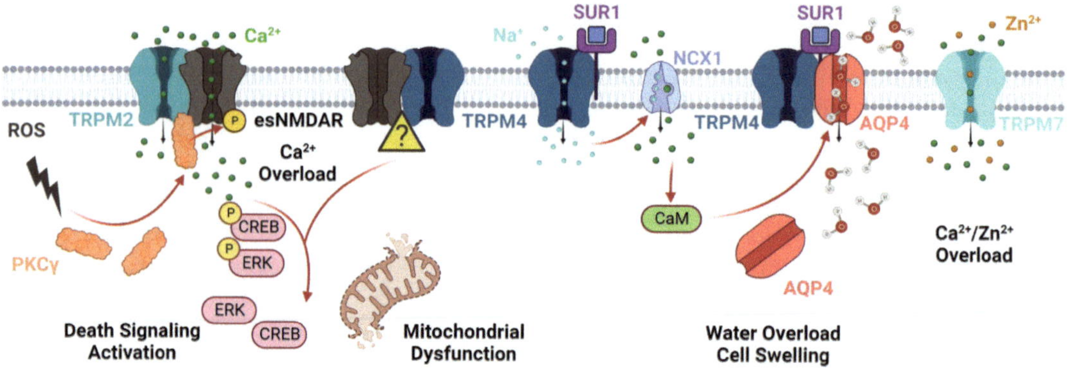

FIGURE 10.2 TRPM2, TRPM4, and TRPM7 in stroke. The EE_3 motif in the NTD of TRPM2 directly interacts with the KKR motif situated in the CTD of GluN2A and GluN2B in esNMDAR. Under oxidative stress, the binding of PKCγ to TRPM2 is increased. This interaction results in PKCγ-mediated phosphorylation of esNMDAR, subsequently augmenting both its surface expression and channel activity, thereby intensifying ischemic excitotoxicity. TRPM2 might act as a crucial mediator between PKCγ and esNMDAR by facilitating Ca^{2+} supply for the activation of PKCγ. The twinF domain in the NTD of TRPM4 associates with the I4 domain in the CTD of GluN2A and GluN2B. This association enhances NMDAR-mediated ischemic excitotoxicity by inhibiting survival signaling pathways and inducing mitochondrial dysfunction. TRPM4 also forms an association with SUR1. Despite being impermeable to Ca^{2+}, TRPM4-mediated Na^+ influx activates the Na^+/Ca^{2+} exchanger NCX1, leading to subsequent Ca^{2+} influx. This influx, in turn, triggers CaM activation, promoting the recruitment of the water channel AQP4 into the TRPM4/SUR1 complex, ultimately causing water overload and cellular swelling. TRPM7 directly induces neuronal cell death due to its high permeability to toxic divalent cations such as Zn^{2+} and Ca^{2+}.

robust protective effect against ischemic stroke in mice.[45] Importantly, our study employed nestin-cre mice, allowing us to demonstrate for the first time that neuron-specific TRPM2 knockout was sufficient to generate a protective effect similar to that seen with global TRPM2 knockout.[45] This novel approach provided compelling evidence showing the direct involvement of TRPM2 in ischemic neuronal death *in vivo.*

The contribution of TRPM2 in other cell types to ischemic stroke has also been well studied. TRPM2 global knockout studies have shown to inhibit microglial proliferation following ischemic stroke in a bilateral common carotid artery stenosis cerebral hypoperfusion mouse model.[67] However, it is difficult to decipher whether the observed reduction in microglial proliferation is a natural response secondary to reduced tissue injury or the reduced brain infarction is due to the inhibited microglia activity by TRPM2 knockout.[67] The absence of a glial cell-specific TRPM2 knockout mouse makes it hard to assess the individual contributions of glial TRPM2 in the intricate brain microenvironment during stroke. Subsequent experiments using isolated microglia and astrocytes from the human brain have revealed that TRPM2 knockdown led to reduced expression of pro-inflammatory cytokines in response to glutathione depletion.[81] This *in vitro* finding suggests a potential role for TRPM2 in facilitating abnormal glial cell activation following a stroke. To precisely evaluate the significance of glial TRPM2 in stroke, future *in vivo* experiments utilizing glial-specific TRPM2 knockout mice will be essential. Apart from its effects on glial cells, TRPM2 has been identified as a promoter of endothelial leakage and facilitator of immune cell invasion during stroke. Its overactivation has been well known for causing endothelial permeability in the peripheral vascular beds.[60,82] A recent study from our group unveiled that TRPM2 triggers dysfunction in cerebral endothelial cells (CECs) by interacting with the scavenger receptor CD36.[68] Our findings revealed that the increased endothelial permeability induced by OGD and the CD36 ligand thrombospondin-1 (TSP1) was significantly reduced by TRPM2 knockout or inhibition.[68]

At the animal level, the specific knockout of TRPM2 in endothelial cells was proven to be sufficient to protect mice against ischemic stroke, which coincided with reduced BBB degradation and mitigated immune cell infiltration.[68] Our data strongly suggest that TRPM2 serves as a major contributor to CEC damage and BBB disruption during ischemic stroke. Additionally, TRPM2-mediated Ca^{2+} influx is essential for the normal activation and migration of immune cells.[59,60,83–89] The impact of TRPM2 on immune cells during stroke has also been demonstrated through bone marrow transplantation experiments in mice. Substituting the bone marrow in wild-type mice with bone marrow from TRPM2 knockout mice attenuated the infiltration of neutrophils and macrophages into the brain after MCAO.[90] Conversely, transplanting bone marrow from wild-type mice into TRPM2 knockout mice abolished the protective effect observed in TRPM2 knockout against ischemic stroke.[90] However, further research using immune cell-specific TRPM2 knockout models is imperative to definitively confirm the direct involvement of immune-TRPM2 in ischemic brain injury.

10.2.1.2 TRPM4 in Ischemic Stroke

Unlike other TRP channels (except TRPM5), TRPM4 is unique due to its monovalent selectivity.[91–93] Similar to TRPM2, the activation of TRPM4 also depends on intracellular Ca^{2+} and phospholipids.[91,92,94] While extracellular nucleotides inhibit TRPM4 activation, phosphorylation by protein kinase C (PKC) promotes it.[95] TRPM4 is highly expressed in the heart and kidneys, and mutations in TRPM4 cause heart block in humans.[57,96] In the brain, TRPM4 is broadly expressed in neurons, glial cells, and endothelial cells, and it has been implicated in exacerbating ischemic brain injury. Despite TRPM4's impermeability to cytotoxic divalent cations like Ca^{2+} and Zn^{2+}, it is unlikely that its detrimental effects in ischemic stroke are directly caused by Na^+ or K^+. Indeed, studies demonstrated that TRPM4 requires an "evil" partner, such as sulfonylurea receptor (SUR1) and NMDAR, as a mediator for inducing cellular dysfunction (Fig. 10.2).

TRPM4 has been found to physically interact with SUR1 in human brain tissue, and both TRPM4 and SUR1 expression levels increase after ischemic stroke.[97,98] Subsequent studies revealed that ischemia enhanced the surface expression and activation of TRPM4-SUR1 specifically in perivascular astrocytic endfeet.[99] This activation led to Na^+ influx, triggering the activation of the Na^+/Ca^{2+} exchanger NCX1 and resulting in Ca^{2+} influx. This Ca^{2+} influx then promoted the calmodulin (CaM)-dependent joining of aquaporin 4 (AQP4) into the TRPM4-SUR1 complex.[99] Activation of AQP4 promotes astrocyte swelling and cerebral edema following an ischemic stroke[99,100] (Fig. 10.2). Pharmacologically inhibiting the SUR1-TRPM4 complex through specific oligodeoxynucleotides or astrocyte-specific deletion of SUR1-TRPM4 or NCX1 has shown promise in reducing cerebral edema and preserving brain function following MCAO in mice.[99,100] This protective effect appears to be independent of infarct size.[99] However, the precise impact of the TRPM4-SUR1-AQP4 coupling on neuronal swelling, neuronal death, or endothelial leakage remains uncertain.[99] To address this, further investigation using in vitro OGD in isolated neurons and CECs using SUR1-TRPM4 complex inhibitors or *in vivo* MCAO with cell-type-specific TRPM4 knockout mice will provide valuable insights.

Recently, a study showed that TRPM4 enhanced ischemic excitotoxicity by physically associating with esNMDAR.[44] Mechanistically, although the coupling with TRPM4 does not enhance the channel function of esNMDAR, it promotes the activation of downstream pro-death CREB and ERK signaling.[44] Uncoupling of TRPM4-esNMDAR association using small chemical compounds protects mice against ischemic stroke.[44] However, the precise mechanism by which TRPM4 influences esNMDAR-CREB/ERK signaling during stroke remains unclear.

Besides functioning in neuronal cells, TRPM4 in endothelial cells also contributes to ischemic brain injury by regulating BBB degradation and angiogenesis.[101,102] In a rat permanent MCAO model, increased expression of TRPM4 was observed in endothelial cells within the penumbra areas as early as 2 hours after ischemia.[101,102] Remarkably, knockdown of TRPM4 by siRNA resulted in several positive outcomes after ischemic stroke, including enhanced angiogenesis and preservation of endothelial integrity.[101,102] These changes were accompanied by a substantial reduction in cerebral edema and brain infarction and improvements in motor function.

The exploration of TRPM4 in immune cell infiltration during stroke is lacking. While TRPM4 itself does not directly facilitate Ca^{2+} influx due to its Ca^{2+}-impermeability, recent research has indicated that TRPM4 may indirectly elevate intracellular Ca^{2+} levels through other mechanisms, such as interaction with SUR1 and the Na^{+}/Ca^{2+} exchangers. Considering TRPM4 has been shown to regulate immune cell functions,[103,104] there is a possibility that TRPM4 might contribute to immune cell invasion during stroke, given that Ca^{2+} influx is crucial for immune cell activation. Further studies are needed to elucidate the potential involvement of TRPM4 in post-stroke immune cell infiltration. Still, additional studies using cell-type-specific TRPM4 deletion mice will be essential to confirm the contribution of endothelial and neuronal TRPM4 in ischemic stroke.

10.2.1.3 TRPM7 in Ischemic Stroke

TRPM7 is a Ca^{2+}-permeable channel-kinase required for early embryonic development and organogenesis.[105–107] TRPM7 has two unique features among all the ion channels. TRPM7 possesses a fully functional α-kinase domain in its CTD, crucial for various cellular processes like insulin production, neutrophil recruitment, T-cell colonization, and thrombus formation.[108–111] Intriguingly, the kinase domain of TRPM7 can be cleaved and translocated to the nucleus, regulating chromatin status.[112] Although having significant biological significance, the possession of a functional kinase domain makes it challenging to separately study the contribution of TRPM7's kinase function and channel function in different animal models, due to the lack of channel-dead TRPM7 knock-in mice. Another distinctive feature of TRPM7 is its high permeability to many trace ions, such as Zn^{2+}.[113] While it plays a crucial role in Mg^{2+} homeostasis, the pathophysiological significance of TRPM7-mediated Zn^{2+} signaling remains less clear.[114–116] The channel function of TRPM7 requires phospholipids and is regulated by intracellular Mg^{2+}, while the regulation of its kinase function remains enigmatic.[117,118] Recent studies have suggested that ARL15, a small G-protein associated with the Mg^{2+} transporter CNNM3, negatively regulates TRPM7, although its biological significance remains unclear.[119–122] Additionally, TRPM7 activity can be influenced by extracellular protons and mechanical stimuli, such as cell swelling.[123,124] Considering the pivotal roles of acidosis and cell swelling in ischemia-related diseases, TRPM7 is very likely to be involved in these conditions. Ubiquitously expressed in different cells, TRPM7 is involved in various cellular functions such as proliferation, differentiation, migration, apoptosis, receptor endocytosis, efferocytosis, chromatin remodeling, and metabolism.[112,125–132]

TRPM7 was the first TRP channel reported in 2003 to be involved in ischemic neuronal death[133] (Fig. 10.2). This pioneering study demonstrated that knockdown of TRPM7 using siRNA mitigated OGD-induced Ca^{2+} overload and reactive oxygen species production in primary cortical neurons.[133] Subsequent investigations revealed that *in vivo* TRPM7 knockdown in CA1 hippocampal neurons by shRNA reduced neuronal death in a rat model of global cerebral ischemia, accompanied by preserved performance in neurobehavior tests.[134] More recently, TRPM7 knockout was found to protect mice against ischemic stroke in a transient MCAO model.[135] Interestingly, selective knockout of TRPM7 in parvalbumin (PV) gamma-aminobutyric acid (GABAergic) neurons was shown to produce a better protective effect compared to the knockout in Ca^{2+}/calmodulin-kinase II (CaMKII) glutamatergic neurons.[135] Initially, TRPM7-induced Ca^{2+} overload was presumed to be the primary cause of its detrimental effects in stroke. However, a subsequent study revealed that TRPM7 activation also enhances Zn^{2+}-induced injury in primary mouse cortical neurons subjected to OGD.[136] This injury was ameliorated by TRPM7 inhibition using Gd^{3+} or 2-aminoethoxydiphenyl borate, or knockdown using siRNA.[136] Despite these findings, whether TRPM7-mediated Zn^{2+} signaling directly contributes to ischemic stroke *in vivo* remains unknown. Additionally, similar to TRPM2, it is challenging to completely differentiate the contributions of Ca^{2+} and Zn^{2+} in ischemic neuronal death.

The early studies primarily investigated TRPM7's channel function in ischemic neuronal death. More recently, the role of TRPM7's kinase function in ischemic brain injury has also been elucidated. Rather than directly inducing neuronal death, TRPM7's kinase activity has been found to promote platelet activation and thrombus formation by influencing the phosphorylation of spleen tyrosine kinase (Syk) and phospholipase C (PLC) $\gamma 2$ and $\beta 3$, thus exacerbating brain injury after MCAO.[111] Mice with kinase-dead TRPM7 knock-in or transplantation of bone marrow from kinase-dead TRPM7 knock-in mice exhibited a significant defect in platelet activation, leading to a reduction in infarct size.[111] Interestingly, TRPM7 kinase promotes Ca^{2+} mobilization by enhancing phosphatidylinositol-4,5-bisphosphate (PIP_2) metabolism.[111] However, the direct phosphorylation of Syk and PLC by TRPM7 kinase remains undetermined. TRPM7 also plays a critical role in the function of endothelial cells and immune cells.[132,137,138] Future *in vivo* experiments using cell-type specific TRPM7 knockout mice will provide useful information regarding the contribution of endothelial and immune-TRPM7 in stroke.

10.2.2 TRPC Channels in Ischemic Stroke

The nomenclature for this family of TRP channels originates from the term "canonical", as TRPC1 being the first identified mammalian TRP channel (originally named TRP1).[31–33] Among mammals, TRPCs are encoded by seven genes, with TRPC2 being a pseudogene in humans.[139] Consequently, the human TRPC family comprises six members (TRPC1, TRPC3, TRPC4, TRPC5, TRPC6, TRPC7).[139] The TRPC family can be categorized into four groups (TRPC1, TRPC2, TRPC3/6/7, and TRPC4/5) based on sequence and functional similarities.[139] A distinguishing feature of TRPCs, setting them apart from other TRP channels, is their inclination to form heterotetramers, especially TRPC1, which is almost non-functional when overexpressed alone in cell lines.[140,141] Under physiological conditions, TRPC1 usually exists in heterotetramers with TRPC4 and TRPC5.[142–145] Other TRPCs can form functional homotetramers when expressed alone, with TRPC3 and TRPC6 appearing to favor the formation of homotetramers *in vivo*.[146] This unique attribute poses a particular challenge in individually characterizing TRPCs in primary cells and animals. TRPCs exhibit broad expression in nearly all cell types and tissues, including neurons, glial cells, immune cells, and endothelial cells, with varying levels of abundance.[34,35,147–150] TRPCs can be activated by a small number of natural and synthesized chemicals, and often function downstream from G protein–coupled receptors (GPCRs) and their effector signaling, especially phospholipase C (PLC)-IP_3/Ca^{2+} pathway.[146,151,152] All TRPC members have exhibited involvement in ischemic stroke, displaying diverse cellular and molecular mechanisms (Fig. 10.3).

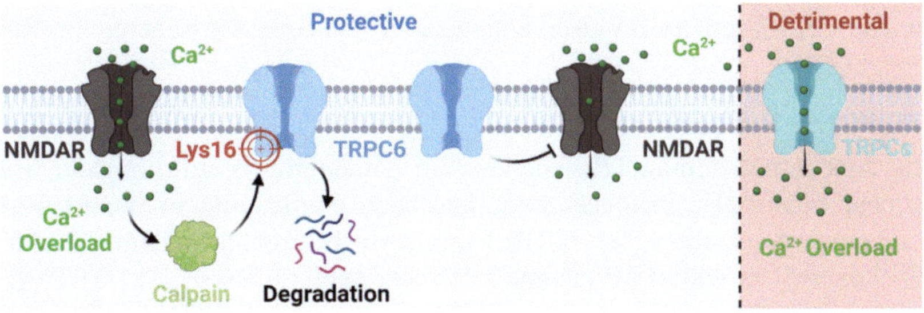

FIGURE 10.3 TRPCs in stroke. TRPC6 has been observed to potentially inhibit NMDAR activity, which could account for the protective effect associated with TRPC6 overexpression in stroke. During stroke, NMDAR activation leads to calcium overload, subsequently activating calpain. This calpain activation cleaves TRPC6 specifically at Lys16. Despite this, the Ca^{2+} influx mediated by TRPCs may contribute to the exacerbation of ischemic neuronal death.

10.2.2.1 TRPC3, TRPC6, and TRPC7 in Ischemic Stroke

TRPC3, TRPC6, and TRPC7 share structural similarity and have the propensity to form functional heterotetramers *in vivo*. These channels play significant roles in regulating neural survival, nerve growth, synaptic transmission, and glutamate signaling.[153–159] The exact involvement of TRPC3, TRPC6, and TRPC3/6/7 in ischemic stroke remains controversial, with studies suggesting both protective and detrimental effects.

Functional interactions between TRPC6 and NMDARs have been extensively investigated.[140,160] In a rat model of transient MCAO and primary cortical neurons subjected to OGD, ischemia triggered rapid TRPC6 degradation in a calpain proteinase-dependent manner, preceding neuronal death.[161] Activation of TRPC3/6 by 1-oleoyl-2-acetyl-sn-glycerol (OAG) mitigated OGD-induced neuronal death, while inhibition of TRPC6 using SKF96365 or TRPC6 knockdown exacerbated it.[161] Blocking NMDARs with MK801 or GluN1A knockdown prevented OGD-induced TRPC6 degradation.[161] Mechanistically, NMDAR-mediated Ca^{2+} influx activated calpain, which cleaved TRPC6 at Lys16.[161] An interfering peptide, TAT-C6, derived from the calpain cleavage site selectively prevented TRPC6 degradation and reduced infarct size, enhancing behavioral performance in rats post-MCAO.[161] The follow-up study demonstrated that TRPC6 transfection protected hippocampal neurons against NMDA and OGD-induced Ca^{2+} overload and cell death, while expression of a dominant-negative TRPC6 mutant abolished this protective effect.[160] TRPC6 overexpression or knockdown did not affect NMDAR expression, suggesting that TRPC6 functions downstream of NMDARs.[160] Furthermore, TRPC6 transgenic mice exhibited resistance to ischemic brain injury.[160] Consistent with these two studies, natural compounds like hyperforin and calycosin were found to reduce brain infarction and ischemic neuronal death by inhibiting NMDAR-induced TRPC6 degradation.[162,163] In another study, TRPC6 was also shown to inhibit NMDAR activity. TRPC6 overexpression or OAG suppressed, while TRPC6 knockdown or SKF96365 enhanced NMDA current in primary hippocampal neurons.[164] Collectively, these findings suggest a mutual regulation between TRPC6 and NMDAR and indicate a protective role of TRPC6 against NMDAR-mediated ischemic excitotoxicity and neuronal death.

Two other studies suggested a detrimental role of TRPC3/6/7 in stroke. A triple knockout of TRPC3, TRPC6, and TRPC7 attenuated ischemic brain injury in mice by inhibiting NF-κB phosphorylation, preventing Ca^{2+} overload, and reducing apoptosis in astrocytes exposed to OGD. Another study found increased TRPC6 expression in cortical neurons after MCAO and primary neurons following OGD.[165] Overexpressing TRPC6 via adenovirus enhanced glutamate-induced Ca^{2+} overload in primary neurons, while TRPC6 knockout attenuated Ca^{2+} overload *in vitro* and ischemic brain injury *in vivo*.[166] There remain two major unresolved problems in the study of TRPC3, TRPC6, and TRPC7 in stroke. Firstly, the absence of TRPC6 expression data in human brain tissue post-stroke diminishes the translational relevance of these findings. Secondly, the lack of ionic current recording in primary cells or brain slices makes it challenging to definitively establish whether TRPC3/6/7 activity is indeed enhanced or reduced after ischemic insults.

10.2.2.2 TRPC1 in Ischemic Stroke

TRPC1 is the first cloned human TRP channel.[31–33] Although it is still not clear whether TRPC1 forms functional homomeric channels as it has been challenging to record TRPC1 currents even in overexpressed cell lines,[140,141] TRPC1 can form heteromeric channels with other TRPC members,[142–145] as well as with TRP channels of other subfamilies such as TRPV4, TRPV6, and TRPP1.[140] Abundantly expressed in neurons, TRPC1 plays a role in growth cone regulation.[153,167,168] Activation of TRPC1 via mGluR1/5 secondary to the (PLC)-IP_3/Ca^{2+} pathway suggests its potential involvement in excitotoxicity.[169–171] Indeed, TRPC1-mediated Ca^{2+} influx has been implicated in modulating neuronal death in various excitotoxicity-related diseases in mice, such as seizures, Alzheimer's disease, and ischemic stroke.[140,172] In hippocampal organotypic slice cultures, glutamate treatment induced a rapid increase in TRPC1 expression and extensive cell death, which was inhibited by TRPC1 knockdown, suggesting a detrimental role of TRPC1 in glutamate excitotoxicity.[173] However, conflicting data emerged that HT22 cells showed reduced TRPC1 expression when subjected to OGD.[172] Knockdown of TRPC1 enhanced reactive oxygen species generation, while TRPC1 overexpression produced an opposite effect.[172] Notably, TRPC1 knockout in mice aggravated ischemic brain injury and resulted in worsened neurobehavioral performance, indicating a protective role of TRPC1 in ischemic brain injury.[172] Nevertheless, the absence of in vitro data using cultured primary neurons makes it challenging to distinguish the direct effect of TRPC1 on ischemic neuronal death.

10.2.2.3 TRPC4 and TRPC5 in Ischemic Stroke

TRPC4 and TRPC5 are close homologues and usually form heterotetramers with TRPC1 in the brain.[174] TRPC4 and TRPC5 are important in many neural activities, such as synaptic transmission, neurite extension, and the normal function of the amygdala.[169–171,174–178] TRPC4 can be activated secondary to NMDAR, suggesting its potential role in excitotoxicity.[169–171,176] However, two recent studies yielded conflicting results regarding the role of TRPC5 in ischemic neuronal death. In a rat model of spinal cord ischemia-reperfusion injury, overexpression of TRPC5 using an adeno-associated virus was observed to reduce inflammatory responses and promote angiogenesis in the spinal cord.[179] Although neuronal injury was not specifically evaluated, this study suggests a protective role of TRPC5 in neuronal survival under ischemic conditions.[179] Conversely, another study revealed that TRPC5 physically associates with phospholipid scramblase 1 (PLSCR1) in mouse cortical neurons, through which TRPC5 promotes phosphatidylserine (PS) externalization in a model of regional cerebral ischemia-reperfusion.[180] Notably, TRPC5 knockout was shown to mitigate ischemic brain injury and neuron apoptosis, leading to a reduction in PS externalization.[180] This study implicates a detrimental role of TRPC5 in ischemic stroke. In a more recent investigation, knockout of TRPC4 or the quadruple knockout of TRPC1/TRPC4/TRPC5/TRPC6 significantly reduced brain infarction in mice following ischemic stroke.[34] TRPC4 knockout exhibited a similarly protective effect against ischemia-induced neuronal death when compared to TRPC1 knockout and the quadruple knockout of TRPC1/TRPC4/TRPC5/TRPC6.[34] This study suggests an intriguing possibility that, rather than forming homotetramers, TRPC1, TRPC4, and TRPC5 may assemble into a heterotetramer complex in the brain, constituting the TRPC1/4/5 channels, and activation of the heterotetrameric TRPC1/4/5 channels exacerbates ischemic brain injury. This finding is pivotal for understanding the physiological constituents of TRPC channels, especially the non-functional TRPC1 homotetramer.[174] However, comprehensive staining and ionic current recording data will be necessary to confirm the presence of the TRPC1/4/5 heterotetramer complex in the brain and the enhanced channel activity in neurons under ischemic conditions.

10.2.3 TRPV Channels in Ischemic Stroke

TRPVs are well known for TRPV1's capability to respond to noxious heat and capsaicin in chili peppers. This unique characteristic led to the nomenclature of this family of TRP channels, as capsaicin is a vanilloid.[181,182] There are six members in the TRPV family (TRPV1-6), which can be subgrouped as TRPV1/V3, TRPV2, TRPV4, and TRPV5/V6. In mammals, TRPVs primarily serve as mechanical, chemical, and thermal sensors. TRPV1, TRPV2, TRPV3, and TRPV4 function as non-selective cation channels with higher permeability to divalent ions, particularly Ca^{2+}, than to monovalent ions, and they are activated by temperature increases with different temperature thresholds. On the other hand, TRPV5 and TRPV6, while not exhibiting thermosensitivity, display the highest selectivity to Ca^{2+} among all TRP channels.[183,184] Like TRPCs and TRPMs, the function of TRPVs is modulated by phospholipids. This regulation not only substantially reduces the temperature threshold, to even below normal body temperature,[185] but also contributes to stabilizing the channel pore, preventing rapid desensitization of TRPVs to various agonists and stimuli.[186,187] Moreover, TRPVs are also regulated by CaM.[188] Additionally, TRPVs can be activated by eicosanoids and arachidonic acid derivatives such as prostaglandins, leukotrienes, and thromboxanes.[188] The production of these compounds and membrane phospholipid levels are elevated during inflammation; TRPVs are expected to exhibit significantly enhanced activity under inflammatory conditions. Expressed prominently in neurons, TRPVs are also broadly present in various other cell types, including glial cells, endothelial cells, and immune cells.[148,189,190] TRPV1, TRPV2, TRPV3, and TRPV4 have demonstrated involvement in ischemic brain injury.

10.2.3.1 TRPV1 in Ischemic Stroke

TRPV1 is primarily recognized for its heat-sensitivity in dorsal root ganglion (DRG) neurons, a discovery that earned a Nobel Prize in 2021 to David Julius.[181,182] Subsequently, TRPV1 was found to be widely expressed in all cell types in the central nervous system.[191–196] Beyond responding to noxious heat, TRPV1 is activated by endogenous eicosanoids and protons, whose production significantly increases during inflammatory conditions, which accompany ischemia.[22,197] Multiple studies have affirmed TRPV1's contribution to ischemic neuronal death. TRPV1 knockout and intracerebroventricular injection of the TRPV1 antagonist capsazepine have demonstrated a reduction in infarct size and improved behavioral scores in mouse transient MCAO models.[198] Meanwhile, TRPV1 knockout did not diminish immune cell invasion and gliosis in the ischemic region, indicating a direct involvement of neuronal TRPV1 in stroke.[198] A subsequent study showed that a more specific TRPV1 inhibitor, AMG9810, exhibited a protective effect against ischemic brain injury in a rat permanent MCAO model.[199] This effect was accompanied by increased levels of the anti-inflammatory IL-10 and pro-inflammatory TNFα, suggesting that TRPV1 activation intensifies local inflammation during stroke.[199] However, the TRPV1 agonist capsaicin did not affect the severity of ischemic brain injury.[199] A recent study also showed that TRPV1 knockout reduces brain damage in a mouse hypoxic-ischemic encephalopathy (HIE) model by decreasing the activation of NLRP3-IL1β inflammasome signaling in astrocytes.[200] Additionally, evidence suggests that TRPV1 promotes microglia activation under varying inflammatory conditions, hinting at the potential contribution of microglial TRPV1 to ischemic brain injury.[201–203]

In contrast to the detrimental effects of TRPV1 in ischemic stroke, several studies strongly suggest its protective role. The thermosensitive nature of TRPV1 positions it as a crucial regulator of core body temperature,[204,205] an aspect that could hold significant relevance in stroke.[206] Body temperature exerts influence over metabolic rates, a pivotal factor in managing metabolic imbalances in brain tissue during stroke episodes,[206] Therapeutic hypothermia has shown promise in treating brain injuries, including stroke.[206] Evidence supporting this protective role comes from studies demonstrating the impact of TRPV1 agonist-induced hypothermia.[207–209] Subcutaneous infusion of the TRPV1 agonist dihydrocapsaicin significantly reduced ischemic brain injury in a mouse transient MCAO model, concurrently inducing a notable reduction in body temperature. Critically, this protection was absent in TRPV1 knockout mice and in mice maintained under normal body temperature conditions, suggesting that TRPV1 activation mitigates ischemic stroke by inducing hypothermia.[207–209] Consistent with these findings, intraperitoneal or intracerebroventricular injection of another TRPV1 agonist, rinvanil, showed a reduction in brain infarction in mice.[208] This protective effect was also absent in normothermic mice.[208] Moreover, administration of dihydrocapsaicin post-stroke at 3.5 hours rapidly induced hypothermia, reduced both primary cortical injury and secondary thalamic injury, and exhibited prolonged benefits in terms of functional recovery.[207–209] The protective effects of TRPV1 activation may also mitigate excitotoxicity, as TRPV1 activator capsaicin inhibits the expression of NMDAR and alleviates glutamate-induced neuronal death in cultured primary neurons.[210] TRPV1 is also important in regulating endothelial permeability and immune cell activation, but the roles of endothelial and immune TRPV1 in stroke remain unexamined.[189,211–213]

10.2.3.2 TRPV4 in Ischemic Stroke

Compared to TRPV1, TRPV4 has a much lower temperature threshold (24~34°C).[214] TRPV4 is abundantly expressed in neurons, astrocytes, microglia, and CECs, and it regulates brain homeostasis by influencing endoplasmic reticulum stress, glial activation, and BBB integrity.[215–219] A significant feature of TRPV4 in the context of stroke is its activation by mechanical stretch due to an increase in cell volume, as cell swelling is an important cause of ischemic brain injury.[220–222] TRPV4 activation via hypotonic stimulation or agonist led to increased NMDAR activity in primary hippocampal CA1 neurons, which was inhibited by the TRPV4 antagonist HC-067047.[223] The potentiation of NMDAR activity by TRPV4 seems contingent upon the phosphorylation of GluN2B by CaMKII, as inhibition of GluN2B and CaMKII abolished this potentiation.[223] Moreover, intracerebroventricular administration of HC-067047 resulted in reduced infarct size after MCAO, implicating the role of TRPV4 in ischemic excitotoxicity.[223] In another study, TRPV4 agonists GSK1016790A, 4α-PDD, and 5,6-EET were found to inhibit GABA receptor activity in CA1 pyramidal neurons.[224] This inhibition on GABA receptors by TRPV4 might contribute to neural hyperexcitability during stroke. Furthermore, TRPV4-mediated Ca^{2+} influx has been shown to promote hypoxia-induced neurotoxicity and apoptosis in a neuronal cell line.[225] These findings collectively suggest that the activation of neuronal TRPV4 could exert detrimental effects in stroke. Glial TRPV4 may also contribute to ischemic brain injury. TRPV4 is expressed in about 30% of astrocytes, and both the expression and activity of TRPV4 in astrocytes are enhanced after ischemia.[226–228] Under normal conditions, TRPV4 promotes glutamate release in astrocytes to regulate synaptic transmission.[226] However, this enhanced release of astrocytic glutamate may further aggravate excitotoxicity during stroke. Moreover, TRPV4 is critical for many normal functions of microglia, such as activation, proliferation, and temperature-dependent motility.[229–231] Furthermore, TRPV4-mediated Ca^{2+} influx regulates the function of peripheral immune cells and CECs.[149,219,232–234] In a spinal cord injury model, TRPV4 activation led to endothelial hyperpermeability and blood-spinal cord barrier degradation.[235] However, more studies are needed to confirm the involvement of glial, endothelial, and immuno-TRPV4 in stroke.

10.2.3.3 TRPV2 and TRPV3 in Ischemic Stroke

In comparison to TRPV1 and TRPV4, the roles of TRPV2 and TRPV3 in stroke remain less understood. TRPV2 exhibits a higher temperature threshold (> 53°C) than TRPV1 and plays a role in mechanical nociception in DRG neurons.[236–238] Within the brain, TRPV2 is involved in regulating the production of nerve growth factor (NGF) and its downstream signaling pathways.[239,240] In astrocytes exposed to OGD, inhibiting TRPV2 has been observed to increase NGF secretion, suggesting a potential benefit in stroke management.[241] Furthermore, cannabidiol, a cannabinoid compound, has been shown to activate TRPV2, promoting microglial phagocytosis and preserving endothelial integrity in human CECs.[242,243] These effects could potentially contribute to the beneficial effects of cannabinoid supplements observed in stroke patients.[244]

TRPV3, with a temperature threshold similar to TRPV4 (ranging from 24 to 34°C) and prominent expression in skin keratinocytes, also exhibits widespread expression in the brain.[214,245–247] Studies exploring TRPV3's role in ischemic brain injury present conflicting observations. On one hand, incensole acetate has been reported to mitigate ischemic brain injury in mice by activating TRPV3.[248] This effect is attributed to the inhibition of NF-κB activation and a subsequent reduction in local inflammation, suggesting a potentially protective role of TRPV3 in stroke.[248] However, contradictory findings emerge from another study where TRPV3 knockout or the TRPV3 inhibitor forsythoside B significantly reduced ischemic stroke.[249] This reduction correlated with decreased neuronal excitability and excitatory synaptic transmissions.[249] Also, the overexpression of TRPV3 in wild-type mice exacerbated ischemic brain injury, while TRPV3 rescue in TRPV3 knockout mice reversed the protective effects.[249] Additional comprehensive investigations will be needed to elucidate the precise roles of TRPV2 and TRPV3 in stroke pathophysiology.

10.2.4 TRPA1 in Ischemic Stroke

TRPA1 (previously known as ANKTM1),[250] the only member of the mammalian ankyrin TRP subfamily, is an oxidant-sensitive TRP channel that has been implicated in nociception[251] and inflammatory diseases.[252] Its expression spans various cell types, including sensory neurons, endothelial cells, and perivascular nerves. A recent study discovered that hypoxia triggers the activation of TRPA1 in cerebral endothelium, leading to TRPA1-dependent vasodilation. This mechanism appears to protect against ischemic stroke by reducing infarct size in wild-type mice. Remarkably, the endothelial cell-specific knockout of TRPA1 blunted the observed reduction in infarct size post-ischemic stroke, underscoring the protective role of endothelial TRPA1.[253]

10.3 CONCLUDING REMARKS AND PERSPECTIVES

Stroke continues to impose a significant public health burden, with a complete lack of effective pharmacological treatments for this debilitating disease. Studies conducted since 2003 on TRP channels in stroke have shown promise, suggesting that the TRP channels could serve as potential therapeutic targets for stroke treatment. While the exploration of TRP channels in hemorrhagic stroke is limited, the mechanism of neuronal death in this type of stroke resembles those observed in ischemic stroke. Consequently, it is plausible to anticipate similar animal phenotypes in hemorrhagic stroke models as observed in ischemic stroke models. Further studies utilizing hemorrhagic stroke mouse models should be conducted to thoroughly investigate the role of TRP channels in this less common but more severe condition.

It appears that the activation of TRPMs during ischemic stroke consistently leads to harmful effects. The lack of specific inhibitors for TRPMs has persisted due to our limited understanding of the channel structure, an issue that is being resolved with the advancement of cryo-EM studies. Given that TRPMs share highly conserved MHR domains not seen in other TRP channels, there is potential to develop a universal TRPM inhibitor by specifically targeting these MHR domains. A comprehensive inhibition strategy that simultaneously targets TRPM2, TRPM4, and TRPM7 may potentially offer a more effective protective effect against ischemic stroke compared to inhibiting these channels individually.

The effects of TRPCs and TRPVs in ischemic stroke are mixed and somewhat controversial, in contrast to the clearly demonstrated detrimental effects of TRPMs. This variability could stem from differences in their expression levels, heteromerization, and diverse roles in various cell types. For instance, while TRPC and TRPV channels might exhibit a protective effect in one cell type, they could potentially be detrimental in another, making it challenging to assess their overall impact on ischemic stroke and resulting in conflicting results. The use of cell-type-specific knockout mice and the application of novel techniques like single-cell RNA sequencing/proteomics hold promise in providing a deeper understanding of the complex roles played by TRPCs and TRPVs in ischemic stroke.

REFERENCES

1. Campbell, B. C. V., De Silva, D. A., Macleod, M. R., Coutts, S. B., Schwamm, L. H., Davis, S. M., Donnan GA. Ischaemic stroke. *Nat Rev Dis Primers*, 2019, 5: 70. doi:10.1038/s41572-019-0118-8
2. Tsao, C. W. et al. Heart disease and stroke statistics-2022 update: A report from the American heart association. *Circulation*, 2022, 145: e153–e639. doi:10.1161/CIR.0000000000001052
3. Campbell, B. C. V., Khatri, P. Stroke. *Lancet*, 2020, 396: 129–142. doi:10.1016/S0140-6736(20)31179-X
4. Lai, T. W., Zhang, S., Wang, Y. T. Excitotoxicity and stroke: Identifying novel targets for neuroprotection. *Prog Neurobiol*, 2014, 115: 157–188. doi:10.1016/j.pneurobio.2013.11.006
5. Hansen, K. B., Wollmuth, L. P., Bowie, D., Furukawa, H., Menniti, F. S., Sobolevsky, A. I., Swanson, G. T., Swanger, S. A., Greger, I. H., Nakagawa, T., McBain, C. J., Jayaraman, V., Low, C. M., Dell'Acqua, M. L., Diamond, J. S., Camp, C. R., Perszyk, R. E., Yuan, H., Traynelis, S. F. Structure, function, and pharmacology of glutamate receptor ion channels. *Pharmacol Rev*, 2021, 73: 298–487. doi:10.1124/pharmrev.120.000131

6. Dong, X. X., Wang, Y., Qin, Z. H. Molecular mechanisms of excitotoxicity and their relevance to pathogenesis of neurodegenerative diseases. *Acta Pharmacol Sin*, 2009, 30: 379–387. doi:10.1038/aps.2009.24
7. Belov Kirdajova, D., Kriska, J., Tureckova, J., Anderova, M. Ischemia-triggered glutamate excitotoxicity from the perspective of glial cells. *Front Cell Neurosci*, 2020, 14: 51. do:10.3389/fncel.2020.00051
8. Parfenova, H., Basuroy, S., Bhattacharya, S., Tcheranova, D., Qu, Y., Regan, R. F., Leffler, C. W. Glutamate induces oxidative stress and apoptosis in cerebral vascular endothelial cells: Contributions of HO-1 and HO-2 to cytoprotection. *Am J Physiol Cell Physiol,* 2006, 290: C1399–C1410. https://doi.org:10.1152/ajpcell.00386.2005
9. Traynelis, S. F., Wollmuth, L. P., McBain, C. J., Menniti, F. S., Vance, K. M., Ogden, K. K., Hansen, K. B., Yuan, H., Myers, S. J., Dingledine, R. Glutamate receptor ion channels: Structure, regulation, and function. *Pharmacol Rev*, 2010, 62: 405–496. doi:10.1124/pr.109.002451
10. Qin, C., Yang, S., Chu, Y. H., Zhang, H., Pang, X. W., Chen, L., Zhou, L. Q., Chen, M., Tian, D. S., Wang, W. Signaling pathways involved in ischemic stroke: Molecular mechanisms and therapeutic interventions. *Signal Transduct Target Ther*, 2022, 7: 215. doi:10.1038/s41392-022-01064-1
11. Hardingham, G. E., Fukunaga, Y., Bading, H. Extrasynaptic NMDARs oppose synaptic NMDARs by triggering CREB shut-off and cell death pathways. *Nat Neurosci*, 2002, 5: 405–414. doi:10.1038/nn835
12. Chen, M., Lu, T. J., Chen, X. J., Zhou, Y., Chen, Q., Feng, X. Y., Xu, L., Duan, W. H., Xiong, Z. Q. Differential roles of NMDA receptor subtypes in ischemic neuronal cell death and ischemic tolerance. *Stroke*, 2008, 39: 3042–3048. doi:10.1161/STROKEAHA.108.521898
13. Ikonomidou, C., Turski, L. Why did NMDA receptor antagonists fail clinical trials for stroke and traumatic brain injury? *Lancet Neurol*, 2002, 1: 383–386. doi:10.1016/s1474-4422(02)00164-3
14. Wu, Q. J., Tymianski, M. Targeting NMDA receptors in stroke: New Hope in neuroprotection. *Mol Brain*, 2018, 11: 15. doi:10.1186/s13041-018-0357-8
15. Xu, S., Lu, J., Shao, A., Zhang, J. H., Zhang, J. Glial cells: Role of the immune response in ischemic stroke. *Front Immunol*, 2020, 11: 294. doi:10.3389/fimmu.2020.00294
16. Bernardo-Castro, S., Sousa, J. A., Brás, A., Cecília, C., Rodrigues, B., Almendra, L., Machado, C., Santo, G., Silva, F., Ferreira, L., Santana, I., Sargento-Freitas, J. Pathophysiology of blood-brain barrier permeability throughout the different stages of ischemic stroke and its implication on hemorrhagic transformation and recovery. *Front Neurol*, 2020, 11: 594672. doi:10.3389/fneur.2020.594672
17. Beuker, C., Strecker, J. K., Rawal, R., Schmidt-Pogoda, A., Ruck, T., Wiendl, H., Klotz, L., Schäbitz, W. R., Sommer, C. J., Minnerup, H., Meuth, S. G., Minnerup, J. Immune cell infiltration into the brain after ischemic stroke in humans compared to mice and rats: A systematic review and meta-analysis. *Transl Stroke Res*, 2021, 12: 976–990. doi:10.1007/s12975-021-00887-4
18. Hollist, M., Morgan, L., Cabatbat, R., Au, K., Kirmani, M. F., Kirmani, B. F. Acute stroke management: Overview and recent updates. *Aging Dis*, 2021, 12: 1000–1009. doi:10.14336/AD.2021.0311
19. Mendelson, S. J., Prabhakaran, S. Diagnosis and management of transient ischemic attack and acute ischemic stroke: A review. *JAMA*, 2021, 325: 1088–1098. doi:10.1001/jama.2020.26867
20. Nilius, B., Owsianik, G. The transient receptor potential family of ion channels. *Genome Biol*, 2011, 12: 218. doi:10.1186/gb-2011-12-3-218
21. Vazquez, E., Valverde, M. A. A review of TRP channels splicing. *Semin Cell Dev Biol*, 2006, 17: 607–617. doi:S1084-9521(06)00118-2
22. Koivisto, A. P., Belvisi, M. G., Gaudet, R., Szallasi, A. Advances in TRP channel drug discovery: From target validation to clinical studies. *Nat Rev Drug Discov*, 2022, 21: 41–59. doi:10.1038/s41573-021-00268-4
23. Cao, E. Structural mechanisms of transient receptor potential ion channels. *J Gen Physiol*, 2020, 152. doi:10.1085/jgp.201811998
24. Clapham, D. E. Calcium signaling. *Cell*, 2007, 131: 1047–1058. doi:10.1016/j.cell.2007.11.028
25. Vangeel, L., Voets, T. Transient receptor potential channels and calcium signaling. *Cold Spring Harb Perspect Biol*, 2019, 11: a035048. doi:10.1101/cshperspect.a035048
26. Nilius, B., Talavera, K., Owsianik, G., Prenen, J., Droogmans, G., Voets, T. Gating of TRP channels: A voltage connection? *J Physiol*, 2005, 567: 35–44. doi:10.1113/jphysiol.2005.088377
27. Xiao, R., Xu, X. Z. S. Temperature sensation: From molecular thermosensors to neural circuits and coding principles. *Annu Rev Physiol*, 2021, 83: 205–230. doi:10.1146/annurev-physiol-031220-095215
28. Venkatachalam, K., Montell, C. TRP channels. *Annu Rev Biochem*, 2007, 76: 387–417. doi:10.1146/annurev.biochem.75.103004.142819
29. Clapham, D. E. TRP channels as cellular sensors. *Nature*, 2003, 426: 517–524. doi:10.1038/nature02196
30. Yue, L., Xu, H. TRP channels in health and disease at a glance. *J Cell Sci*, 2021, 134. doi:10.1242/jcs.258372

31. Wes, P. D. TRPC1, a human homolog of a drosophila store-operated channel. *Proc Natl Acad Sci U S A*, 1995, 92: 9652–9656. doi:10.1073/pnas.92.21.9652
32. Zhu, X., Chu, P. B., Peyton, M., Birnbaumer, L. Molecular cloning of a widely expressed human homologue for the drosophila trp Gene. *FEBS Lett*, 1995, 373: 193–198. doi:10.1016/0014-5793(95)01038-g
33. Zitt, C., Zobel, A., Obukhov, A. G., Harteneck, C., Kalkbrenner, F., Lückhoff, A., Schultz, G. Cloning and functional expression of a human Ca^{2+}-permeable cation channel activated by calcium store depletion. *Neuron*, 1996, 16: 1189–1196. doi:10.1016/s0896-6273(00)80145-2
34. Jeon, J., Bu, F., Sun, G., Tian, J. B., Ting, S. M., Li, J., Aronowski, J., Birnbaumer, L., Freichel, M., Zhu, M. X. Contribution of TRPC channels in neuronal excitotoxicity associated with neurodegenerative disease and ischemic stroke. *Front Cell Dev Biol*, 2020, 8: 618663. doi:10.3389/fcell.2020.618663
35. Partida-Sanchez, S., Desai, B. N., Schwab, A., Zierler, S. Editorial: TRP channels in inflammation and immunity. *Front Immunol*, 2021, 12: 684172. doi:10.3389/fimmu.2021.684172
36. Zong, P., Lin, Q., Feng, J., Yue, L. A systemic review of the integral role of TRPM2 in ischemic stroke: From upstream risk factors to ultimate neuronal death. *Cells*, 2022, 11: 491. doi:10.3390/cells11030491
37. Rosenbaum, T., Morales-Lazaro, S. L., Islas, L. D. TRP channels: A journey towards A molecular understanding of pain. *Nat Rev Neurosci*, 2022, 23: 596–610. doi:10.1038/s41583-022-00611-7
38. Wen, H., Gwathmey, J. K., Xie, L. H. Role of transient receptor potential canonical channels in heart physiology and pathophysiology. *Front Cardiovasc Med*, 2020, 7: 24. doi:10.3389/fcvm.2020.00024
39. Uchida, K., Tominaga, M. The role of TRPM2 in pancreatic beta-cells and the development of diabetes. *Cell Calcium*, 2014, 56: 332–339. doi:10.1016/j.ceca.2014.07.001
40. Miller, B. A. TRPM2 in cancer. *Cell Calcium*, 2019, 80: 8–17. doi:10.1016/j.ceca.2019.03.002
41. Liu, L., Gu, L., Chen, M., Zheng, Y., Xiong, X., Zhu, S. Novel targets for stroke therapy: Special focus on TRPC channels and TRPC6. *Front Aging Neurosci*, 2020, 12: 70. doi:10.3389/fnagi.2020.00070
42. Xu, Q., Zou, Y., Miao, Z., Jiang, L., Zhao, X. Transient receptor potential ion channels and cerebral stroke. *Brain Behav*, 2023, 13: e2843. doi:10.1002/brb3.2843
43. Zong, P., Li, C. X., Feng, J., Cicchetti, M., Yue, L. TRP channels in stroke. *Neurosci Bull*, 2024, 40: 1141-1159. doi:10.1007/s12264-023-01151-5
44. Yan, J., Bengtson, C. P., Buchthal, B., Hagenston, A. M., Bading, H. Coupling of NMDA receptors and TRPM4 guides discovery of unconventional neuroprotectants. *Science*, 2020, 370: eaay3302. doi:10.1126/science.aay3302
45. Zong, P., Feng, J., Yue, Z., Li, Y., Wu, G., Sun, B., He, Y., Miller, B., Yu, A. S., Su, Z., Xie, J., Mori, Y., Hao, B., Yue, L. Functional coupling of TRPM2 and extrasynaptic NMDARs exacerbates excitotoxicity in ischemic brain injury. *Neuron*, 2022. doi:10.1016/j.neuron.2022.03.021
46. Zong, P., Li, C. X., Feng, J., Yue, L. Targeting TRPM2- and TRPM4-extrasynaptic N-methyl-D-aspartate receptor coupling in ischemic stroke. *Neural Regen Res*, 2023, 18: 2383–2384. doi:10.4103/1673-5374.371354
47. Kashio, M., Tominaga, M. TRP channels in thermosensation. *Curr Opin Neurobiol*, 2022, 75: 102591. doi:10.1016/j.conb.2022.102591
48. Talavera, K., Yasumatsu, K., Voets, T., Droogmans, G., Shigemura, N., Ninomiya, Y., Margolskee, R. F., Nilius, B. Heat activation of TRPM5 underlies thermal sensitivity of sweet taste. *Nature*, 2005, 438: 1022–1025. doi:10.1038/nature04248
49. Guo, J., She, J., Zeng, W., Chen, Q., Bai, X. C., Jiang, Y. Structures of the calcium-activated, non-selective cation channel TRPM4. *Nature*, 2017, 552: 205–209. doi:10.1038/nature24997
50. Duan, J., Li, Z., Li, J., Santa-Cruz, A., Sanchez-Martinez, S., Zhang, J., Clapham, D. E. Structure of full-length human TRPM4. *Proc Natl Acad Sci U S A*, 2018, 115: 2377–2382. doi:10.1073/pnas.1722038115
51. Wang, L., Fu, T. M., Zhou, Y., Xia, S., Greka, A., Wu, H. Structures and gating mechanism of human TRPM2. *Science*, 2018, 362: eaav4809. doi: 10.1126/science.aav4809.
52. Perraud, A. L., Fleig, A., Dunn, C. A., Bagley, L. A., Launay, P., Schmitz, C., Stokes, A. J., Zhu, Q., Bessman, M. J., Penner, R., Kinet, J. P., Scharenberg, A. M. ADP-ribose gating of the calcium-permeable LTRPC2 channel revealed by Nudix motif homology. *Nature*, 2001, 411: 595–599. doi:10.1038/35079100
53. Iordanov, I., Toth, B., Szollosi, A., Csanady, L. Enzyme activity and selectivity filter stability of ancient TRPM2 channels were simultaneously lost in early vertebrates. *Elife*, 2019, 8: e44556. doi:10.7554/eLife.44556
54. Shen, B. W., Perraud, A. L., Scharenberg, A., Stoddard, B. L. The crystal structure and mutational analysis of human NUDT9. *J Mol Biol*, 2003, 332: 385–398.
55. Huang, Y., Winkler, P. A., Sun, W., Lu, W., Du, J. Architecture of the TRPM2 channel and its activation mechanism by ADP-ribose and calcium. *Nature*, 2018, 562: 145–149. doi:10.1038/s41586-018-0558-4

56. Huang, Y., Roth, B., Lu, W., Du, J. Ligand recognition and gating mechanism through three ligand-binding sites of human TRPM2 channel. *Elife*, 2019, 8: e50175. doi:10.7554/eLife.50175
57. Huang, Y., Fliegert, R., Guse, A. H., Lu, W., Du, J. A structural overview of the ion channels of the TRPM family. *Cell Calcium*, 2020, 85: 102111. doi:10.1016/j.ceca.2019.102111
58. Lange, I., Yamamoto, S., Partida-Sanchez, S., Mori, Y., Fleig, A., Penner, R. TRPM2 functions as a lysosomal Ca^{2+}-release channel in beta cells. *Sci Signal*, 2009, 2: ra23. doi:10.1126/scisignal.2000278
59. Zhong, Z., Zhai, Y., Liang, S., Mori, Y., Han, R., Sutterwala, F. S., Qiao, L. TRPM2 links oxidative stress to NLRP3 inflammasome activation. *Nat Commun* , 2013, 4: 1611. doi: 10.1038/ncomms2608.
60. Mittal, M., Nepal, S., Tsukasaki, Y., Hecquet, C. M., Soni, D., Rehman, J., Tiruppathi, C., Malik, A. B. Neutrophil activation of endothelial cell-expressed TRPM2 mediates transendothelial neutrophil migration and vascular injury. *Circ Res*, 2017, 121: 1081–1091. doi:10.1161/CIRCRESAHA.117.311747
61. Almasi, S., Kennedy, B. E., El-Aghil, M., Sterea, A. M., Gujar, S., Partida-Sánchez, S., El Hiani, Y. TRPM2 channel-mediated regulation of autophagy maintains mitochondrial function and promotes gastric cancer cell survival via the JNK-signaling pathway. *J Biol Chem*, 2018, 293: 3637–3650. doi:10.1074/jbc.M117.817635
62. Zhang, W., Chu, X., Tong, Q., Cheung, J. Y., Conrad, K., Masker, K., Miller, B. A. A novel TRPM2 isoform inhibits calcium influx and susceptibility to cell death. *J Biol Chem*, 2003. 278: 16222–16229. doi: 10.1074/jbc.M300298200
63. Chen, S. J., Hoffman, N. E., Shanmughapriya, S., Bao, L., Keefer, K., Conrad, K., Merali, S., Takahashi, Y., Abraham, T., Hirschler-Laszkiewicz, I., Wang, J., Zhang, X. Q., Song, J., Barrero, C., Shi, Y., Kawasawa, Y. I., Bayerl, M., Sun, T., Barbour, M., Wang, H. G., Madesh, M., Cheung, J. Y., Miller, B. A. A splice variant of the human ion channel TRPM2 modulates neuroblastoma tumor growth through hypoxia-inducible factor (HIF)-1/2alpha. *J Biol Chem*, 2014, 289: 36284–36302. doi:10.1074/jbc.M114.620922
64. Uemura, T., Kudoh, J., Noda, S., Kanba, S., Shimizu, N. Characterization of human and mouse TRPM2 genes: Identification of a novel N-terminal truncated protein specifically expressed in human striatum. *Biochem Biophys Res Commun*, 2005, 328: 1232–1243. doi:S0006-291X(05)00112-9
65. Wehage, E., Eisfeld, J., Heiner, I., Jüngling, E., Zitt, C., Lückhoff, A. Activation of the cation channel long transient receptor potential channel 2 (LTRPC2) by hydrogen peroxide. A splice variant reveals a mode of activation independent of ADP-ribose. *J Biol Chem*, 2002. 277: 23150–23156. doi: 10.1074/jbc.M112096200
66. Orfanelli, U., Wenke, A. K., Doglioni, C., Russo, V., Bosserhoff, A. K., Lavorgna, G. Identification of novel sense and antisense transcription at the TRPM2 locus in cancer. *Cell Res*, 2008, 18: 1128–1140. doi:10.1038/cr.2008.296
67. Miyanohara, J., Kakae, M., Nagayasu, K., Nakagawa, T., Mori, Y., Arai, K., Shirakawa, H., Kaneko, S. TRPM2 channel aggravates CNS inflammation and cognitive impairment via activation of microglia in chronic cerebral hypoperfusion. *J Neurosci*, 2018, 38: 3520–3533. doi:10.1523/JNEUROSCI.2451-17.2018
68. Zong, P., Feng, J., Li, C. X., Jellison, E. R., Yue, Z., Miller, B., Yue, L. Activation of endothelial TRPM2 exacerbates blood-brain barrier degradation in ischemic stroke. *Cardiovasc Res*, 2024, 120: 188–202. doi:10.1093/cvr/cvad126
69. Hara, Y., Wakamori, M., Ishii, M., Maeno, E., Nishida, M., Yoshida, T., Yamada, H., Shimizu, S., Mori, E., Kudoh, J., Shimizu, N., Kurose, H., Okada, Y., Imoto, K., Mori, Y. LTRPC2 Ca^{2+}-permeable channel activated by changes in redox status confers susceptibility to cell death. *Mol Cell*, 2002, 9: 163–173. doi:10.1016/s1097-2765(01)00438-5
70. Fonfria, E., Marshall, I. C., Boyfield, I., Skaper, S. D., Hughes, J. P., Owen, D. E., Zhang, W., Miller, B. A., Benham, C. D., McNulty, S. Amyloid β-peptide(1–42) and hydrogen peroxide-induced toxicity are mediated by TRPM2 in rat primary striatal cultures. *J Neurochem*, 2005, 95: 715–723. doi:doi:10.1111/j.1471-4159.2005.03396.x
71. Kaneko, S., Kawakami, S., Hara, Y., Wakamori, M., Itoh, E., Minami, T., Takada, Y., Kume, T., Katsuki, H., Mori, Y., Akaike, A. A critical role of TRPM2 in neuronal cell death by hydrogen peroxide. *J Pharmacol Sci*, 2006, 101: 66–76. doi:JST.JSTAGE/jphs/FP0060128 [pii]
72. Jia, J., Verma, S., Nakayama, S., Quillinan, N., Grafe, M. R., Hurn, P. D., Herson, P. S. Sex differences in neuroprotection provided by inhibition of TRPM2 channels following experimental stroke. *J Cereb Blood Flow Metab*, 2011, 31: 2160–2168. doi:10.1038/jcbfm.2011.77
73. Ostapchenko, V. G., Chen, M., Guzman, M. S., Xie, Y. F., Lavine, N., Fan, J., Beraldo, F. H., Martyn, A. C., Belrose, J. C., Mori, Y., MacDonald, J. F., Prado, V. F., Prado, M. A., Jackson, M. F. The transient receptor potential melastatin 2 (TRPM2) channel contributes to beta-amyloid oligomer-related neurotoxicity and memory impairment. *J Neurosci*, 2015, 35: 15157–15169. doi:10.1523/JNEUROSCI.4081-14.2015

74. Verma, S., Quillinan, N., Yang, Y. F., Nakayama, S., Cheng, J., Kelley, M. H., Herson, P. S. TRPM2 channel activation following in vitro ischemia contributes to male hippocampal cell death. *Neurosci Lett*, 2012, 530: 41–46. doi:10.1016/j.neulet.2012.09.044
75. Nakayama, S., Vest, R., Traystman, R. J., Herson, P. S. Sexually dimorphic response of TRPM2 inhibition following cardiac arrest-induced global cerebral ischemia in mice. *J Mol Neurosci*, 2013, 51: 92–98. doi:10.1007/s12031-013-0005-9
76. Alim, I., Teves, L., Li, R., Mori, Y., Tymianski, M. Modulation of NMDAR subunit expression by TRPM2 channels regulates neuronal vulnerability to ischemic cell death. *J Neurosci*, 2013, 33: 17264–17277. doi:10.1523/JNEUROSCI.1729-13.2013
77. Ye, M., Yang, W., Ainscough, J. F., Hu, X. P., Li, X., Sedo, A., Zhang, X. H., Zhang, X., Chen, Z., Li, X. M., Beech, D. J., Sivaprasadarao, A., Luo, J. H., Jiang, L. H. TRPM2 channel deficiency prevents delayed cytosolic Zn^{2+} accumulation and CA1 pyramidal neuronal death after transient global ischemia. *Cell Death Dis*, 2014, 5: e1541. doi:10.1038/cddis.2014.494
78. Huang, S., Turlova, E., Li, F., Bao, M. H., Szeto, V., Wong, R., Abussaud, A., Wang, H., Zhu, S., Gao, X., Mori, Y., Feng, Z. P., Sun, H. S. Transient receptor potential melastatin 2 channels (TRPM2) mediate neonatal hypoxic-ischemic brain injury in mice. *Exp Neurol*, 2017, 296: 32–40. doi:10.1016/j.expneurol.2017.06.023
79. Liu, H. W., Gong, L. N., Lai, K., Yu, X. F., Liu, Z. Q., Li, M. X., Yin, X. L., Liang, M., Shi, H. S., Jiang, L. H., Yang, W., Shi, H. B., Wang, L. Y., Yin, S. K. Bilirubin gates the TRPM2 channel as a direct agonist to exacerbate ischemic brain damage. *Neuron*, 2023, 111: 1609–1625.e6. doi:10.1016/j.neuron.2023.02.022
80. Xie, Y. F., Belrose, J. C., Lei, G., Tymianski, M., Mori, Y., Macdonald, J. F., Jackson, M. F. Dependence of NMDA/GSK-3β mediated metaplasticity on TRPM2 channels at hippocampal CA3-CA1 synapses. *Mol Brain*, 2011, 4: 44. doi:10.1186/1756-6606-4-44
81. Lee, M., Cho, T., Jantaratnotai, N., Wang, Y. T., McGeer, E., McGeer, P. L. Depletion of GSH in glial cells induces neurotoxicity: Relevance to aging and degenerative neurological diseases. *FASEB J*, 2010, 24: 2533–2545. doi:10.1096/fj.09-149997
82. Hecquet, C. M., Ahmmed, G. U., Vogel, S. M., Malik, A. B. Role of TRPM2 channel in mediating H_2O_2-induced Ca^{2+} entry and endothelial hyperpermeability. *Circ Res*, 2008, 102: 347–355. doi:10.1161/CIRCRESAHA.107.160176
83. Zhang, Z., Zhang, W., Jung, D. Y., Ko, H. J., Lee, Y., Friedline, R. H., Lee, E., Jun, J., Ma, Z., Kim, F., Tsitsilianos, N., Chapman, K., Morrison, A., Cooper, M. P., Miller, B. A., Kim, J. K. TRPM2 Ca^{2+} channel regulates energy balance and glucose metabolism. *Am J Physiol Endocrinol Metab*, 2012, 302: E807–E816. doi:10.1152/ajpendo.00239.2011
84. Zong, P., Feng, J., Yue, Z., Yu, A. S., Mori, Y., Yue, L. TRPM2 deficiency protects against atherosclerosis by inhibiting TRPM2-CD36 inflammatory axis in macrophages. *BioRxiv*, 2021.07.29.454234; doi:10.1101/2021.07.29.454234.
85. Beceiro, S., Radin, J. N., Chatuvedi, R., Piazuelo, M. B., Horvarth, D. J., Cortado, H., Gu, Y., Dixon, B., Gu, C., Lange, I., Koomoa, D. L., Wilson, K. T., Algood, H. M., Partida-Sánchez, S. TRPM2 ion channels regulate macrophage polarization and gastric inflammation during Helicobacter pylori infection. *Mucosal Immunol*, 2017, 10: 493–507. doi:10.1038/mi.2016.60
86. Knowles H, Heizer JW, Li Y, Chapman K, Ogden CA, Andreasen K, Shapland E, Kucera G, Mogan J, Humann J, Lenz LL, Morrison AD, Perraud AL. Transient receptor potential melastatin 2 (TRPM2) ion channel is required for innate immunity against listeria monocytogenes. *Proc Natl Acad Sci U S A*, 2011, 108: 11578–11583. doi:10.1073/pnas.1010678108
87. Kashio M, Sokabe T, Shintaku K, Uematsu T, Fukuta N, Kobayashi N, Mori Y, Tominaga M. Redox signal-mediated sensitization of transient receptor potential melastatin 2 (TRPM2) to temperature affects macrophage functions. *Proc Natl Acad Sci U S A*, 2012, 109: 6745–6750. doi:10.1073/pnas.1114193109
88. Zou J, Ainscough JF, Yang W, Sedo A, Yu SP, Mei ZZ, Sivaprasadarao A, Beech DJ, Jiang LH. A differential role of macrophage TRPM2 channels in Ca^{2+} signaling and cell death in early responses to H_2O_2. *Am J Physiol Cell Physiol*, 2013, 305: C61–C69. doi:10.1152/ajpcell.00390.2012
89. Zong P, Feng J, Yue Z, Yu AS, Vacher J, Jellison ER, Miller B, Mori Y, Yue L. TRPM2 deficiency in mice protects against atherosclerosis by inhibiting TRPM2-CD36 inflammatory axis in macrophages. *Nat Cardiovasc Res*, 2022, 1: 344–360. doi:10.1038/s44161-022-00027-7
90. Gelderblom M, Melzer N, Schattling B, Göb E, Hicking G, Arunachalam P, Bittner S, Ufer F, Herrmann AM, Bernreuther C, Glatzel M, Gerloff C, Kleinschnitz C, Meuth SG, Friese MA, Magnus T. Transient receptor potential melastatin subfamily member 2 cation channel regulates detrimental immune cell invasion in ischemic stroke. *Stroke*, 2014, 45: 3395–3402. doi:10.1161/STROKEAHA.114.005836

91. Launay P, Fleig A, Perraud AL, Scharenberg AM, Penner R, Kinet JP. TRPM4 is a Ca^{2+}-activated nonselective cation channel mediating cell membrane depolarization. *Cell*, 2002. 109: 397–407. doi: 10.1016/s0092-8674(02)00719-5
92. Hofmann, T., Chubanov, V., Gudermann, T., Montell, C. TRPM5 is a voltage-modulated and $Ca2^{+}$-activated monovalent selective cation channel. *Curr Biol*, 2003, 13: 1153–1158. doi: 10.1016/s0960-9822(03)00431-7
93. Zhang Y, Hoon MA, Chandrashekar J, Mueller KL, Cook B, Wu D, Zuker CS, Ryba NJ. Coding of sweet, bitter, and umami tastes: Different receptor cells sharing similar signaling pathways. *Cell*, 2003, 112: 293–301. doi: 10.1016/s0092-8674(03)00071-0
94. Nilius B, Mahieu F, Prenen J, Janssens A, Owsianik G, Vennekens R, Voets T. The Ca^{2+}-activated cation channel TRPM4 is regulated by phosphatidylinositol 4,5-biphosphate. *EMBO J*, 2006, 25: 467–478. doi: 10.1038/sj.emboj.7600963
95. Nilius B, Prenen J, Tang J, Wang C, Owsianik G, Janssens A, Voets T, Zhu MX. Regulation of the Ca^{2+} sensitivity of the nonselective cation channel TRPM4. *J Biol Chem*, 2005, 280: 6423–6433. doi:10.1074/jbc.M411089200
96. Xu, X. Z., Moebius, F., Gill, D. L., Montell, C. Regulation of melastatin, a TRP-related protein, through interaction with a cytoplasmic isoform. *Proc Natl Acad Sci U S A*, 2001, 98: 10692–10697. doi: 10.1073/pnas.191360198
97. Mehta RI, Tosun C, Ivanova S, Tsymbalyuk N, Famakin BM, Kwon MS, Castellani RJ, Gerzanich V, Simard JM. Surl-Trpm4 cation channel expression in human cerebral infarcts. *J Neuropathol Exp Neurol*, 2015, 74, 835–8:49. doi:10.1097/NEN.0000000000000223
98. Sala-Rabanal, M., Wang, S., Nichols, C. G. On potential interactions between non-selective cation channel TRPM4 and sulfonylurea receptor SUR1. *J Biol Chem*, 2012, 287: 8746–8756. doi:10.1074/jbc.M111.336131
99. Stokum JA, Shim B, Negoita S, Tsymbalyuk N, Tsymbalyuk O, Ivanova S, Keledjian K, Bryan J, Blaustein MP, Jha RM, Kahle KT, Gerzanich V, Simard JM. Cation flux through SUR1-TRPM4 and NCX1 in astrocyte endfeet induces water influx through AQP4 and brain swelling after ischemic stroke. *Sci Signal*, 2023, 16: eadd6364. doi:10.1126/scisignal.add6364
100. Woo SK, Tsymbalyuk N, Tsymbalyuk O, Ivanova S, Gerzanich V, Simard JM. SUR1-TRPM4 channels, not K(ATP), mediate brain swelling following cerebral ischemia. *Neurosci Lett*, 2020, 718: 134729. doi:10.1016/j.neulet.2019.134729
101. Loh KP, Ng G, Yu CY, Fhu CK, Yu D, Vennekens R, Nilius B, Soong TW, Liao P. TRPM4 inhibition promotes angiogenesis after ischemic stroke. *Pflugers Arch*, 2014, 466: 563–576. doi:10.1007/s00424-013-1347-4
102. Chen B, Ng G, Gao Y, Low SW, Sandanaraj E, Ramasamy B, Sekar S, Bhakoo K, Soong TW, Nilius B, Tang C, Robins EG, Goggi J, Liao P. Non-invasive multimodality imaging directly shows TRPM4 inhibition ameliorates stroke reperfusion injury. *Transl Stroke Res*, 2019, 10: 91–103. doi:10.1007/s12975-018-0621-3
103. Serafini N, Dahdah A, Barbet G, Demion M, Attout T, Gautier G, Arcos-Fajardo M, Souchet H, Jouvin MH, Vrtovsnik F, Kinet JP, Benhamou M, Monteiro RC, Launay P. The TRPM4 channel controls monocyte and macrophage, but not neutrophil, function for survival in sepsis. *J Immunol*, 2012, 189: 3689–3699. doi:10.4049/jimmunol.1102969
104. Launay P, Cheng H, Srivatsan S, Penner R, Fleig A, Kinet JP. TRPM4 regulates calcium oscillations after T cell activation. *Science*, 2004, 306: 1374–1377. doi:10.1126/science.1098845
105. Jin J, Desai BN, Navarro B, Donovan A, Andrews NC, Clapham DE. Deletion of Trpm7 disrupts embryonic development and thymopoiesis without altering Mg^{2+} homeostasis. *Science*, 2008, 322: 756–760. doi: 10.1126/science.1163493
106. Jin J, Wu LJ, Jun J, Cheng X, Xu H, Andrews NC, Clapham DE. The channel kinase, TRPM7, is required for early embryonic development. *Proc Natl Acad Sci U S A*, 2012, 109: E225–E233. doi: 10.1073/pnas.1120033109
107. Sah R, Mesirca P, Mason X, Gibson W, Bates-Withers C, Van den Boogert M, Chaudhuri D, Pu WT, Mangoni ME, Clapham DE. Timing of myocardial trpm7 deletion during cardiogenesis variably disrupts adult ventricular function, conduction, and repolarization. *Circulation*, 2013, 128: 101–114. doi:10.1161/CIRCULATIONAHA.112.000768
108. Khajavi N, Beck A, Riçku K, Beyerle P, Jacob K, Syamsul SF, Belkacemi A, Reinach PS, Schreier PC, Salah H, Popp T, Novikoff A, Breit A, Chubanov V, Müller TD, Zierler S, Gudermann T. TRPM7 kinase is required for insulin production and compensatory islet responses during obesity. *JCI Insight*, 2023, 8: e163397. doi:10.1172/jci.insight.163397

109. Nadolni W, Immler R, Hoelting K, Fraticelli M, Ripphahn M, Rothmiller S, Matsushita M, Boekhoff I, Gudermann T, Sperandio M, Zierler S. TRPM7 kinase is essential for neutrophil recruitment and function via regulation of Akt/mTOR signaling. *Front Immunol*, 2020, 11: 606893. doi:10.3389/fimmu.2020.606893
110. Romagnani A, Vettore V, Rezzonico-Jost T, Hampe S, Rottoli E, Nadolni W, Perotti M, Meier MA, Hermanns C, Geiger S, Wennemuth G, Recordati C, Matsushita M, Muehlich S, Proietti M, Chubanov V, Gudermann T, Grassi F, Zierler S. TRPM7 kinase activity is essential for T cell colonization and alloreactivity in the gut. *Nat Commun*, 2017, 8: 1917. doi:10.1038/s41467-017-01960-z
111. Gotru SK, Chen W, Kraft P, Becker IC, Wolf K, Stritt S, Zierler S, Hermanns HM, Rao D, Perraud AL, Schmitz C, Zahedi RP, Noy PJ, Tomlinson MG, Dandekar T, Matsushita M, Chubanov V, Gudermann T, Stoll G, Nieswandt B, Braun A. TRPM7 kinase controls calcium responses in arterial thrombosis and stroke in mice. *Arterioscler Thromb Vasc Biol,* 2018, 38: 344–352. doi:10.1161/ATVBAHA.117.310391
112. Krapivinsky, G., Krapivinsky, L., Manasian, Y., Clapham, D. E. The TRPM7 chanzyme is cleaved to release a chromatin-modifying kinase. *Cell*, 2014, 157: 1061–1072. doi:10.1016/j.cell.2014.03.046
113. Monteilh-Zoller MK, Hermosura MC, Nadler MJ, Scharenberg AM, Penner R, Fleig A. TRPM7 provides an ion channel mechanism for cellular entry of trace metal ions. *J Gen Physiol*, 2003, 121: 49–60. doi:10.1085/jgp.20028740
114. Nadler MJ, Hermosura MC, Inabe K, Perraud AL, Zhu Q, Stokes AJ, Kurosaki T, Kinet JP, Penner R, Scharenberg AM, Fleig A. LTRPC7 is a Mg.ATP-regulated divalent cation channel required for cell viability. *Nature*, 2001, 411: 590–595. doi:10.1038/35079092
115. Chubanov V, Waldegger S, Mederos y Schnitzler M, Vitzthum H, Sassen MC, Seyberth HW, Konrad M, Gudermann T. Disruption of TRPM6/TRPM7 complex formation by a mutation in the TRPM6 gene causes hypomagnesemia with secondary hypocalcemia. *Proc Natl Acad Sci U S A*, 2004. 101: 2894–2899. doi: 10.1073/pnas.0305252101
116. Schmitz C, Perraud AL, Johnson CO, Inabe K, Smith MK, Penner R, Kurosaki T, Fleig A, Scharenberg AM. Regulation of vertebrate cellular Mg^{2+} homeostasis by TRPM7. *Cell*, 2003. 114: 191–200. doi: 10.1016/s0092-8674(03)00556-7
117. Runnels, L. W., Yue, L., Clapham, D. E. The TRPM7 channel is inactivated by PIP(2) hydrolysis. *Nat Cell Biol*, 2002, 4: 329–336. doi:10.1038/ncb781
118. Penner, R., Fleig, A. The Mg^{2+} and Mg^{2+}-nucleotide-regulated channel-kinase TRPM7. *Handb Exp Pharmacol*, 2007, 313–328. doi:10.1007/978-3-540-34891-7_19
119. Mahbub L, Kozlov G, Zong P, Tetteh S, Nethramangalath T, Knorn C, Jiang J, Shahsavan A, Lee E, Yue L, Runnels LW, Gehring K. Structural insights into regulation of TRPM7 divalent cation uptake by the small GTPase ARL15. *BioRxiv*, 2023 Jan 20:2023.01.19.524765. doi:10.1101/2023.01.19.524765
120. Bai, Z., Feng, J., Franken, G. A. C., Al'Saadi, N., Cai, N., Yu, A. S., Lou, L., Komiya, Y., Hoenderop, J. G. J., de Baaij, J. H. F., Yue, L., Runnels, L. W. CNNM proteins regulate the TRPM7 channel to control divalent cation entry. *PLOS Biol*, 2021, 19: e3001496. doi: 10.1371/journal.pbio.3001496.
121. Kollewe A, Chubanov V, Tseung FT, Correia L, Schmidt E, Rössig A, Zierler S, Haupt A, Müller CS, Bildl W, Schulte U, Nicke A, Fakler B, Gudermann T. The molecular appearance of native TRPM7 channel complexes identified by high-resolution proteomics. *Elife*, 2021, 10: e68544. doi:10.7554/eLife.68544
122. Hardy S, Zolotarov Y, Coleman J, Roitman S, Khursheed H, Aubry I, Uetani N, Tremblay ML. PRL-1/2 phosphatases control TRPM7 magnesium-dependent function to regulate cellular bioenergetics. *Proc Natl Acad Sci U S A*, 2023, 120: e2221083120. doi:10.1073/pnas.2221083120
123. Jiang, J., Li, M., Yue, L. Potentiation of TRPM7 inward currents by protons. *J Gen Physiol*, 2005, 126: 137–150. doi:10.1085/jgp.200409185
124. Numata, T., Shimizu, T., Okada, Y. TRPM7 is a stretch- and swelling-activated cation channel involved in volume regulation in human epithelial cells. *Am J Physiol Cell Physiol*, 2007, 292: C460–467. doi:10.1152/ajpcell.00367.2006
125. Qian N, Ichimura A, Takei D, Sakaguchi R, Kitani A, Nagaoka R, Tomizawa M, Miyazaki Y, Miyachi H, Numata T, Kakizawa S, Nishi M, Mori Y, Takeshima H. TRPM7 channels mediate spontaneous Ca^{2+} fluctuations in growth plate chondrocytes that promote bone development. *Sci Signal*, 2019, 12: eaaw4847. doi: 10.1126/scisignal.aaw4847.
126. Du J, Xie J, Zhang Z, Tsujikawa H, Fusco D, Silverman D, Liang B, Yue L. TRPM7-mediated Ca^{2+} signals confer fibrogenesis in human atrial fibrillation. *Circ Res,* 2010, 106: 992–1003. doi:CIRCRESAHA.109.206771
127. Wei C, Wang X, Chen M, Ouyang K, Song LS, Cheng H. Calcium flickers steer cell migration. *Nature*, 2009, 457: 901–905. doi: 10.1038/nature07577

128. Desai BN, Krapivinsky G, Navarro B, Krapivinsky L, Carter BC, Febvay S, Delling M, Penumaka A, Ramsey IS, Manasian Y, Clapham DE. Cleavage of TRPM7 releases the kinase domain from the ion channel and regulates its participation in Fas-induced apoptosis. *Dev Cell,* 2012, 22: 1149–1162. doi:10.1016/j.devcel.2012.04.006
129. Schappe MS, Szteyn K, Stremska ME, Mendu SK, Downs TK, Seegren PV, Mahoney MA, Dixit S, Krupa JK, Stipes EJ, Rogers JS, Adamson SE, Leitinger N, Desai BN. Chanzyme TRPM7 mediates the Ca^{2+} influx essential for lipopolysaccharide-induced toll-like receptor 4 endocytosis and macrophage activation. *Immunity*, 2018, 48: 59–74.e5. doi:10.1016/j.immuni.2017.11.026
130. Schappe MS, Stremska ME, Busey GW, Downs TK, Seegren PV, Mendu SK, Flegal Z, Doyle CA, Stipes EJ, Desai BN. Efferocytosis requires periphagosomal Ca^{2+}-signaling and TRPM7-mediated electrical activity. *Nat Commun*, 2022, 13: 3230. doi:10.1038/s41467-022-30959-4
131. Sahni, J., Tamura, R., Sweet, I. R., Scharenberg, A. M. TRPM7 regulates quiescent/proliferative metabolic transitions in lymphocytes. *Cell Cycle*, 2010, 9: 3565–3574. doi:12798
132. Wu W, Wang X, Liao L, Chen J, Wang Y, Yao M, Zhu L, Li J, Wang X, Chen AF, Zhang G, Zhang Z, Bai Y. The TRPM7 channel reprograms cellular glycolysis to drive tumorigenesis and angiogenesis. *Cell Death Dis*, 2023, 14: 183. doi:10.1038/s41419-023-05701-7
133. Aarts M, Iihara K, Wei WL, Xiong ZG, Arundine M, Cerwinski W, MacDonald JF, Tymianski M. A key role for TRPM7 channels in anoxic neuronal death. *Cell*, 2003, 115: 863–877. doi:10.1016/s0092-8674(03)01017-1
134. Sun HS, Jackson MF, Martin LJ, Jansen K, Teves L, Cui H, Kiyonaka S, Mori Y, Jones M, Forder JP, Golde TE, Orser BA, Macdonald JF, Tymianski M. Suppression of hippocampal TRPM7 protein prevents delayed neuronal death in brain ischemia. *Nat Neurosci*, 2009, 12: 1300–1307. doi: 10.1038/nn.2395
135. Zhang, P., Li, W., Liu, Y., Gao, Y., Abumaria, N. Neuroprotective effects of TRPM7 deletion in parvalbumin GABAergic vs. glutamatergic neurons following ischemia. *Cells*, 2022, 11. doi:10.3390/cells11071178
136. Inoue, K., Branigan, D., Xiong, Z. G. Zinc-induced neurotoxicity mediated by transient receptor potential melastatin 7 channels. *J Biol Chem*, 2010, 285: 7430–7439. doi:10.1074/jbc.M109.040485
137. Paravicini, T. M., Yogi, A., Mazur, A., Touyz, R. M. Dysregulation of vascular TRPM7 and annexin-1 is associated with endothelial dysfunction in inherited hypomagnesemia. *Hypertension*, 2009, 53: 423–429. doi:10.1161/HYPERTENSIONAHA.108.124651
138. Nadolni, W., Zierler, S. The channel-kinase TRPM7 as novel regulator of immune system homeostasis. *Cells*, 2018, 7. doi:10.3390/cells7080109
139. Wang H, Cheng X, Tian J, Xiao Y, Tian T, Xu F, Hong X, Zhu MX. TRPC channels: Structure, function, regulation and recent advances in small molecular probes. *Pharmacol Ther*, 2020, 209: 107497. doi:10.1016/j.pharmthera.2020.107497
140. Dietrich, A., Fahlbusch, M., Gudermann, T. Classical transient receptor potential 1 (TRPC1): channel or channel regulator? *Cells*, 2014, 3: 939–962. doi:10.3390/cells3040939
141. Beech, D. J., Muraki, K., Flemming, R. Non-selective cationic channels of smooth muscle and the mammalian homologues of drosophila TRP. *J Physiol*, 2004, 559: 685–706. doi:10.1113/jphysiol.2004.068734
142. Lintschinger B, Balzer-Geldsetzer M, Baskaran T, Graier WF, Romanin C, Zhu MX, Groschner K. Coassembly of Trp1 and Trp3 proteins generates diacylglycerol- and Ca^{2+}-sensitive cation channels. *J Biol Chem*, 2000, 275: 27799–27805. doi:10.1074/jbc.M002705200
143. Hofmann, T., Schaefer, M., Schultz, G., Gudermann, T. Subunit composition of mammalian transient receptor potential channels in living cells. *Proc Natl Acad Sci U S A*, 2002, 99: 7461–7466. doi: 10.1073/pnas.102596199
144. Strubing, C., Krapivinsky, G., Krapivinsky, L., Clapham, D. E. TRPC1 and TRPC5 form a novel cation channel in mammalian brain. *Neuron*, 2001, 29: 645–655. doi: 10.1016/s0896-6273(01)00240-9
145. Storch, U., Forst, A. L., Philipp, M., Gudermann, T., Mederos y Schnitzler, M. Transient receptor potential channel 1 (TRPC1) reduces calcium permeability in heteromeric channel complexes. *J Biol Chem*, 2012, 287: 3530–3540. doi:10.1074/jbc.M111.283218
146. Bon, R. S., Wright, D. J., Beech, D. J., Sukumar, P. Pharmacology of TRPC channels and its potential in cardiovascular and metabolic medicine. *Annu Rev Pharmacol Toxicol*, 2022, 62: 427–446. doi:10.1146/annurev-pharmtox-030121-122314
147. Tai, Y., Jia, Y. TRPC channels and neuron development, plasticity, and activities. *Adv Exp Med Biol*, 2017, 976: 95–110. doi:10.1007/978-94-024-1088-4_9

148. Verkhratsky, A., Reyes, R. C., Parpura, V., TRP channels coordinate ion signalling in astroglia. *Rev Physiol Biochem Pharmacol*, 2014, 166: 1–22. doi:10.1007/112_2013_15
149. Negri, S., Faris, P., Berra-Romani, R., Guerra, G., Moccia, F. Endothelial transient receptor potential channels and vascular remodeling: Extracellular Ca^{2+} entry for angiogenesis, arteriogenesis and vasculogenesis. *Front Physiol*, 2019, 10: 1618. doi:10.3389/fphys.2019.01618
150. Formoso, K., Susperreguy, S., Freichel, M., Birnbaumer, L. RNA-seq analysis reveals TRPC genes to impact an unexpected number of metabolic and regulatory pathways. *Sci Rep*, 2020, 10: 7227. doi:10.1038/s41598-020-61177-x
151. Veldhuis, N. A., Poole, D. P., Grace, M., McIntyre, P., Bunnett, N. W. The G protein-coupled receptor-transient receptor potential channel axis: Molecular insights for targeting disorders of sensation and inflammation. *Pharmacol Rev*, 2015, 67: 36–73. doi:10.1124/pr.114.009555
152. Venkatachalam, K., Zheng, F., Gill, D. L. Regulation of canonical transient receptor potential (TRPC) channel function by diacylglycerol and protein kinase C. *J Biol Chem*, 2003, 278: 29031–29040. doi:10.1074/jbc.M302751200
153. Riccio A, Medhurst AD, Mattei C, Kelsell RE, Calver AR, Randall AD, Benham CD, Pangalos MN. mRNA distribution analysis of human TRPC family in CNS and peripheral tissues. *Brain Res Mol Brain Res*, 2002, 109: 95–104. doi:10.1016/s0169-328x(02)00527-2
154. Li Y, Jia YC, Cui K, Li N, Zheng ZY, Wang YZ, Yuan XB. Essential role of TRPC channels in the guidance of nerve growth cones by brain-derived neurotrophic factor. *Nature*, 2005, 434: 894–898. doi:10.1038/nature03477
155. Becker EB, Oliver PL, Glitsch MD, Banks GT, Achilli F, Hardy A, Nolan PM, Fisher EM, Davies KE. A point mutation in TRPC3 causes abnormal Purkinje cell development and cerebellar ataxia in moonwalker mice. *Proc Natl Acad Sci U S A*, 2009, 106: 6706–6711. doi:10.1073/pnas.0810599106
156. Jia, Y., Zhou, J., Tai, Y., Wang, Y. TRPC channels promote cerebellar granule neuron survival. *Nat Neurosci*, 2007, 10: 559–567. doi:10.1038/nn1870
157. Hartmann J, Dragicevic E, Adelsberger H, Henning HA, Sumser M, Abramowitz J, Blum R, Dietrich A, Freichel M, Flockerzi V, Birnbaumer L, Konnerth A. TRPC3 channels are required for synaptic transmission and motor coordination. *Neuron*, 2008, 59: 392–398. doi:10.1016/j.neuron.2008.06.009
158. Berg, A. P., Sen, N., Bayliss, D. A. TrpC3/C7 and Slo2.1 are molecular targets for metabotropic glutamate receptor signaling in rat striatal cholinergic interneurons. *J Neurosci*, 2007, 27: 8845–8856. doi:10.1523/JNEUROSCI.0551-07.2007
159. Zhou J, Du W, Zhou K, Tai Y, Yao H, Jia Y, Ding Y, Wang Y. Critical role of TRPC6 channels in the formation of excitatory synapses. *Nat Neurosci*, 2008, 11: 741–743. doi:10.1038/nn.2127
160. Li H, Huang J, Du W, Jia C, Yao H, Wang Y. TRPC6 inhibited NMDA receptor activities and protected neurons from ischemic excitotoxicity. *J Neurochem*, 2012, 123: 1010–1018. doi:10.1111/jnc.12045
161. Du W, Huang J, Yao H, Zhou K, Duan B, Wang Y. Inhibition of TRPC6 degradation suppresses ischemic brain damage in rats. *J Clin Invest*, 2010, 120: 3480–3492. doi: 10.1172/JCI43165
162. Lin Y, Zhang JC, Fu J, Chen F, Wang J, Wu ZL, Yuan SY. Hyperforin attenuates brain damage induced by transient middle cerebral artery occlusion (MCAO) in rats via inhibition of TRPC6 channels degradation. *J Cereb Blood Flow Metab*, 2013, 33: 253–262. doi:10.1038/jcbfm.2012.164
163. Guo C, Ma Y, Ma S, Mu F, Deng J, Duan J, Xiong L, Yin Y, Wang Y, Xi M, Wen A. The role of TRPC6 in the neuroprotection of calycosin against cerebral ischemic injury. *Sci Rep*, 2017, 7: 3039. doi:10.1038/s41598-017-03404-6
164. Shen, H., Pan, J., Pan, L., Zhang, N. TRPC6 inhibited NMDA current in cultured hippocampal neurons. *Neuromolecular Med*, 2013, 15: 389–395. doi:10.1007/s12017-013-8226-1
165. Chen X, Lu M, He X, Ma L, Birnbaumer L, Liao Y. TRPC3/6/7 knockdown protects the brain from cerebral ischemia injury via astrocyte apoptosis inhibition and effects on NF-κB translocation. *Mol Neurobiol*, 2017, 54: 7555–7566. doi:10.1007/s12035-016-0227-2
166. Chen J, Li Z, Hatcher JT, Chen QH, Chen L, Wurster RD, Chan SL, Cheng Z. Deletion of TRPC6 attenuates NMDA receptor-mediated Ca^{2+} entry and Ca^{2+}-induced neurotoxicity following cerebral ischemia and oxygen-glucose deprivation. *Front Neurosci,* 2017, 11: 138. doi:10.3389/fnins.2017.00138
167. Wang, G. X., Poo, M. M. Requirement of TRPC channels in netrin-1-induced chemotropic turning of nerve growth cones. *Nature*, 2005, 434: 898–904. doi:10.1038/nature03478
168. Shim S, Yuan JP, Kim JY, Zeng W, Huang G, Milshteyn A, Kern D, Muallem S, Ming GL, Worley PF. Peptidyl-prolyl isomerase FKBP52 controls chemotropic guidance of neuronal growth cones via regulation of TRPC1 channel opening. *Neuron*, 2009, 64: 471–483. doi:10.1016/j.neuron.2009.09.025
169. Kim SJ, Kim YS, Yuan JP, Petralia RS, Worley PF, Linden DJ., Activation of the TRPC1 cation channel by metabotropic glutamate receptor mGluR1. *Nature*, 2003. 426: 285–291. doi: 10.1038/nature02162

170. Reiner, A., Levitz, J. Glutamatergic signaling in the Central nervous system: Ionotropic and metabotropic receptors in concert. *Neuron*, 2018, 98: 1080–1098. doi:10.1016/j.neuron.2018.05.018
171. Lepannetier S, Gualdani R, Tempesta S, Schakman O, Seghers F, Kreis A, Yerna X, Slimi A, de Clippele M, Tajeddine N, Voets T, Bon RS, Beech DJ, Tissir F, Gailly P. Activation of TRPC1 channel by metabotropic glutamate receptor mGluR5 modulates synaptic plasticity and spatial working memory. *Front Cell Neurosci*, 2018, 12: 318. doi:10.3389/fncel.2018.00318
172. Xu N, Meng H, Liu T, Feng Y, Qi Y, Wang H. TRPC1 deficiency exacerbates cerebral Ischemia/Reperfusion-induced neurological injury by potentiating Nox4-derived reactive oxygen species generation. *Cell Physiol Biochem*, 2018, 51: 1723–1738. doi:10.1159/000495676
173. Narayanan, K. L., Irmady, K., Subramaniam, S., Unsicker, K., von Bohlen und Halbach, O. Evidence that TRPC1 is involved in hippocampal glutamate-induced cell death. *Neurosci Lett*, 2008, 446: 117–122. doi:10.1016/j.neulet.2008.09.034
174. Riccio A, Li Y, Moon J, Kim KS, Smith KS, Rudolph U, Gapon S, Yao GL, Tsvetkov E, Rodig SJ, Van't Veer A, Meloni EG, Carlezon WA Jr, Bolshakov VY, Clapham DE. Essential role for TRPC5 in amygdala function and fear-related behavior. *Cell*, 2009, 137: 761–772. doi:10.1016/j.cell.2009.03.039
175. Weick, J. P., Austin Johnson, M., Zhang, S. C. Developmental regulation of human embryonic stem cell-derived neurons by calcium entry via transient receptor potential channels. *Stem Cells*, 2009, 27: 2906–2916. doi:10.1002/stem.212
176. Stroh O, Freichel M, Kretz O, Birnbaumer L, Hartmann J, Egger V. NMDA receptor-dependent synaptic activation of TRPC channels in olfactory bulb granule cells. *J Neurosci*, 2012, 32: 5737–5746. doi:10.1523/JNEUROSCI.3753-11.2012
177. Greka, A., Navarro, B., Oancea, E., Duggan, A., Clapham, D. E. TRPC5 is a regulator of hippocampal neurite length and growth cone morphology. *Nat Neurosci*, 2003, 6: 837–845. doi:10.1038/nn1092
178. Riccio A, Li Y, Tsvetkov E, Gapon S, Yao GL, Smith KS, Engin E, Rudolph U, Bolshakov VY, Clapham DE. Decreased anxiety-like behavior and Galphaq/11-dependent responses in the amygdala of mice lacking TRPC4 channels. *J Neurosci*, 2014, 34: 3653–3667. doi:10.1523/JNEUROSCI.2274-13.2014
179. Shen N, Wang L, Wu Y, Liu Y, Pei H, Xiang H. Adeno-associated virus packaged TRPC5 gene therapy alleviated spinal cord ischemic reperfusion injury in rats. *Neuroreport*, 2020, 31: 29–36. doi:10.1097/WNR.0000000000001359
180. Guo J, Li J, Xia L, Wang Y, Zhu J, Du J, Lu Y, Liu G, Yao X, Shen B. Transient receptor potential canonical 5-scramblase signaling complex mediates neuronal phosphatidylserine externalization and apoptosis. *Cells*, 2020, 9. doi:10.3390/cells9030547
181. Cesare, P., McNaughton, P. A novel heat-activated current in nociceptive neurons and its sensitization by bradykinin. *Proc Natl Acad Sci U S A*, 1996, 93: 15435–15439. doi:10.1073/pnas.93.26.15435
182. Caterina MJ, Schumacher MA, Tominaga M, Rosen TA, Levine JD, Julius D. The capsaicin receptor: A heat-activated ion channel in the pain pathway. *Nature*, 1997. 389: 816–824. doi: 10.1038/39807
183. Nilius B, Vennekens R, Prenen J, Hoenderop JG, Droogmans G, Bindels RJ. The single pore residue Asp542 determines Ca^{2+} permeation and Mg^{2+} block of the epithelial Ca^{2+} channel. *J Biol Chem*, 2001. 276: 1020–1025. doi: 10.1074/jbc.M006184200
184. Saotome, K., Singh, A. K., Yelshanskaya, M. V., Sobolevsky, A. I. Crystal structure of the epithelial calcium channel TRPV6. *Nature*, 2016, 534: 506–511. doi:10.1038/nature17975
185. Sun, X., Zakharian, E. Regulation of the temperature-dependent activation of transient receptor potential vanilloid 1 (TRPV1) by phospholipids in planar lipid bilayers. *J Biol Chem*, 2015, 290: 4741–4747. doi:10.1074/jbc.M114.611459
186. Prescott, E. D., Julius, D. A modular PIP2 binding site as a determinant of capsaicin receptor sensitivity. *Science*, 2003, 300: 1284–1288. doi:10.1126/science.1083646
187. Doerner, J. F., Hatt, H., Ramsey, I. S. Voltage- and temperature-dependent activation of TRPV3 channels is potentiated by receptor-mediated PI(4,5)P2 hydrolysis. *J Gen Physiol*, 2011, 137: 271–288. doi: 10.1085/jgp.200910388
188. Rosenbaum, T., Islas, L. D. Molecular physiology of TRPV channels: Controversies and future challenges. *Annu Rev Physiol*, 2023, 85: 293–316. doi:10.1146/annurev-physiol-030222-012349
189. Froghi S, Grant CR, Tandon R, Quaglia A, Davidson B, Fuller B. New insights on the role of TRP channels in calcium signalling and immunomodulation: Review of pathways and implications for clinical practice. *Clin Rev Allergy Immunol*, 2021, 60: 271–292. doi:10.1007/s12016-020-08824-3
190. Thakore, P., Earley, S. Transient receptor potential channels and endothelial cell calcium signaling. *Compr Physiol*, 2019, 9: 1249–1277. doi:10.1002/cphy.c180034
191. Marrone MC, Morabito A, Giustizieri M, Chiurchiù V, Leuti A, Mattioli M, Marinelli S, Riganti L, Lombardi M, Murana E, Totaro A, Piomelli D, Ragozzino D, Oddi S, Maccarrone M, Verderio C,

Marinelli S. TRPV1 channels are critical brain inflammation detectors and neuropathic pain biomarkers in mice. *Nat Commun*, 2017, 8: 15292. doi:10.1038/ncomms15292

192. Menigoz, A., Boudes, M. The expression pattern of TRPV1 in brain. *J Neurosci*, 2011, 31: 13025–13027. doi:10.1523/JNEUROSCI.2589-11.2011
193. Shuba, Y. M. Beyond neuronal heat sensing: Diversity of TRPV1 heat-capsaicin receptor-channel functions. *Front Cell Neurosci*, 2020, 14: 612480. doi:10.3389/fncel.2020.612480
194. Molinas AJR, Desmoulins LD, Hamling BV, Butcher SM, Anwar IJ, Miyata K, Enix CL, Dugas CM, Satou R, Derbenev AV, Zsombok A. Interaction between TRPV1-expressing neurons in the hypothalamus. *J Neurophysiol*, 2019, 121: 140–151. doi:10.1152/jn.00004.2018
195. Balleza-Tapia H, Crux S, Andrade-Talavera Y, Dolz-Gaiton P, Papadia D, Chen G, Johansson J, Fisahn A. TrpV1 receptor activation rescues neuronal function and network gamma oscillations from abeta-induced impairment in mouse hippocampus in vitro. *Elife*, 2018, 7: e37703 . doi:10.7554/eLife.37703
196. Hurtado-Zavala JI, Ramachandran B, Ahmed S, Halder R, Bolleyer C, Awasthi A, Stahlberg MA, Wagener RJ, Anderson K, Drenan RM, Lester HA, Miwa JM, Staiger JF, Fischer A, Dean C. TRPV1 regulates excitatory innervation of OLM neurons in the hippocampus. *Nat Commun*, 2017, 8: 15878. doi:10.1038/ncomms15878
197. Aneiros E, Cao L, Papakosta M, Stevens EB, Phillips S, Grimm C. The biophysical and molecular basis of TRPV1 proton gating. *EMBO J*, 2011, 30: 994–1002. doi:10.1038/emboj.2011.19
198. Miyanohara, J., Shirakawa, H., Sanpei, K., Nakagawa, T., Kaneko, S. A pathophysiological role of TRPV1 in ischemic injury after transient focal cerebral ischemia in mice. *Biochem Biophys Res Commun*, 2015, 467: 478–483. doi:10.1016/j.bbrc.2015.10.027
199. Hakimizadeh E, Shamsizadeh A, Roohbakhsh A, Arababadi MK, Hajizadeh MR, Shariati M, Rahmani MR, Allahtavakoli M. Inhibition of transient receptor potential vanilloid-1 confers neuroprotection, reduces tumor necrosis factor-alpha, and increases IL-10 in a rat stroke model. *Fundam Clin Pharmacol*, 2017, 31: 420–428. doi:10.1111/fcp.12279
200. Yang XL, Wang X, Shao L, Jiang GT, Min JW, Mei XY, He XH, Liu WH, Huang WX, Peng BW. TRPV1 mediates astrocyte activation and interleukin-1beta release induced by hypoxic ischemia (HI). *J Neuroinflammation*, 2019, 16: 114. doi:10.1186/s12974-019-1487-3
201. Zhang Y, Hou B, Liang P, Lu X, Wu Y, Zhang X, Fan Y, Liu Y, Chen T, Liu W, Peng B, Yin J, Han S, He X. TRPV1 channel mediates NLRP3 inflammasome-dependent neuroinflammation in microglia. *Cell Death Dis*, 2021, 12: 1159. doi:10.1038/s41419-021-04450-9
202. Wang, C., Huang, W., Lu, J., Chen, H., Yu, Z. TRPV1-mediated microglial autophagy attenuates Alzheimer's disease-associated pathology and cognitive decline. *Front Pharmacol*, 2021, 12: 763866. doi:10.3389/fphar.2021.763866
203. Zhu, R., Luo, Y., Li, S., Wang, Z. The role of microglial autophagy in Parkinson's disease. *Front Aging Neurosci*, 2022, 14: 1039780. doi:10.3389/fnagi.2022.1039780
204. Gavva NR, Bannon AW, Surapaneni S, Hovland DN Jr, Lehto SG, Gore A, Juan T, Deng H, Han B, Klionsky L, Kuang R, Le A, Tamir R, Wang J, Youngblood B, Zhu D, Norman MH, Magal E, Treanor JJ, Louis JC. The vanilloid receptor TRPV1 is tonically activated in vivo and involved in body temperature regulation. *J Neurosci*, 2007, 27: 3366–3374. doi:10.1523/JNEUROSCI.4833-06.2007
205. Yue, W. W. S., Yuan, L., Braz, J. M., Basbaum, A. I., Julius, D. TRPV1 drugs alter core body temperature via central projections of primary afferent sensory neurons. *Elife*, 2022, 11: e80139. doi:10.7554/eLife.80139
206. Yenari, M. A., Hemmen, T. M. Therapeutic hypothermia for brain ischemia: Where have we come and where do we go? *Stroke*, 2010, 41: S72–74. doi:10.1161/STROKEAHA.110.595371
207. Cao, Z., Balasubramanian, A., Marrelli, S. P. Pharmacologically induced hypothermia via TRPV1 channel agonism provides neuroprotection following ischemic stroke when initiated 90 min after reperfusion. *Am J Physiol Regul Integr Comp Physiol*, 2014, 306: R149–156. doi:10.1152/ajpregu.00329.2013
208. Muzzi M, Felici R, Cavone L, Gerace E, Minassi A, Appendino G, Moroni F, Chiarugi A. Ischemic neuroprotection by TRPV1 receptor-induced hypothermia. *J Cereb Blood Flow Metab*, 2012, 32: 978–982. doi:10.1038/jcbfm.2012.36
209. Cao Z, Balasubramanian A, Pedersen SE, Romero J, Pautler RG, Marrelli SP. TRPV1-mediated pharmacological hypothermia promotes improved functional recovery following ischemic stroke. *Sci Rep*, 2017, 7: 17685. doi:10.1038/s41598-017-17548-y
210. Huang M, Cheng G, Tan H, Qin R, Zou Y, Wang Y, Zhang Y. Capsaicin protects cortical neurons against ischemia/reperfusion injury via down-regulating NMDA receptors. *Exp Neurol*, 2017, 295: 66–76. doi:10.1016/j.expneurol.2017.05.001

211. Ching LC, Chen CY, Su KH, Hou HH, Shyue SK, Kou YR, Lee TS. Implication of AMP-activated protein kinase in transient receptor potential vanilloid type 1-mediated activation of endothelial nitric oxide synthase. *Mol Med*, 2012, 18: 805–815. doi:10.2119/molmed.2011.00461
212. Ching LC, Kou YR, Shyue SK, Su KH, Wei J, Cheng LC, Yu YB, Pan CC, Lee TS. Molecular mechanisms of activation of endothelial nitric oxide synthase mediated by transient receptor potential vanilloid type 1. *Cardiovasc Res*, 2011, 91: 492–501. doi:10.1093/cvr/cvr104
213. Lv Z, Xu X, Sun Z, Yang YX, Guo H, Li J, Sun K, Wu R, Xu J, Jiang Q, Ikegawa S, Shi D. TRPV1 alleviates osteoarthritis by inhibiting M1 macrophage polarization via Ca^{2+}/CaMKII/Nrf2 signaling pathway. *Cell Death Dis*, 2021, 12: 504. doi:10.1038/s41419-021-03792-8
214. Peng, S., Poole, D. P., Veldhuis, N. A. Mini-review: Dissecting receptor-mediated stimulation of TRPV4 in nociceptive and inflammatory pathways. *Neurosci Lett*, 2022, 770: 136377. https://doi.org:10.1016/j.neulet.2021.136377
215. Toft-Bertelsen, T. L., MacAulay, N. TRPing to the point of clarity: Understanding the function of the complex TRPV4 ion channel. *Cells*, 2021, 10: 165. doi:10.3390/cells10010165
216. Willette RN, Bao W, Nerurkar S, Yue TL, Doe CP, Stankus G, Turner GH, Ju H, Thomas H, Fishman CE, Sulpizio A, Behm DJ, Hoffman S, Lin Z, Lozinskaya I, Casillas LN, Lin M, Trout RE, Votta BJ, Thorneloe K, Lashinger ES, Figueroa DJ, Marquis R, Xu X. Systemic activation of the transient receptor potential vanilloid subtype 4 channel causes endothelial failure and circulatory collapse: Part 2. *J Pharmacol Exp Ther*, 2008, 326: 443–452. doi:10.1124/jpet.107.134551
217. Malczyk M, Veith C, Fuchs B, Hofmann K, Storch U, Schermuly RT, Witzenrath M, Ahlbrecht K, Fecher-Trost C, Flockerzi V, Ghofrani HA, Grimminger F, Seeger W, Gudermann T, Dietrich A, Weissmann N. Classical transient receptor potential channel 1 in hypoxia-induced pulmonary hypertension. *Am J Respir Crit Care Med*, 2013, 188: 1451–1459. doi:10.1164/rccm.201307-1252OC
218. Yang XR, Lin AH, Hughes JM, Flavahan NA, Cao YN, Liedtke W, Sham JS. Upregulation of osmo-mechanosensitive TRPV4 channel facilitates chronic hypoxia-induced myogenic tone and pulmonary hypertension. *Am J Physiol Lung Cell Mol Physiol*, 2012, 302: L555–L568. doi:10.1152/ajplung.00005.2011
219. Kumar, H., Lee, S. H., Kim, K. T., Zeng, X., Han, I. TRPV4: A sensor for homeostasis and pathological events in the CNS. *Mol Neurobiol*, 2018, 55: 8695–8708. doi:10.1007/s12035-018-0998-8
220. Toft-Bertelsen, T. L., MacAulay, N. TRPing on cell swelling – TRPV4 senses it. *Front Immunol*, 2021, 12: 730982. doi:10.3389/fimmu.2021.730982
221. Rungta RL, Choi HB, Tyson JR, Malik A, Dissing-Olesen L, Lin PJC, Cain SM, Cullis PR, Snutch TP, MacVicar BA. The cellular mechanisms of neuronal swelling underlying cytotoxic edema. *Cell*, 2015, 161: 610–621. doi:10.1016/j.cell.2015.03.029
222. Fricker, M., Tolkovsky, A. M., Borutaite, V., Coleman, M., Brown, G. C. Neuronal cell death. *Physiol Rev*, 2018, 98: 813–880. doi:10.1152/physrev.00011.2017
223. Li L, Qu W, Zhou L, Lu Z, Jie P, Chen L, Chen L. Activation of transient receptor potential vanilloid 4 increases NMDA-activated current in hippocampal pyramidal neurons. *Front Cell Neurosci*, 2013, 7: 17. doi:10.3389/fncel.2013.00017
224. Hong Z, Tian Y, Qi M, Li Y, Du Y, Chen L, Liu W, Chen L. Transient receptor potential vanilloid 4 inhibits gamma-aminobutyric acid-activated current in hippocampal pyramidal neurons. *Front Mol Neurosci*, 2016, 9: 77. doi:10.3389/fnmol.2016.00077
225. Ozsimsek, A., Naziroglu, M. The involvement of TRPV4 on the hypoxia-induced oxidative neurotoxicity and apoptosis in a neuronal cell line: Protective role of melatonin. *Neurotoxicology*, 2021, 87: 136–148. doi:10.1016/j.neuro.2021.09.003
226. Shibasaki, K., Ikenaka, K., Tamalu, F., Tominaga, M., Ishizaki, Y. A novel subtype of astrocytes expressing TRPV4 (transient receptor potential vanilloid 4) regulates neuronal excitability via release of gliotransmitters. *J Biol Chem*, 2014, 289: 14470–14480. doi:10.1074/jbc.M114.557132
227. Benfenati V, Amiry-Moghaddam M, Caprini M, Mylonakou MN, Rapisarda C, Ottersen OP, Ferroni S. Expression and functional characterization of transient receptor potential vanilloid-related channel 4 (TRPV4) in rat cortical astrocytes. *Neuroscience*, 2007, 148: 876–892. doi:10.1016/j.neuroscience.2007.06.039
228. Butenko O, Dzamba D, Benesova J, Honsa P, Benfenati V, Rusnakova V, Ferroni S, Anderova M. The increased activity of TRPV4 channel in the astrocytes of the adult rat hippocampus after cerebral hypoxia/ischemia. *PLoS One*, 2012, 7: e39959. doi:10.1371/journal.pone.0039959
229. Hu X, Du L, Liu S, Lan Z, Zang K, Feng J, Zhao Y, Yang X, Xie Z, Wang PL, Ver Heul AM, Chen L, Samineni VK, Wang YQ, Lavine KJ, Gereau RW 4th, Wu GF, Hu H. A TRPV4-dependent neuroimmune axis in the spinal cord promotes neuropathic pain. *J Clin Invest*, 2023, 133: e161507. doi:10.1172/JCI161507

230. Nishimoto R, Derouiche S, Eto K, Deveci A, Kashio M, Kimori Y, Matsuoka Y, Morimatsu H, Nabekura J, Tominaga M. Thermosensitive TRPV4 channels mediate temperature-dependent microglia movement. *Proc Natl Acad Sci U S A*, 2021, 118: e2012894118. doi:10.1073/pnas.2012894118
231. Beeken J, Mertens M, Stas N, Kessels S, Aerts L, Janssen B, Mussen F, Pinto S, Vennekens R, Rigo JM, Nguyen L, Brône B, Alpizar YA. Acute inhibition of transient receptor potential vanilloid-type 4 cation channel halts cytoskeletal dynamism in microglia. *Glia,* 2022, 70: 2157–2168. doi:10.1002/glia.24243
232. Dutta B, Arya RK, Goswami R, Alharbi MO, Sharma S, Rahaman SO. Role of macrophage TRPV4 in inflammation. *Lab Invest*, 2020, 100: 178–185. doi:10.1038/s41374-019-0334-6
233. Luo J, Qian A, Oetjen LK, Yu W, Yang P, Feng J, Xie Z, Liu S, Yin S, Dryn D, Cheng J, Riehl TE, Zholos AV, Stenson WF, Kim BS, Hu H. TRPV4 channel signaling in macrophages promotes gastrointestinal motility via direct effects on smooth muscle cells. *Immunity*, 2018, 49: 107–119.e4. doi:10.1016/j.immuni.2018.04.021
234. Yin J, Michalick L, Tang C, Tabuchi A, Goldenberg N, Dan Q, Awwad K, Wang L, Erfinanda L, Nouailles G, Witzenrath M, Vogelzang A, Lv L, Lee WL, Zhang H, Rotstein O, Kapus A, Szaszi K, Fleming I, Liedtke WB, Kuppe H, Kuebler WM. Role of transient receptor potential vanilloid 4 in neutrophil activation and acute lung injury. *Am J Respir Cell Mol Biol*, 2016, 54: 370–383. doi:10.1165/rcmb.2014-0225OC
235. Kumar H, Lim CS, Choi H, Joshi HP, Kim KT, Kim YH, Park CK, Kim HM, Han IB. Elevated TRPV4 levels contribute to endothelial damage and scarring in experimental spinal cord injury. *J Neurosci*, 2020, 40: 1943–1955. doi:10.1523/JNEUROSCI.2035-19.2020
236. Pumroy RA, Protopopova AD, Fricke TC, Lange IU, Haug FM, Nguyen PT, Gallo PN, Sousa BB, Bernardes GJL, Yarov-Yarovoy V, Leffler A, Moiseenkova-Bell VY. Structural insights into TRPV2 activation by small molecules. *Nat Commun*, 2022, 13: 2334. doi:10.1038/s41467-022-30083-3
237. Katanosaka, K., Takatsu, S., Mizumura, K., Naruse, K., Katanosaka, Y. TRPV2 is required for mechanical nociception and the stretch-evoked response of primary sensory neurons. *Sci Rep*, 2018, 8: 16782. doi:10.1038/s41598-018-35049-4
238. Nedungadi, T. P., Dutta, M., Bathina, C. S., Caterina, M. J., Cunningham, J. T. Expression and distribution of TRPV2 in rat brain. *Exp Neurol*, 2012, 237: 223–237. doi:10.1016/j.expneurol.2012.06.017
239. Shibasaki, K., Murayama, N., Ono, K., Ishizaki, Y., Tominaga, M. TRPV2 enhances axon outgrowth through its activation by membrane stretch in developing sensory and motor neurons. *J Neurosci*, 2010, 30: 4601–4612. doi:10.1523/JNEUROSCI.5830-09.2010
240. Cohen MR, Johnson WM, Pilat JM, Kiselar J, DeFrancesco-Lisowitz A, Zigmond RE, Moiseenkova-Bell VY. Nerve growth factor regulates transient receptor potential vanilloid 2 via extracellular signal-regulated kinase signaling to enhance neurite outgrowth in developing neurons. *Mol Cell Biol*, 2015, 35: 4238–4252. doi:10.1128/MCB.00549-15
241. Zhang H, Xiao J, Hu Z, Xie M, Wang W, He D. Blocking transient receptor potential vanilloid 2 channel in astrocytes enhances astrocyte-mediated neuroprotection after oxygen-glucose deprivation and reoxygenation. *Eur J Neurosci*, 2016, 44: 2493–2503. doi:10.1111/ejn.13352
242. Hassan S, Eldeeb K, Millns PJ, Bennett AJ, Alexander SP, Kendall DA. Cannabidiol enhances microglial phagocytosis via transient receptor potential (TRP) channel activation. *Br J Pharmacol*, 2014, 171: 2426–2439. doi:10.1111/bph.12615
243. Luo H, Rossi E, Saubamea B, Chasseigneaux S, Cochois V, Choublier N, Smirnova M, Glacial F, Perrière N, Bourdoulous S, Smadja DM, Menet MC, Couraud PO, Cisternino S, Declèves X. Cannabidiol increases proliferation, migration, tubulogenesis, and integrity of human brain endothelial cells through TRPV2 activation. *Mol Pharm*, 2019, 16: 1312–1326. doi:10.1021/acs.molpharmaceut.8b01252
244. Choi, S. H., Mou, Y., Silva, A. C. Cannabis and cannabinoid biology in stroke. *Stroke*, 2019, 50: 2640–2645. doi:10.1161/STROKEAHA.118.023587
245. Peier AM, Reeve AJ, Andersson DA, Moqrich A, Earley TJ, Hergarden AC, Story GM, Colley S, Hogenesch JB, McIntyre P, Bevan S, Patapoutian A. A heat-sensitive TRP channel expressed in keratinocytes. *Science*, 2002, 296: 2046–2049. doi:10.1126/science.1073140
246. Smith GD, Gunthorpe MJ, Kelsell RE, Hayes PD, Reilly P, Facer P, Wright JE, Jerman JC, Walhin JP, Ooi L, Egerton J, Charles KJ, Smart D, Randall AD, Anand P, Davis JB. TRPV3 is a temperature-sensitive vanilloid receptor-like protein. *Nature*, 2002. 418: 186–190. doi: 10.1038/nature00894
247. Xu H, Ramsey IS, Kotecha SA, Moran MM, Chong JA, Lawson D, Ge P, Lilly J, Silos-Santiago I, Xie Y, DiStefano PS, Curtis R, Clapham DE. TRPV3 is a calcium-permeable temperature-sensitive cation channel. *Nature*, 2002. 418: 181–186. doi: 10.1038/nature00882
248. Moussaieff A, Yu J, Zhu H, Gattoni-Celli S, Shohami E, Kindy MS. Protective effects of incensole acetate on cerebral ischemic injury. *Brain Res*, 2012, 1443: 89–97. doi:10.1016/j.brainres.2012.01.001

249. Chen, X., Zhang, J., Wang, K. Inhibition of intracellular proton-sensitive Ca^{2+}-permeable TRPV3 channels protects against ischemic brain injury. *Acta Pharm Sin B*, 2022, 12: 2330–2347. doi:10.1016/j.apsb.2022.01.001
250. Jordt SE, Bautista DM, Chuang HH, McKemy DD, Zygmunt PM, Högestätt ED, Meng ID, Julius D. Mustard oils and cannabinoids excite sensory nerve fibres through the TRP channel ANKTM1. *Nature*, 2004. 427: 260–265. doi: 10.1038/nature02282
251. Bandell M, Story GM, Hwang SW, Viswanath V, Eid SR, Petrus MJ, Earley TJ, Patapoutian A. Noxious cold ion channel TRPA1 is activated by pungent compounds and bradykinin. *Neuron*, 2004, 41: 849–857. doi:S0896627304001503 [pii]
252. Bautista, D. M., Pellegrino, M., Tsunozaki, M. TRPA1: A gatekeeper for inflammation. *Annu Rev Physiol*, 2013, 75: 181–200. doi:10.1146/annurev-physiol-030212-183811
253. Pires, P. W., Earley, S. Neuroprotective effects of TRPA1 channels in the cerebral endothelium following ischemic stroke. *Elife*, 2018, 7: e35316. https://doi.org:10.7554/eLife.35316

11 Evolution of Thermosensitive TRP Channels as a Source of Sensory Diversification and Environmental Adaptation

Shigeru Saito and Makoto Tominaga

11.1 INTRODUCTION

Sensory perception, including vision, olfaction, taste, hearing, and somatosensation, provides information about ambient conditions that enables animals to respond behaviorally and physiologically to changing environments and is essential for survival. Considering this important role in survival, evolutionary changes in sensory perception would be expected to play pivotal roles in adaptive evolution. Among the various sensory perceptions, mechanical, thermal, and chemical stimuli activate primary afferent neurons in the somatosensory system. Such activation is essential for detecting noxious stimuli that can damage tissues.[1,2] Several ion channels belonging to the transient receptor potential (TRP) channel family are expressed in skin and primary sensory neurons, where they initiate somatosensory signal transduction.[1,3] TRP channels are activated by critical environmental factors like temperature as well as chemical and other physical stimuli.[1,3] Alterations in the functional properties of TRP channels directly affect somatosensory perception. Thus, these channels are likely to be involved in evolutionary adaptation to diverse environments.

Most TRP channels are non-selective cation channels that, upon activation, are permeable to sodium and calcium ions. This ion movement causes membrane depolarization that leads to action potential generation in peripheral sensory neurons derived from the dorsal root ganglion (DRG) and trigeminal ganglion (TG).[1,3] Mammalian species such as mice possess 28 different TRP channels, of which 11 members are temperature-sensititive.[4,5] The temperature range that activates each TRP channel differs, and collectively, they roughly cover the entire range of temperatures that mammals sense. In addition to temperature, these channels are also activated by various physical and chemical stimuli and thus act as multimodal receptors. Expression of TRP channels in various tissues allows their involvement in diverse physiological functions.[1–4,6–10]

TRP channels form a gene family that can be divided into several subfamilies according to sequence similarity and domain structure. Vertebrate TRP channels are classified into seven subfamilies, including TRP canonical (TRPC), TRP vanilloid (TRPV), TRP melastatin (TRPM), TRP ankyrin (TRPA), TRP polycystin (TRPP), and TRP mucolipin (TRPML), and TRP NOMP (TRPN, not present in amniotes).[11] TRP channels are composed of four subunits, each possessing six transmembrane domains and relatively large intracellular N- and C-terminal regions.[5,11–13] The ion permeation pathway lies at the center of the channel formed by the fifth and sixth transmembrane domains, with a selectivity filter located in the loop region between these two domains. Domain structures of TRP channels vary among different subfamilies. Channels belonging to the TRPV, TRPC, and TRPA subfamilies have multiple ankyrin repeat domains. TRPA1 has a long stretch of 16 ankyrin repeat domains,[14] whereas some TRPC, TRPM, and TRPV channels have a short sequence motif termed the "TRP box," located near the sixth transmembrane domain in the C-terminal region. Intercellular components such as ATP and calmodulin regulate TRP channel

 DOI: 10.1201/9781003402732-11

activity by interacting with these ankyrin repeat domains or the part of the C-terminal region.[1,2] Phosphatidylinositol 4,5-bisphosphate (PIP2) also interacts with transmembrane regions of TRP channels to regulate activity.[15]

Our understanding of TRP channels in various animal species has expanded over the past two decades.[16–19] Comparative analyses using various animal models have revealed that functional alterations in TRP channels led to diversification of somatosensation among species and, in some species, were involved in the acquisition of unique sensory systems. Recent studies have revealed that the functional properties of TRP channels vary in relation to thermal niche selection. This chapter summarizes the results of studies that focused on the evolutionary aspect of thermosensitive TRP channels, and particularly their roles in environmental adaptation.

11.2 EVOLUTIONARY CHANGES IN THERMOSENSITIVE TRP CHANNELS

TRP channels are widely shared among eukaryotes and have diversified into multiple subfamilies.[20] Although TRP channels belonging to different subfamilies are found across the animal kingdom, the gene repertoires showed species-specific variations with substantial differences between invertebrates and vertebrates.[21–24] As mentioned above, vertebrates possess TRP channels belonging to seven subfamilies: TRPC, TRPV, TRPM, TRPA, TRPP, TRPML, and TRPN.[11] Rodents have 11 thermosensitive TRP channels, with four belonging to the TRPV subfamily (TRPV1-TRPV4), five to the TRPM subfamily (TRPM2-TRPM5, and TRPM8), one to the TRPA subfamily (TRPA1), and one to the TRPC subfamily (TRPC5).[5] Thermal perception by insect species like fruit flies involves two channels belonging to the TRPC subfamily (TRP and TRPL), three to the TRPA subfamily (TRPA1, painless, and pyrexia), and three to the TRPP subfamily (biribido-1, -2, and -3), although whether all of these proteins function as ion channels is unclear.[25] Therefore, TRP channels involved in thermal perception largely differ between invertebrates and vertebrates.

Phylogenetic analysis of TRP channels revealed that TRPA1 is likely the only orthologous gene shared between vertebrates and invertebrates.[8] Most thermosensitive TRP channels found in mammals newly emerged at the latest in the most recent common ancestor between teleosts and tetrapods, suggesting that a pivotal rearrangement of thermal receptors occurred during that evolutionary period.[22] Subsequent gene duplication and loss events in the respective vertebrate lineages led to the diversification of thermosensitive TRP channels, even among extant vertebrate classes such as teleosts, amphibians, reptiles, birds, and mammals.[21] Especially, the gene repertoires of thermosensitive TRP channels diverged between teleosts and tetrapods. For instance, TRPV2, TRPV3, and TRPM8 genes were lost from teleost species, whereas the number of channels belonging to the TRPM subfamily increased.[17] Moreover, some teleost species possess two copies of TRPV1 and TRPA1 paralogs that likely arose from gene duplication events that occurred in the most recent common ancestors of teleosts. The diversification of TRP channel repertoires between teleosts and tetrapods may be related to adaptation to aquatic or terrestrial environments, respectively.

11.3 FUNCTIONAL DIVERSITY IN THERMOSENSITIVE TRP CHANNELS

In addition to variations in the gene repertoire, functional diversification of TRP channels among species is of particular interest, as function directly influences sensory perception. The ion channel properties of TRP channels can be investigated using electrophysiological approaches with in vitro expression systems that enable comparison of the functional properties of TRP channels from diverse animal species. Among the thermosensitive TRP channels, TRPA1s from various species, including vertebrate and invertebrate species, have been cloned and functionally characterized.[16,18]

TRPA1 cloned from rats was first reported to be activated by cold stimulus and was later shown to be activated by pungent chemicals and environmental irritants like mustard oil, cinnamaldehyde, and acrolein.[26–32] Despite its established role as a receptor for chemical irritants, debate continues about the role of TRPA1 in thermosensation in humans and rodents.[8,26,33] Some studies reported that TRPA1 does not respond to cold stimulation, whereas others suggested that TRPA1 functions as a bimodal thermal sensor activated by both cold and heat. Further characterization has been made for TRPA1 across various classes of vertebrates, including reptiles (e.g., lizards and snakes), birds (e.g., chickens), amphibians (e.g., frogs and salamanders), and teleost fish (such as zebrafish, medaka, and pufferfish; Fig. 11.1). TRPA1 in reptiles, birds, and amphibians appears to be primarily activated by heat rather than cold stimulation.[34,35] Two paralogous TRPA1 genes, TRPA1a and TRPA1b, are present in zebrafish and likely arose through gene duplication.[22] Chemical, but not thermal, stimulation activates zebrafish TRPA1a, whereas TRPA1b is predominantly activated by thermal stimulation, including both heat and cold. These different properties suggest that subfunctionalization occurred after gene duplication in zebrafish.[36] Pufferfish TRPA1 also exhibits dual activation by both cold and heat stimuli, while TRPA1 in medaka is solely activated by heat.[37,38] In invertebrates like starfish and insects, TRPA1 is activated by warmth or heat as well as chemical compounds that also activate vertebrate TRPA1.[39–41] Collectively, these observations suggest that TRPA1 emerged as a thermal and chemical sensor to help animals avoid harmful environmental stimuli, and likely originated in common ancestors of animals (Fig. 11.1).

Vertebrate species possess another heat-sensitive channel, TRPV1, that likely emerged in a common ancestor of tetrapods and teleosts (Fig. 11.1).[21,22] TRPV1 is activated by heat, a feature consistent in various vertebrate species, including rodents, chickens, snakes, amphibians, and teleosts.[42–45] TRPV1 is often co-expressed with TRPA1 in subsets of DRG and TG neurons in vertebrate species such as rodents, chickens, clawed frogs, and zebrafish,[35,46] suggesting that TRPV1 was likely acquired as a heat-sensitive channel in the most recent common ancestor of tetrapods and teleosts, leading to redundancy in heat receptors.[17,18]

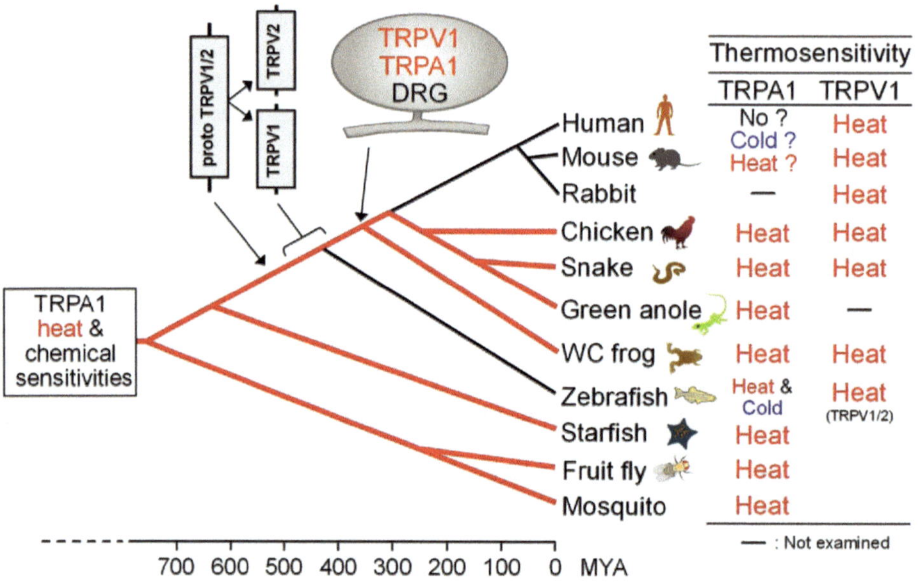

FIGURE 11.1 Evolutionary trajectory of TRPA1 and TRPV1. Summary of thermal sensitivity of TRPA1 and TRPV1 in representative animal species. TRPA1 is likely to have possessed chemical and heat sensitivity at an early stage of animal evolution. The timing of the gene duplication that produced TRPV1 has not been precisely determined but occurred, at the latest, in the most recent common ancestor of tetrapods. MYA: Million years ago.

11.4 SPECIES DIFFERENCES IN TRPA1 AND TRPV1 CHEMICAL SENSITIVITY

TRPA1 is widely conserved among diverse animal species due to its functional importance as a nociceptive receptor, but some species differences in sensitivity to specific chemical compounds have been reported. Methyl anthranilate (MA), a chemical compound found in Concord grapes and used as a food additive to impart grape flavor, is known to elicit aversive responses in birds and thus has industrial applications as a safe bird repellent.[47] TG neurons are responsible for the aversive response of birds to MA, and the activation of chicken TRPA1 by MA suggests that noxious stimulus is perceived via TRPA1.[34] Comparative analysis revealed variations in TRPA1 response to MA among vertebrate species. TRPA1 from chickens, mice, and humans is potently activated by MA, whereas frog and lizard TRPA1 show considerably reduced activity. A comparison of the amino acid sequences of vertebrate TRPA1 together with mutagenesis analysis identified three amino acid residues that are required for TRPA1 activation by MA.[34] These amino acids are located close to each other around the loop region that connects the cytosolic ankyrin repeat domains and the first transmembrane domain in TRPA1.

Various TRPA1 inhibitors have been developed as it is one of the potential targets for developing analgesia. These chemical compounds are typically screened using mammalian TRPA1, and the inhibitory effects vary among TRPA1 from non-mammalian species. For example, A-967079 is a potent inhibitor of mouse and human TRPA1 but does not inhibit TRPA1 from chickens, frogs, or fruit flies.[48] Detailed chimeric and mutagenesis analyses revealed that A-967079 interacts with amino acids located in the pore loop region of TRPA1.[14,48] On the other hand, HC-030031 shows inhibitory effects toward TRPA1 of humans, mice, chickens, and lizards but does not inhibit frog and zebrafish TRPA1. The amino acid responsible for the inhibitory effect of HC-030031 is located in the loop between the fourth and fifth transmembrane domains, a region that is distant from the pore loop region involved in A-967079 inhibition.[49] These observations suggest that HC-030031 and A-967079 act on TRPA1 in different manners, due to the structural differences between these two chemical compounds. Interestingly, A-967079 acts as an agonist to chicken TRPA1. Mutagenesis analysis revealed that a single amino acid change in the binding pocket in the pore loop region is responsible for the opposite effect of A-967079 on TRPA1 between chickens and humans,[50] suggesting that even slight structural modifications in the binding pocket can drastically affect the conformational movements of the channel gate.

Species diversity in TRPV1 sensitivity to capsaicin is also well documented. Mammalian species, including humans, rodents, and dogs, possess TRPV1 that has high sensitivity to capsaicin, whereas rabbit TRPV1 shows reduced sensitivity.[51,52] Furthermore, several vertebrate species, like tropical clawed frogs, chickens, and zebrafish, also possess TRPV1 channels that have reduced or no sensitivity to capsaicin.[42,43,53] The amino acid residues responsible for these species differences in TRPV1 capsaicin sensitivity have been localized to the third and fourth transmembrane domains (positions 512, 550, and 570 in rat TRPV1).[17,43,51,53,54] These amino acids form hydrogen bonds with capsaicin molecules,[12,55] again indicating that alterations in the binding pocket lead to diverse TRPV1 sensitivity to capsaicin.

Although species differences in TRPV1 capsaicin sensitivity appear to be largely unrelated to environmental adaptation, the reduced sensitivity of TRPV1 to capsaicin seen for tree shrews is one exception. Tree shrews (*Tupaia belangeri chinensis*) are reported to consume chili peppers and have reduced TRPV1-mediated sensitivity to capsaicin.[56] *T. belangeri chinensis* inhabits south Asia, where *Piper boehmeriaefolium*, a spicy plant that produces a capsaicin analog, is distributed. Thus, the reduced capsaicin sensitivity of TRPV1 in *T. belangeri chinensis* could confer a selective dietary advantage to these animals. It has also been hypothesized that the reduced sensitivity of bird TRPV1 to capsaicin could facilitate seed dispersal of capsicum species (chili peppers), while such sensitivity helps deter consumption by herbivorous mammals like rodents.[57]

11.5 INVOLVEMENT OF THERMOSENSITIVE TRP CHANNELS IN A UNIQUE PHYSIOLOGICAL SYSTEM

Some species have developed unique physiological systems associated with thermal adaptation during the course of evolution. Several snake species, including pit vipers, boas, and pythons, acquired the ability to sense infrared radiation via a specialized organ in the head called the pit organ that is innervated by TG sensory neurons. To investigate the involvement of TRP channels in pit organs, RNA-seq analysis was carried out to compare gene expression levels between TG and DRG,[44] the latter of which is involved in somatosensory perception in the trunk. This transcriptome analysis revealed that TRPA1 expression is higher in the TG than in the DRG in species possessing a pit organ. In contrast, snake species without a pit organ had similar TRPA1 expression levels in the TG and DRG. Further characterization of TRPA1 in snakes revealed that species with a pit organ had considerably lower TRPA1 thermal activation thresholds than snakes without a pit organ. Detailed molecular evolutionary analysis detected signatures of positive selection in terms of an increased evolutionary rate of non-synonymous substitutions in the evolutionary lineages of snakes with pit organs.[58,59] These results suggest that alterations in channel expression and functional properties played important roles in the acquisition of infrared detection by pit organs in snake lineages.

Infrared detection has also been acquired in mammalian lineages. Vampire bats are obligated blood feeders and locate hotspots on endothermic prey by detecting infrared radiation using leaf pits that surround the nose. Heat-sensitive neurons innervating leaf pits respond to temperatures above 29°C. Infrared detection by vampire bats is thought to involve TRPV1, which in these bats is expressed as two alternatively spliced (AS) variants. In the novel AS variant termed TRPV1-S, the C-terminal region is truncated relative to the canonical AS variant termed TRPV1-L. TRPV1-S is predominately expressed in the TG and is activated by temperatures around 30°C, whereas TRPV1-L is predominantly expressed in the DRG and is activated by temperatures around 40°C. Therefore, vampire bats likely rely on the TRPV1 AS variant with its higher thermal sensitivity for infrared detection.

TRPV1-S is also found in TG in fruit bats, which do not have the ability to detect infrared radiation; however, its expression level is considerably low. These observations suggest that the gain in the expression level of TRPV1-S is a key innovation to acquire infrared detection in vampire bats. It is intriguing that functional changes in TRP channels were likely independently involved in the acquisition of infrared detection, although the types of channels and molecular basis are different in snake and bat lineages.[44,60]

TRPA1 is also involved in seasonal adaptation of silk moths (*Bombix mori*), which lay two types of eggs, dormant and non-diapausing eggs. Production of dormant eggs is essential for silk moths to survive adverse winter seasons. The type of eggs laid by female adult moths is determined by the environmental temperatures to which they were exposed during their embryonic stage. Adult female silk moths (Kosetsu type) exposed to 25°C during the embryonic stage lay dormant eggs, but those exposed to 15°C lay non-diapause eggs. TRPA1, which is expressed in the embryonic stage, has been identified as a thermal sensor that determines the type of eggs laid by females.[41] Activation of *B. mori* TRPA1 by temperatures around 25°C induces expression of a dormancy-inducing peptide hormone. *B. mori* having TRPA1 knockout cannot respond to environmental temperature and instead the photoperiod determines the type of eggs laid.[61]

Silk moths are thought to have been domesticated from *Bombyx mandarina* 10,000–5,000 years ago. *B. mandarina* also produces dormant and non-diapausing eggs, and the photoperiod determines the type of eggs laid by the moths. Examination of *B. mandarina* TRPA1 showed activation by 25°C and expression in embryos, suggesting that TRPA1 might have functioned as a thermal sensor in moths that could have been exploited during the domestication process to control the egg-laying phenotype.[61]

In most animal species, sex is genetically determined. However, in some reptilian species, sex is determined by environmental temperature. For example, in American alligators (*Alligator mississippiensis*), incubation of eggs at 33°C results in all male offspring, whereas only females hatch from eggs incubated at 30°C. TRPV4 is reported to be expressed during a critical temperature-sensitive period that determines the sexual fate. Activation of TRPV4 by warm temperatures may thus play a role in temperature-dependent sex determination in American alligators. Moreover, treatment of eggs with pharmacological agents that affect TRPV4 function also affects gene expression patterns associated with male differentiation, further supporting the suggestion that TRPV4 is involved in temperature-dependent sex determination in American alligators.[62]

11.6 ASSOCIATION BETWEEN EVOLUTIONARY CHANGES IN TRP CHANNELS AND ENVIRONMENTAL ADAPTATION

Elucidating the roles of TRP channels in environmental adaptation is of particular interest in the context of evolutionary biology, as temperature is a critical environmental factor for most organisms. The temperature range suitable for each species diverged as a consequence of adaptation to habitats that have different thermal conditions. Considering that animals evolved the ability to behaviorally respond to ambient temperature, thermal perception that is essential for exerting proper behavioral responses would have been altered during adaptation processes. Comparative analyses among species inhabiting diverse thermal niches can provide informative clues to understand the association between evolutionary changes in thermal receptors and environmental adaptation.

To address this issue, a comparative analysis of tadpoles of frog species was performed to elucidate the role of TRP channels in environmental adaptation.[63] Water environments are essential for amphibians, especially for breeding, as embryos and larvae generally develop in water. Due to the variation in spawning seasons and water environments, larvae grow under different thermal conditions even among species with similar distributions. *Rana japonica, Glandirana rugosa, Rhacophorus schlegelii*, and *Buergeria buergeri* are four species of frogs that are widely distributed across Japan. On the other hand, *Buergeria japonica* is distributed in the Ryukyu Archipelago in Japan. *B. japonica* tadpoles are found in various water environments. Field observations in shallow paddles revealed that water temperatures can reach as high as 40°C on sunny days, suggesting that *B. japonica* tadpoles are frequently exposed to heat. Tadpoles of this species inhabit even in natural geothermal hot springs where water temperatures reach up to 44°C.[64] In contrast, *R. japonica* tadpoles grow in relatively cool conditions since parental frogs lay eggs in late winter and tadpoles metamorphose and move to terrestrial environments before summer.[63]

To compare the heat tolerance, the critical thermal maximum (CTmax), defined as the temperature at which loss of righting reflex upon heating occurs, was determined for tadpoles of the five frog species mentioned above. The CTmax values ranged from 38°C to 43°C among the five species (Fig. 11.2A). As expected from their ecological characteristics, *R. japonica* had the lowest CTmax and *B. japonica* had the highest; *B. buergeria* had an intermediate value. These results show that the heat tolerance of tadpoles is likely tuned by the thermal conditions of the habitats.[63] Avoidance behaviors of these five species had a similar trend as that seen for CTmax, suggesting that heat tolerance and avoidance behaviors cooperatively evolved during adaptation to diverse thermal niches. Interestingly, the species difference in avoidance temperatures was 2.6 times larger than that seen for CTmax, indicating that avoidance behavior can be flexibly altered with the thermal niches of the species.

To clarify the association between TRP channel properties and thermal niches of the different species, TRPA1 channels from *R. japonica*, *B. buergeri*, and *B. japonica* were cloned and functionally characterized. These species harbor a novel alternatively spliced (AS) variant having a single valine insertion between the sixth and seventh ankyrin repeat domains compared to the canonical AS variant shared among vertebrate species.[63] The novel AS variant and canonical version are termed TRPA1(V+) and TRPA1(V-), respectively. Both TRPA1 AS variants in *R. japonica* potently respond to heat. In *B. buergeri*, only TRPA1(V+) responds to

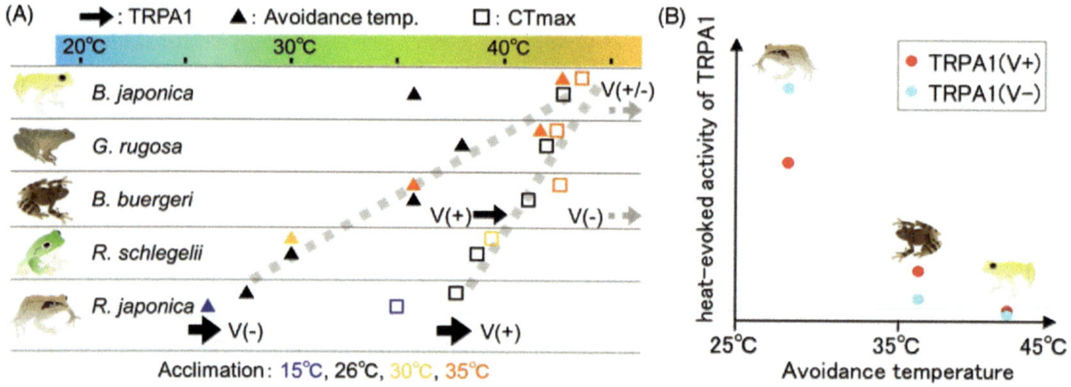

FIGURE 11.2 Heat tolerance, avoidance temperature, and TRPA1 properties of five frog species. (A) Summary of heat tolerance (CTmax), avoidance temperature, and TRPA1 properties of five frog species. Tadpoles were used for behavioral assays. CTmax and avoidance temperature values are indicated by open squares and filled triangles, respectively. Colors indicate the acclimation temperatures of tadpoles shown below. Each arrow indicates the thermal activation threshold of a TRPA1 AS variant, and the activity in response to heat corresponds to the thickness of the arrow. Gray dashed arrows depict near loss of heat-evoked activity of TRPA1 AS variants. (B) Comparison of the heat-evoked activity of TRPA1 AS variants in three frog species. Heat-evoked activity is estimated from the ionic currents obtained from *Xenopus laevis* oocytes expressing TRPA1 from each species.

heat, while TRPA1(V-) exhibits only a faint heat-evoked response. Furthermore, in *B. japonica*, both AS variants almost completely lack heat sensitivity. In all three species, both TRPA1 AS variants could be dose-dependently activated by cinnamaldehyde, a well-known TRPA1 agonist. Normalization of heat-evoked TRPA1 activity using responses elicited by cinnamaldehyde revealed that it declines in the order of *R. japonica*, *B. buergeri*, and *B. japonica* for both AS variants (Fig. 11.2B).[63] These results suggest that heat-evoked TRPA1 activity is maintained in species adapted to relatively cool environments, such as *R. japonica*, as they are susceptible to heat exposure. In contrast, heat-evoked TRPA1 activity has been nearly abolished in species that have increased heat resistance that allow adaptation to harsh thermal niches, such as *B. japonica*.

A shift in the thermal sensitivity of TRPA1 was observed between *R. japonica* and *B. buergeri*. The average thermal activation thresholds of TRPA1(V+) and TRPA1(V-) in *R. japonica* are approximately 37°C and 26°C, respectively, while that of *B. buergeri* TRPA1(V+) is approximately 39°C (Fig. 11.2A). *R. japonica* possesses TRPA1 AS variants with high and low thermal sensitivity, while *B. buergeri* only possesses a TRPA1 AS variant that has low thermal sensitivity. Due to the near absence of the heat-evoked responses, the thermal activation threshold of *B. buergeri* TRPA1(V-) and both AS variants of *B. japonica* cannot be estimated.

An association between TRPA1 channel properties and thermal niches has also been found in clawed frogs from Africa. *Xenopus tropicalis* and *Xenopus muelleri* are distributed in relatively hot regions, while *Xenopus laevis* and *Xenopus borealis* are found in relatively cool areas. A behavioral assay comparing *X. laevis* and *X. tropicalis* showed that the former species exhibit nocifensive heat responses (i.e., jumping) from lower temperatures.[65] The thermal activation threshold was lower in dissociated DRG neurons from *X. laevis* relative to those from *X. tropicalis*. Comparison of TRPA1 channel properties revealed that its heat-evoked activity is significantly higher in *X. laevis* and *X. borealis*, which inhabit cooler regions, compared to that in *X. tropicalis* and *X. muelleri*. Together, these results suggest that TRPA1 activity changed according to the thermal conditions of their habitats.[66] The thermal activation threshold of TRPA1 also showed species-specific variations, although these differences were not consistent with the thermal conditions of habitats.

Closely related lizard species in Caribbean islands are reported to have alterations in TRPA1 thermal response. Three sympatric lizard species (*Anolis allogus, Anolis homolechis*, and *Anolis sagrei*) occupy microhabitats with different thermal conditions. Behavioral assays showed that the avoidance temperatures of *A. homolechis* and *A. sagrei* are higher than those of *A. allogus.*[67] Similarly, the thermal activation thresholds of TRPA1 in *A. homolechis* and *A. sagrei* are significantly higher than those of *A. allogus*, suggesting that TRPA1 thermal sensitivity shifted with the thermal niches of the three lizard species.

In addition to TRPA1, TRPV1 is likely to be involved in thermal adaptation. Investigation of thirteen-lined ground squirrels and camels, which both possess extreme heat resistance, revealed that these species have TRPV1 channels with nearly no heat sensitivity.[68] TRPV1 channels in these two species exhibit similar sensitivity to capsaicin and acid, suggesting that the channel function is maintained. On the other hand, TRPV1 in axolotl and several species of salamanders, which inhabit cool aquatic environments, is activated by considerably lower temperatures (28–34°C) compared to the orthologs in mammals and clawed frogs, which are activated by temperatures higher than 40°C.[69,70] Axolotl TRPA1 is reported to be activated by 40°C; thus, TRPV1, which is activated at 31°C, potentially plays a crucial role in noxious heat detection in this species.[69–71] Meanwhile, masu salmon, which also prefers cool temperatures, possesses TRPV1 with a thermal activation threshold around 28°C.[45] These observations suggest that shifts in the thermal sensitivity of TRPV1 might also be related to thermal adaptation in some species. In Antarctic notothenioid fish living in extreme cold habitats, TRPV1 and TRPA1 have unexpected thermal activation thresholds that are considerably higher than temperatures experienced in the wild.[72] Detailed functional assays revealed that the thermal activation threshold of TRPA1, but not of TRPV1, declines with alterations in cellular conditions such as oxidation and reduction in cholesterol and becomes close to the expected temperature range.

Variations in thermal activity of a cold-activated channel have also been reported. A study comparing mammalian and avian species (seven species from mammals and one from avians) that inhabit different thermal niches showed that the cold-evoked activity of TRPM8 decreased as the temperature of the species' habitat declined.[73] In this study, TRPM8 of African elephants was shown to exhibit the highest cold-evoked activity, while that of emperor penguins, which live in Antarctica, had the lowest activity in response to cold. The TRPM8 of all species compared in the study maintained menthol-evoked responses, suggesting that the observed difference in cold-evoked activity is unrelated to the expression levels of the channel protein or chemical gating. Interestingly, the introduction of TRPM8 from emperor penguins into TRPM8 knockout mice failed to restore the behavioral responses to cold, implying that a reduction in the cold-evoked activity of TRPM8 dampens behavioral responses. Cold-evoked activity of TRPM8 in thirteen-lined ground squirrels and Syrian hamsters, which both hibernate, is also considerably lower than other species,[74] but whether the reduced activity of TRPM8 is indeed related to hibernation awaits further comparative studies between hibernating and non-hibernating species.

Taken together, comparative analysis of TRP channels among various animal species revealed that changes in thermal activity likely served as a common mechanism of adaptation to different thermal niches, as consistent associations between species differences and thermal niches are found for both heat- and cold-activated channels and across different groups of vertebrates.[63,66,73] Although the association of TRP channel thermal sensitivity (activation threshold) and thermal niche of species appears to be weak,[66] in some cases changes in sensitivity do play a crucial role in adaptation.[75] In many instances, alterations in TRP channel properties are specifically associated with thermal responses, but sensitivities to chemical stimulations are maintained.[63,66] For example, the thermal sensitivity of both TRPV1 and TRPM8 in thirteen-lined ground squirrels is nearly absent, but sensitivity to chemical stimuli (capsaicin for TRPV1 and menthol for TRPM8) remains similar to that of rat orthologs.[68,74] For TRPA1 in frog species like *B. japonica* and *X. tropicalis*, TRPA1 retains sensitivity to cinnamaldehyde despite having considerable reductions in heat-evoked activity.[63,65,66] These findings suggest that selective pressure during the evolutionary processes specifically acts on channel function related to thermal responses. Further comparative analyses using diverse animal species may deepen our understanding of the role of TRP channels in thermal adaptation.

11.7 STRUCTURAL BASIS FOR THE FUNCTIONAL DIVERSITY OF THERMOSENSITIVE TRP CHANNELS RELATED TO ENVIRONMENTAL ADAPTATION

Species diversity in TRP channel properties provides an opportunity to understand the structural basis for functional evolution. Chimeric and mutagenesis analyses of TRPV1 between rats and thirteen-lined ground squirrels identified two amino acids located within the ankyrin repeat domains that are associated with the species difference in thermal activity (Fig. 11.3A, B).[68] Additionally, three amino acid substitutions between African and tropical clawed frogs are reported to partially contribute to species differences in TRPV1 responses to repeated heat stimulation (sensitization vs desensitization).[65] Comparative and mutagenesis experiments using salamander TRPV1 revealed that two single amino acids located in the first ankyrin repeat domain are responsible for its species differences in heat sensitivity.[70] Further detailed analysis showed that the thermal stability of ankyrin repeat domains is indeed corelated with the thermal sensitivity of TRPV1 channels.

Ankyrin repeat domains are also involved in different thermal responses of TRPA1. Thermal activation thresholds of TRPA1 are altered between mosquito species that occupy different thermal habitats (*Culex pipiens* and *Anopheles stephensi*). Chimeric and point mutagenesis analyses revealed that two charged residues in the N-terminus ankyrin repeat are important for determining temperature thresholds in these insects. Identified charged amino acids likely participate in salt bridge networks for heat-evoked structural changes.[76,77] Furthermore, the TRPA1 AS variants in frogs (*R. japonica*, *B. buergeri*, and *B. japonica*) have a single amino acid insertion between the sixth and seventh ankyrin repeat domains that greatly affects both the activity and sensitivity of TRPA1 to heat stimulation.[63] Together, these findings suggest that the ankyrin repeat area is an important target domain that affects thermal responses of TRP channels during the evolutionary process.

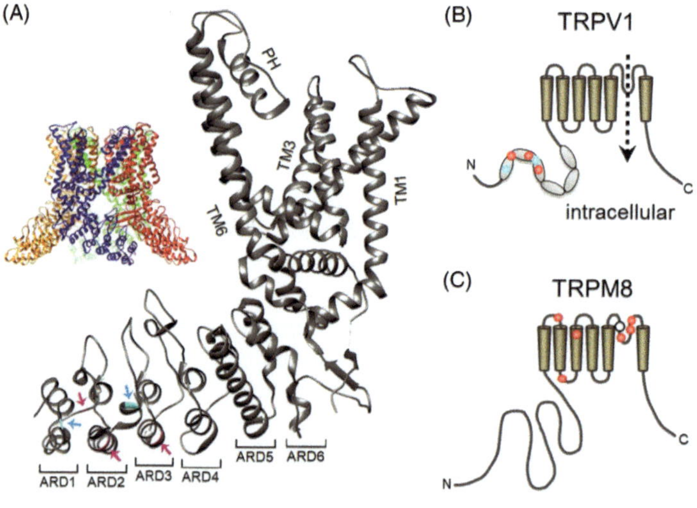

FIGURE 11.3 Amino acids associated with species differences in TRPV1 and TRPM8 thermal responses. (A) 3D structure of rat TRPV1. Structures of a single subunit of rat TRPV1 (right) and a tetrameric channel (left). The position of amino acids associated with species differences of TRPV1 between rats and thirteen-lined ground squirrels, or *X. laevis* and *X. tropicalis*, is indicated with cyan and magenta, respectively (A and B). (C) The position of amino acids responsible for species differences of TRPM8 between rats and thirteen-lined ground squirrels, or emperor penguins and African elephants are indicated by magenta and open circles, respectively. Figure adapted and modified from fig. 4 in Saito (2023).[79]

The structural basis for cold-activated TRPM8 channels has been identified. In the loop and transmembrane domains, six amino acids were shown to be involved in the difference in TRPM8 cold responses of rats and thirteen-lined ground squirrels (Fig. 11.3C).[68] On the other hand, a single amino acid substitution located in the pore domain can explain nearly all species differences in the cold-evoked responses of TRPM8 between African elephants and emperor penguins.[73] Furthermore, comparison of TRPM8 among various species revealed the tendency that an increase in the pore domain hydrophobicity enhances the cold-evoked activity of TRPM8.

Overall, identification of amino acids responsible for species differences in TRP channel responses revealed that their thermal sensitivity can be altered by a small number of amino acid substitutions, and in some cases, even a single amino acid substitution. This structural feature of TRP channels might be related to the flexible nature of channel function that enables participation in environmental adaptation during evolutionary processes.

11.8 FUTURE PERSPECTIVES

Given that TRP channels serve as peripheral thermal sensors initiating signal transduction, they likely played crucial roles in changing thermal perception during environmental adaptation. Indeed, recent studies focusing on species adapted to diverse thermal niches revealed that the temperature-related function of TRP channels shifted during the evolutionary process. However, in many studies, whether the observed species differences of TRP channels revealed by in vitro functional assays are associated with sensory perception and alter behavioral responses has not been investigated.[44,60,63,65–68,74] Such investigations may be challenging since the comparative analyses largely involved non-model species. Making genetic modifications in such species is difficult, although recent advances in genome editing technologies may allow manipulation of TRP channel genes to investigate the physiological role of functional divergence even in non-model species.

The molecular mechanisms for functional alterations in TRP channels have been extensively investigated, and the amino acids responsible for the species differences in TRP channels have been identified through chimeric and mutagenesis analyses using extant species. However, comparative analyses of these extant species did not always highlight amino acid changes involved in functional alterations of proteins in an evolutionary context. For example, regarding the difference in TRPV1 heat-evoked activity of thirteen-lined ground squirrels and rats, two amino acids were identified; however, these amino acids altered in the lineage leading to rats,[68] suggesting that they may not have played a role in TRPV1 functional shift in these squirrels. Similarly, comparative analysis of TRPV1 heat responses between African and tropical clawed frogs identified three amino acid substitutions responsible for the species difference, but only one of the three occurred in the lineage that led to tropical clawed frogs in which functional shift took place.[66] Thus, identification of bona fide amino acid substitutions that played critical roles in the functional alteration of TRP channels must consider the evolutionary trajectory. Reconstruction and functional characterization of ancestral proteins are effective approaches to identify amino acid changes that actually occurred in the evolutionary lineage in which the functional alteration took place.[78] Novel sequencing technologies and artificial gene synthesis will contribute to more approaches for elucidating the molecular basis of functional shifts in TRP channels in an evolutionary context.

REFERENCES

1. Julius, D. TRP channels and pain. *Annu Rev Cell Dev Biol*, 2013, 29: 355–384. doi:10.1146/annurev-cellbio-101011-155833.
2. Laing, R. J., Dhaka, A. ThermoTRPs and pain. *Neuroscientist*, 2016, 22: 171–187. doi:10.1177/1073858414567884.
3. Patapoutian, A., Peier, A. M., Story, G. M., Viswanath, V. ThermoTRP channels and beyond: Mechanisms of temperature sensation. *Nat Rev Neurosci*, 2003, 4: 529–539. doi:10.1038/nrn1141.

4. Kashio, M., Tominaga, M. TRP channels in thermosensation. *Curr Opin Neurobiol*, 2022, 75: 102591. doi:10.1016/j.conb.2022.102591.
5. García-Ávila, M., Islas, L. D. What is new about mild temperature sensing? A review of recent findings. *Temperature*, 2019, 6: 132–141. doi:10.1080/23328940.2019.1607490.
6. Bandell, M., Macpherson, L. J., Patapoutian, A. From chills to chilis: Mechanisms for thermosensation and chemesthesis via thermoTRPs. *Curr Opin Neurobiol*, 2007, 17: 490–497. doi:10.1016/j.conb.2007.07.014.
7. Dhaka, A., Viswanath, V., Patapoutian, A. Trp ion channels and temperature sensation. *Annu Rev Neurosci*, 2006, 29: 135–161. doi:10.1146/annurev.neuro.29.051605.112958.
8. Nilius, B., Appendino, G., Owsianik, G. The transient receptor potential channel TRPA1: From gene to pathophysiology. *Pflugers Arch*, 2012, 464: 425–458. doi:10.1007/s00424-012-1158-z.
9. Uchida, K., Tominaga, M. The role of thermosensitive TRP (transient receptor potential) channels in insulin secretion. *Endocr J*, 2011, 58: 1021–1028. doi: 10.1507/endocrj.ej11-0130.
10. Wang, H., Siemens, J. TRP ion channels in thermosensation, thermoregulation and metabolism. *Temperature (Austin)*, 2015, 2: 178–187. doi:10.1080/23328940.2015.1040604.
11. Pedersen, S. F., Owsianik, G., Nilius, B. TRP channels: An overview. *Cell Calcium*, 2005, 38: 233–252. doi:10.1016/j.ceca.2005.06.028.
12. Cao, E., Liao, M., Cheng, Y., Julius, D. TRPV1 structures in distinct conformations reveal activation mechanisms. *Nature*, 2013, 504: 113–118. doi:10.1038/nature12823.
13. Liao, M., Cao, E., Julius, D., Cheng, Y. Structure of the TRPV1 ion channel determined by electron cryo-microscopy. *Nature*, 2013, 504: 107–112. doi:10.1038/nature12822.
14. Paulsen, C. E., Armache, J. P., Gao, Y., Cheng, Y., Julius, D. Structure of the TRPA1 ion channel suggests regulatory mechanisms. *Nature*, 2015, 520: 511–517. doi:10.1038/nature14367.
15. Rohacs, T. 2023. Phosphoinositide regulation of TRP channels: A functional overview in the structural era. *Annu Rev Physiol*. doi:10.1146/annurev-physiol-042022-013956.
16. Laursen, W. J., Anderson, E. O., Hoffstaetter, L. J., Bagriantsev, S. N., Gracheva, E. O. Species-specific temperature sensitivity of TRPA1. *Temperature (Austin)*, 2015, 2: 214–226. doi:10.1080/23328940.2014.1000702.
17. Saito, S., Tominaga, M. Functional diversity and evolutionary dynamics of thermoTRP channels. *Cell Calcium*, 2015, 57: 214–221. doi:10.1016/j.ceca.2014.12.001.
18. Saito, S., Tominaga, M. Evolutionary tuning of TRPA1 and TRPV1 thermal and chemical sensitivity in vertebrates. *Temperature (Austin)*, 2017, 4: 141–152. doi:10.1080/23328940.2017.1315478.
19. Bagriantsev, S. N., Gracheva, E. O. Molecular mechanisms of temperature adaptation. *J Physiol*, 2015, 593: 3483–3491. doi:10.1113/jphysiol.2014.280446.
20. Himmel, N. J., Cox, D. N. Transient receptor potential channels: Current perspectives on evolution, structure, function and nomenclature. *Proc Biol Sci*, 2020, 287: 20201309. doi:10.1098/rspb.2020.1309.
21. Saito, S., Fukuta, N., Shingai, R., Tominaga, M. Evolution of vertebrate transient receptor potential vanilloid 3 channels: Opposite temperature sensitivity between mammals and Western clawed frogs. *PLoS Genet*, 2011, 7: e1002041. doi:10.1371/journal.pgen.1002041.
22. Saito, S., Shingai, R. Evolution of thermoTRP ion channel homologs in vertebrates. *Physiol Genomics*, 2006, 27: 219–230. doi:10.1152/physiolgenomics.00322.2005.
23. Matsuura, H., Sokabe, T., Kohno, K., Tominaga, M., Kadowaki, T. Evolutionary conservation and changes in insect TRP channels. *BMC Evol Biol*, 2009, 9: 228. doi:10.1186/1471-2148-9-228.
24. Peng, G., Shi, X., Kadowaki, T. Evolution of TRP channels inferred by their classification in diverse animal species. *Mol Phylogenet Evol*, 2015, 84: 145–157. doi:10.1016/j.ympev.2014.06.016.
25. Bellemer, A. Thermotaxis, circadian rhythms, and TRP channels in *Drosophila*. *Temperature (Austin)*, 2015, 2: 227–243. doi:10.1080/23328940.2015.1004972.
26. Bautista, D. M. et al. TRPA1 mediates the inflammatory actions of environmental irritants and proalgesic agents. *Cell*, 2006, 124: 1269–1282. doi:10.1016/j.cell.2006.02.023.
27. Bandell, M., Story, G. M., Hwang, S. W., Viswanath, V., Eid, S. R., Petrus, M. J., Earley, T. J., Patapoutian, A. Noxious cold ion channel TRPA1 is activated by pungent compounds and bradykinin. *Neuron*, 2004, 41: 849–857. doi: 10.1016/s0896-6273(04)00150-3.
28. Jordt, S. E., Bautista, D. M., Chuang, H. H., McKemy, D. D., Zygmunt, P. M., Högestätt, E. D., Meng, I. D., Julius, D. Mustard oils and cannabinoids excite sensory nerve fibres through the TRP channel ANKTM1. *Nature*, 2004, 427: 260–265. doi:10.1038/nature02282.
29. Macpherson, L. J., Geierstanger, B. H., Viswanath, V., Bandell, M., Eid, S. R., Hwang, S., Patapoutian, A. The pungency of garlic: Activation of TRPA1 and TRPV1 in response to allicin. *Curr Biol*, 2005, 15: 929–934. doi:10.1016/j.cub.2005.04.018.

30. Bautista, D. M., Movahed, P., Hinman, A., Axelsson, H. E., Sterner, O., Högestätt, E. D., Julius, D., Jordt, S. E., Zygmunt, P. M. Pungent products from garlic activate the sensory ion channel TRPA1. *Proc Natl Acad Sci U S A*, 2005, 102: 12248–12252. doi:10.1073/pnas.0505356102.
31. McNamara, C. R., Mandel-Brehm, J., Bautista, D. M., Siemens, J., Deranian, K. L., Zhao, M., Hayward, N. J., Chong. J. A., Julius, D., Moran, M. M., Fanger, C. M. TRPA1 mediates formalin-induced pain. *Proc Natl Acad Sci U S A*, 2007, 104: 13525–13530. doi:10.1073/pnas.0705924104.
32. Story, G. M., Peier, A. M., Reeve, A. J., Eid, S. R., Mosbacher, J., Hricik, T. R., Earley, T. J., Hergarden, A. C., Andersson, D. A., Hwang, S. W., McIntyre, P., Jegla, T., Bevan, S., Patapoutian, A. ANKTM1, a TRP-like channel expressed in nociceptive neurons, is activated by cold temperatures. *Cell*, 2003, 112: 819–829. doi: 10.1016/s0092-8674(03)00158-2.
33. Talavera, K., Startek, J. B., Alvarez-Collazo, J., Boonen, B., Alpizar, Y. A., Sanchez, A., Naert, R., Nilius, B. Mammalian transient receptor potential TRPA1 channels: From structure to disease. *Physiol Rev*, 2020, 100: 725–803. doi:10.1152/physrev.00005.2019.
34. Saito, S., Banzawa, N., Fukuta, N., Saito, C. T., Takahashi, K., Imagawa, T., Ohta, T., Tominaga, M. Heat and noxious chemical sensor, chicken TRPA1, as a target of bird repellents and identification of its structural determinants by multispecies functional comparison. *Mol Biol Evol*, 2014, 31: 708–722. doi:10.1093/molbev/msu001.
35. Saito, S., Nakatsuka, K., Takahashi, K., Fukuta, N., Imagawa, T., Ohta, T., Tominaga, M. Analysis of transient receptor potential ankyrin 1 (TRPA1) in frogs and lizards illuminates both nociceptive heat and chemical sensitivities and coexpression with TRP vanilloid 1 (TRPV1) in ancestral vertebrates. *J Biol Chem*, 2012, 287: 30743–30754. doi:10.1074/jbc.M112.362194.
36. Oda, M., Kurogi, M., Kubo, Y., Saitoh, O. Sensitivities of two zebrafish TRPA1 paralogs to chemical and thermal stimuli analyzed in heterologous expression systems. *Chem. Senses*, 2016, 41: 261–272. doi:10.1093/chemse/bjv091.
37. Oda, M., Kubo, Y., Saitoh, O. Sensitivity of TRPA1 to thermal stimulations analyzed in oocytes expression system. *Neuroreport*, 2018, 29: 280–285. doi:10.1097/WNR.0000000000000939.
38. Oda, M., Saito, K., Hatta, S., Kubo, Y., Saitoh, O. Chemical and thermal sensitivity of medaka TRPA1 analyzed in heterologous expression system. *Biochem Biophys Res Commun*, 2017, 494: 194–201. doi:10.1016/j.bbrc.2017.10.057.
39. Saito, S., Hamanaka, G., Kawai, N., Furukawa, R., Gojobori, J., Tominaga, M., Kaneko, H., Satta, Y. Characterization of TRPA channels in the starfish *Patiria pectinifera*: Involvement of thermally activated TRPA1 in thermotaxis in marine planktonic larvae. *Sci Rep*, 2017, 7: 2173. doi:10.1038/s41598-017-02171-8.
40. Viswanath, V., Story, G. M., Peier, A. M., Petrus, M. J., Lee, V. M., Hwang, S. W., Patapoutian, A., Jegla, T. Opposite thermosensor in fruitfly and mouse. *Nature*, 2003, 423: 822–823. doi:10.1038/423822a.
41. Sato, A., Sokabe, T., Kashio, M., Yasukochi, Y., Tominaga, M., Shiomi, K. Embryonic thermosensitive TRPA1 determines transgenerational diapause phenotype of the silkworm, *Bombyx mori*. *Proc Natl Acad Sci U S A*, 2014, 111: E1249–E1255. doi:10.1073/pnas.1322134111.
42. Gau, P., Poon, J., Ufret-Vincenty, C., Snelson, C. D., Gordon, S. E., Raible, D. W., Dhaka, A. The zebrafish ortholog of TRPV1 is required for heat-induced locomotion. *J Neurosci*, 2013, 33: 5249–5260. doi:10.1523/JNEUROSCI.5403-12.2013.
43. Jordt, S. E., Julius, D. Molecular basis for species-specific sensitivity to "hot" chili peppers. *Cell*, 2002, 108: 421–430. doi:10.1016/s0092-8674(02)00637-2.
44. Gracheva, E. O., Ingolia, N. T., Kelly, Y. M., Cordero-Morales, J. F., Hollopeter, G., Chesler, A. T., Sánchez, E. E., Perez, J. C., Weissman, J. S., Julius, D. Molecular basis of infrared detection by snakes. *Nature*, 2010, 464: 1006–1011. doi:10.1038/nature08943.
45. Yoshimura, A., Saito, S., Saito, C. T., Takahashi, K., Tominaga, M., Ohta, T. Functional analysis of thermo-sensitive TRPV1 in an aquatic vertebrate, masu salmon (Oncorhynchus masou ishikawae). *Biochem Biophys Rep*, 2022, 31: 101315. doi:10.1016/j.bbrep.2022.101315.
46. Prober, D. A., Zimmerman, S., Myers, B. R., McDermott, B. M. Jr., Kim, S. H., Caron, S., Rihel, J., Solnica-Krezel, L., Julius, D., Hudspeth, A. J., Schier, A. F. Zebrafish TRPA1 channels are required for chemosensation but not for thermosensation or mechanosensory hair cell function. *J Neurosci*, 2008, 28: 10102–10110. doi:10.1523/JNEUROSCI.2740-08.2008.
47. Mason, J. R., Adams, M. A., Clark, L. Anthranilate repellency to starlings - chemical correlates and sensory perception. *J Wildl Manag*, 1989, 53: 55–64. doi:10.2307/3801306.
48. Nakatsuka, K., Gupta, R., Saito, S., Banzawa, N., Takahashi, K., Tominaga, M., Ohta, T. Identification of molecular determinants for a potent mammalian TRPA1 antagonist by utilizing species differences. *J Mol Neurosci*, 2013, 51: 754–762. doi:10.1007/s12031-013-0060-2.

49. Gupta, R., Saito, S., Mori, Y., Itoh, S. G., Okumura, H., Tominaga, M. Structural basis of TRPA1 inhibition by HC-030031 utilizing species-specific differences. *Sci Rep*, 2016, 6: 37460. doi:10.1038/srep37460.
50. Banzawa, N., Saito, S., Imagawa, T., Kashio, M., Takahashi, K., Tominaga, M., Ohta, T. Molecular basis determining Inhibition/Activation of nociceptive receptor TRPA1: A single amino acid dictates species-specific actions of the most potent mammalian TRPA1 antagonists. *J Biol Chem*, 2014, 289: 31927-31939. doi:10.1074/jbc.M114.586891.
51. Gavva, N. R., Klionsky, L., Qu, Y., Shi, L., Tamir, R., Edenson, S., Zhang, T. J., Viswanadhan, V. N., Toth, A., Pearce, L. V., Vanderah, T. W., Porreca, F., Blumberg, P. M., Lile, J., Sun, Y., Wild, K., Louis, J. C., Treanor, J. J. Molecular determinants of vanilloid sensitivity in TRPV1. *J Biol Chem*, 2004, 279: 20283–20295. doi:10.1074/jbc.M312577200.
52. Phelps, P. T., Anthes, J. C., Correll, C. C. Cloning and functional characterization of dog transient receptor potential vanilloid receptor-1 (TRPV1). *Eur J Pharmacol*, 2005, 513: 57–66. doi:10.1016/j.ejphar.2005.02.045.
53. Ohkita, M., Saito, S., Imagawa, T., Takahashi, K., Tominaga, M., Ohta, T. Molecular cloning and functional characterization of *Xenopus tropicalis* frog transient receptor potential vanilloid 1 reveal its functional evolution for heat, acid, and capsaicin sensitivities in terrestrial vertebrates. *J Biol Chem*, 2012, 287: 2388–2397. doi:10.1074/jbc.M111.305698.
54. Chu, Y., Cohen, B. E., Chuang, H. H. A single TRPV1 amino acid controls species sensitivity to capsaicin. *Sci Rep*, 2020, 10: 8038. doi:10.1038/s41598-020-64584-2.
55. Gao, Y., Cao, E., Julius, D., Cheng, Y. TRPV1 structures in nanodiscs reveal mechanisms of ligand and lipid action. *Nature*, 2016, 534: 347–351. doi:10.1038/nature17964.
56. Han, Y., Li, B., Yin, T. T., Xu, C., Ombati, R., Luo, L., Xia, Y., Xu, L., Zheng, J., Zhang, Y., Yang, F., Wang, G. D., Yang, S., Lai, R. Molecular mechanism of the tree shrew's insensitivity to spiciness. *PLoS Biol*, 2018, 16: e2004921. doi:10.1371/journal.pbio.2004921.
57. Tewksbury, J. J., Nabhan, G. P. Seed dispersal. Directed deterrence by capsaicin in chilies. *Nature*, 2001, 412: 403–404. doi:10.1038/35086653.
58. Geng, J., Liang, D., Jiang, K., Zhang, P. Molecular evolution of the infrared sensory gene TRPA1 in snakes and implications for functional studies. *PLoS One*, 2011, 6: e28644. doi:10.1371/journal.pone.0028644.
59. Yokoyama, S., Altun, A., DeNardo, D. F. Molecular convergence of infrared vision in snakes. *Mol Biol Evol*, 2011, 28: 45–48. doi:10.1093/molbev/msq267.
60. Gracheva, E. O., Cordero-Morales, J. F., González-Carcacía, J. A., Ingolia, N. T., Manno, C., Aranguren, C. I., Weissman, J. S., Julius, D. Ganglion-specific splicing of TRPV1 underlies infrared sensation in vampire bats. *Nature*, 2011, 476: 88–91. doi:10.1038/nature10245.
61. Yokoyama, T., Saito, S., Shimoda, M., Kobayashi, M., Takasu, Y., Sezutsu, H., Kato, Y., Tominaga, M., Mizoguchi, A., Shiomi, K. Comparisons in temperature and photoperiodic-dependent diapause induction between domestic and wild mulberry silkworms. *Sci Rep*, 2021, 11: 8052. doi:10.1038/s41598-021-87590-4.
62. Yatsu, R., Miyagawa, S., Kohno, S., Saito, S., Lowers, R. H., Ogino, Y., Fukuta, N., Katsu, Y., Ohta, Y., Tominaga, M., Guillette, L. J. Jr., Iguchi, T. TRPV4 associates environmental temperature and sex determination in the American alligator. *Sci Rep*, 2015, 5: 18581. doi:10.1038/srep18581.
63. Saito, S., Saito, C. T., Igawa, T., Takeda, N., Komaki, S., Ohta, T., Tominaga, M. Evolutionary tuning of transient receptor potential ankyrin 1 underlies the variation in heat avoidance behaviors among frog species inhabiting diverse thermal niches. *Mol Biol Evol*, 2022, 39: msac180. doi:10.1093/molbev/msac180.
64. Komaki, S., Lau, Q., Igawa, T. Living in a Japanese onsen: Field observations and physiological measurements of hot spring amphibian tadpoles, *Buergeria japonica. Amphibia-Reptilia*, 2016, 37: 311. doi: 10.1163/15685381-00003052.
65. Saito, S., Ohkita, M., Saito, C. T., Takahashi, K., Tominaga, M., Ohta, T. Evolution of heat sensors drove shifts in thermosensation between Xenopus species adapted to different thermal niches. *J Biol Chem*, 2016, 291: 11446–11459. doi:10.1074/jbc.M115.702498.
66. Saito, S., Saito, C. T., Nozawa, M., Tominaga, M. Elucidating the functional evolution of heat sensors among *Xenopus* species adapted to different thermal niches by ancestral sequence reconstruction. *Mol Ecol*, 2019, 28: 3561–3571. doi:10.1111/mec.15170.
67. Akashi, H.D., Saito, S., Cádiz Díaz, A., Makino, T., Tominaga, M., Kawata, M. Comparisons of behavioural and TRPA1 heat sensitivities in three sympatric Cuban Anolis lizards. *Mol Ecol*, 2018, 27: 2234–2242. doi:10.1111/mec.14572.
68. Laursen, W. J., Schneider, E. R., Merriman, D. K., Bagriantsev, S. N., Gracheva, E. O. Low-cost functional plasticity of TRPV1 supports heat tolerance in squirrels and camels. *Proc Natl Acad Sci U S A*, 2016, 113: 11342–11347. doi:10.1073/pnas.1604269113.

69. Hori, S., Saitoh, O. Unique high sensitivity to heat of axolotl TRPV1 revealed by the heterologous expression system. *Biochem Biophys Res Commun*, 2020, 521: 914–920. doi:10.1016/j.bbrc.2019.10.203.
70. Hori, S., Tateyama, M., Shirai, T., Kubo, Y., Saitoh, O. Two single-point mutations in ankyrin repeat one drastically change the threshold temperature of TRPV1. *Nat Commun*, 2023, 14: 2415. doi:10.1038/s41467-023-38051-1.
71. Oda, M., Ogino, H., Kubo, Y., Saitoh, O. Functional properties of axolotl transient receptor potential ankyrin 1 revealed by the heterologous expression system. *Neuroreport*, 2019, 30: 323–330. doi:10.1097/WNR.0000000000001197.
72. York, J. M. Temperature activated transient receptor potential ion channels from Antarctic fishes. *Open Biol*, 2023, 13: 230215. doi:10.1098/rsob.230215.
73. Yang, S., Lu, X., Wang, Y., Xu, L., Chen, X., Yang, F., Lai, R. A paradigm of thermal adaptation in penguins and elephants by tuning cold activation in TRPM8. *Proc Natl Acad Sci U S A*, 2020, 117: 8633–8638. doi:10.1073/pnas.1922714117.
74. Matos-Cruz, V., Schneider, E. R., Mastrotto, M., Merriman, D. K., Bagriantsev, S. N., Gracheva, E. O. Molecular prerequisites for diminished cold sensitivity in ground squirrels and hamsters. *Cell Rep*, 2017, 21: 3329–3337. doi:10.1016/j.celrep.2017.11.083.
75. Akashi, H. Thermal sensitivity of heat sensor TRPA1 correlates with temperatures inducing heat avoidance behavior in terrestrial ectotherms. *Front Ecol Evol*, 2021, 9: 583837. doi:10.3389/fevo.2021.583837.
76. Li, T., Saito, C. T., Hikitsuchi, T., Inoguchi, Y., Mitsuishi, H., Saito, S., Tominaga, M. Diverse sensitivities of TRPA1 from different mosquito species to thermal and chemical stimuli. *Sci Rep*, 2019, 9: 20200. doi:10.1038/s41598-019-56639-w.
77. Nguyen, T. H. D., Chapman, S., Kashio, M., Saito, C., Strom, T., Yasui, M., Tominaga, M. Single amino acids set apparent temperature thresholds for heat-evoked activation of mosquito transient receptor potential channel TRPA1. *J Biol Chem*, 2022, 298: 102271. doi:10.1016/j.jbc.2022.102271.
78. Yokoyama, S. Synthetic biology of phenotypic adaptation in vertebrates: The next frontier. *Mol Biol Evol*, 2013, 30: 1495–1499. doi:10.1093/molbev/mst075.
79. Saito, S. The association between the evolutionary changes of the thermosensory system and environmental adaptation in vertebrates. *Low Temperature Science*, 2023, 81: 71–80. doi:10.14943/lowtemsci.81.71.

Index

Note: Page numbers in **bold** and *italics* refer to tables and figures, respectively.

D

E

F

G

H

P

Q

R

S

T